*Energy Resources
and Supply*

Energy Resources and Supply

J. T. McMullan
R. Morgan
R. B. Murray

School of Physical Sciences, New University of Ulster,
Coleraine, Northern Ireland

A Wiley–Interscience Publication

JOHN WILEY & SONS
London · New York · Sydney · Toronto

Library of Congress Cataloging in Publication Data:

McMullan, John T
  Energy resources and supply.

  'A Wiley–Interscience publication.'
  1. Power resources. 2. Power (Mechanics)
I. Morgan, R., joint author. II. Murray, R. B., joint author.
III. Title.
TJ153.M183   333.7   75–6973

ISBN 0 471 58975 6

Photosetting by Thomson Press (India) Limited, New Delhi
and Printed in Great Britain by J. W. Arrowsmith Ltd., Bristol.

To
Cara, Daniel, Astrid
Helen, David
Sarah-Jane
who will inherit our mistakes

Preface

The aim of this book is to present a comprehensive survey of available energy resources and the technologies by which they can be exploited. It is intended for graduate and advanced undergraduate students of physical sciences, engineering and environmental studies, and also for the specialists within the energy industries who wish to become acquainted with sectors other than their own. We hope that it will also be a useful reference source for scientific writers and journalists.

The first three chapters form a general introduction. The problems are outlined and natural energy flows on the Earth are related to the incoming solar flux. This demonstrates our ultimate dependence on the Sun for most of our energy resources and all of our food. Next, fossil fuels and their exploitation, electricity generation and nuclear power are discussed and the treatment of resources ends with a consideration of natural power supplies. We then change emphasis from resources to the associated problems of waste, transmission and storage, and finish with a chapter on the impact of the energy industries on people and a brief discussion of energy policy for the future.

Although no subject is covered in great detail, the essential features are discussed in some depth and there is a list of further reading material at the end of each chapter to enable the reader to pursue his studies further. Throughout we have tried to eliminate personal opinion and to present an objective account of the advantages and limitations of each source of power, although we have allowed ourselves the luxury of a little personal freedom of expression in the last two chapters.

Acknowledgments to specific sources are made in the text, but we would like to record here our appreciation and thanks to all those who have so willingly given us information and assistance. We have learnt a great deal in the course of writing this book, and we hope that our readers will learn as much from reading it.

J. T. McM.
R. M.
R. B. M.

Coleraine, January 1975.

Contents

1 Introduction: Man and Energy

Energy crisis! Fuel shortage! The bottom of the oil barrel has been reached! We must cut our standard of living if posterity is to be saved! These warnings and many others have been common in recent years, and have begun to influence official policy. This is especially so since the artificial crisis at the beginning of 1974 when the Arab oil producing states reduced the supply of crude oil and simultaneously raised their prices substantially. This had a dramatic effect on industrialized economies, and in the United Kingdom the situation was exacerbated by a national strike in the coal mines. The western world suddenly became aware that it depends on fuel for its continuance, and that easy access to fuel supplies may not continue to be available.

These warnings are not new; Jules Verne refers to future shortages of fuel in *The Mysterious Island*, and in an admirable little book published in 1913, A. H. Gibson has in his preface, 'Much attention has been paid, in recent years, to the depletion of the coal resources of our own and other actively industrial countries, and to the problems which this depletion must ultimately involve.'

What is new is that modern western society has made itself heavily industrialized and machine-dependent, even in the home, and overwhelmingly committed to one energy source, oil. The crisis has been precipitated by the abrupt realization that oil supplies are limited since oil is a finite resource. Even worse as far as an industrial country is concerned, most of the oil used in the West is imported, and every western country faces an increasing import bill for its fuel supplies. As reserves are depleted, or as oil becomes more expensive to extract, the price will increase and it will become more difficult for the oil importing countries to sell enough goods overseas to pay for their fuel. This is the immediate nature of the problem, but in reality there is no energy crisis as yet. There is, rather, a chronic situation.

To be strictly correct, there is no energy shortage, there is a fuel shortage. Fuel is the material we consume, either by burning or by some other means, in order to generate energy which we then use to suit our needs. A rule of thumb definition could be: Energy *is*. If its entropy is low it is fuel; if its entropy is high it is pollution. Misuse of language in recent years has led to a situation in which the three terms energy, fuel and power have become interchangeable in popular articles and in some technical journals. This is unfortunate but little can now be done except to accept their new roles.

How has the energy problem come about? To see this, it is easiest to examine

1

briefly the history of fuel use in western society. Until the beginning of the 17th century, man's productivity was primarily determined by his own strength and industry and that of his animals. He used fire for cooking, for keeping warm, for smelting and for firing kilns. These latter two activities were on a much smaller scale than today, of course, and were dependent on wood for fuel. Smelting was also limited by supplies of charcoal. With the development of the steam engine in 1663 by Somerset, and its refinement by Papin, Savery and eventually Newcomen in 1712, the possibilities for expansion of industrial output were improved. The steam engine also increased the demand for fuel and precipitated the first energy crisis in the form of an acute shortage of trees. There was a long period when iron was in extremely short supply in Britain due to this timber shortage, and the situation was not alleviated until coal replaced wood as fuel and coke replaced charcoal as the reducing agent in the furnaces. Parts of Britain still show the scars of the denudation of trees that occurred during this period.

The improvement in productivity made possible by the steam engine laid the foundations for the industrial revolution in Britain and later in the rest of the World. In 1824, Sadi Carnot wrote, 'To rob Britain of her steam engines would be to rob her of her coal and iron, to deprive her of her sources of wealth, to ruin her prosperity, to annihilate that colossal power'. The industrial revolution produced a completely new life style. Cities became larger and fewer people were needed to farm the land. The mixture of new farming techniques and new industries could support more people and the population expanded rapidly from the old limits that could be supported by an agrarian society. All of this expansion placed new stresses on the supplies of fuel and raw materials. Coal became the primary fuel material for the industrial society and mining expanded. By the late 19th century, industrialization based solely upon coal and steam had reached a limit and newer energy sources were being developed. Electricity had appeared as a means of supplying energy to domestic and industrial users in a clean and convenient way, and domestic gas supplies were common in cities. Unfortunately, the conversion efficiencies from coal to electricity were low, about ten per cent, making electricity an expensive form of energy.

At approximately the same time, oil appeared as a primary fuel. The first modern oil well, Drake's Well, Pennsylvania, was drilled in 1859. Since then the story is well known. The United States, with its own domestic supplies of oil and natural gas, became highly dependent on these fuels quite early in the 20th century. European countries, without large domestic oil supplies, continued to expand their coal based economies but imported oil as they needed it. After the Second World War, however, oil was significantly cheaper than coal, and Europe's dependence on it as a primary fuel increased dramatically. Not only is it the basic feedstock of the petrochemical industry, but also deliberate government decisions were made to change from coal to oil for electricity generation. The consequences were inevitable. Sooner or later the producing countries were certain to increase prices or to use their enormous

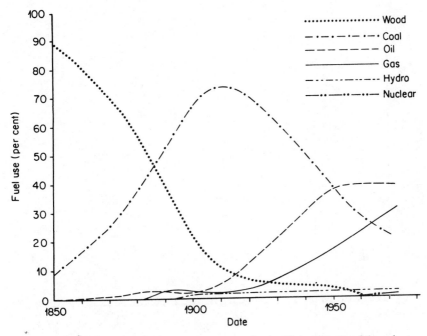

Figure 1.1. Variable pattern of fuel use in the United States of America from 1850

economic power over industrial nations for other ends. This is what happened in the upheaval at the beginning of 1974.

The changing pattern of fuel use is easy to see. The statistics for the United States are shown diagrammatically in Figure 1.1. There is a noticeable transition from the economy of 1850, in which wood formed 90 per cent of the fuel supply, to the present time. Wood has all but vanished as a fuel. Coal has dropped from its peak of 75 per cent in 1910 to about 25 per cent today, with oil and natural gas supplying most of the remaining needs. Hydroelectric power is a small contributor and nuclear power hardly shows on the diagram at 2 per cent in 1971. The pattern of fuel use in the United Kingdom, shown in Figure 1.2, is interesting as a comparison. While showing a not unrelated trend, there are important differences. By 1850, coal had already become established as the primary source, and, whereas oil and natural gas appeared quite early in the United States, oil did not figure significantly in the United Kingdom until about 1900 and natural gas entered only in the 1960's with the discoveries of gas under the North Sea. Another noteworthy comparison is that, while oil and gas had replaced coal as the dominant fuel in the United States by 1950, this point was not reached in the United Kingdom until 1970, when it was becoming evident that oil reserves were limited. This phenomenon is paralleled by the development of large sailing ships to their ultimate performance in the clipper ships at exactly the time that they were becoming

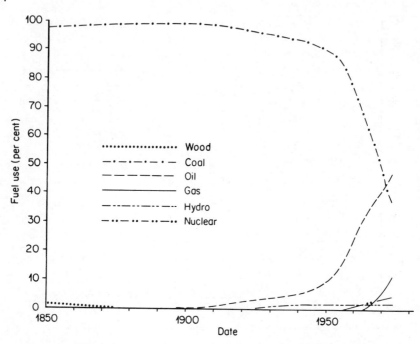

Figure 1.2. Pattern of fuel use in the United Kingdom from 1850

redundant. The curves also clearly indicate the degree of industrialization of the United Kingdom relative to the United States. The industrial revolution began in Britain and the early dependence on coal reflects this. By contrast, the America of the late 19th century was still largely agrarian and depended on local timber as its primary fuel source. As the United States became more industrialized its dependence on coal as a primary fuel increased, reaching its peak as a percentage at about 1910. After this the other fossil fuels steadily replaced coal in importance until by 1970 it supplied only about 23 per cent of American fuel needs.

This then was the background against which the oil crisis developed and the warnings of previous years began to be appreciated. The Arab–Israeli war of 1973 led to an embargo by the Arab oil-producing states on supplies to those countries that they felt had been overly cooperative with Israel. In addition, they dramatically increased the price of crude oil and simultaneously reduced output. The whole of western society suddenly became aware of the true situation regarding fuel supplies. In particular, the situation in the United Kingdom was exacerbated by the coincidence of the Arab action and a national strike by coal miners. Consequently, great strains were placed on stocks of all the fossil fuels. The irony is that the abrupt crisis would not have been possible if more diverse energy policies had been followed.

What is the position regarding fuel supplies for the future? Are they as seriously limited as it would appear from the hue and cry that has been raised?

Is there a better future for our children, or have we, through our own greed and ignorance, condemned them to a much lower standard of living than our own? Would we be better to switch off all our electric appliances and our oil-fired heating systems, and return to a much simpler society in which the horse again supplied the power? This suggestion has been seriously made from several quarters and has been the philosophy behind the formation of some of the 'back-to-nature' communes in the United States and Europe. That it is untenable is easily demonstrated. In 1918 there were 25 million horses and mules in the United States. Almost all of the good arable land was already under cultivation, but one quarter of the total harvested crop acreage had to be allocated to provide feed for these power animals. To revert to such a system would cause total chaos with world food supplies. Today's population densities and large city groups can only be supported by an automated agricultural system and this automation depends on there being a supply of suitable fuel. Fuel is also necessary to maintain the complicated commercial and communication systems required to support our society even at present population levels.

At the moment there are three dominant sources of energy supply: fossil fuels (coal, oil and gas), nuclear fuels and hydroelectric power. As a future energy source hydroelectricity can be taken as having only a limited potential for development. This is reflected in Figures 1.1 and 1.2 where its contributions to both American and British energy supplies are shown to be small and also fairly steady.

Estimates of the reserves of fossil fuels have been established. Coal is totally dominant with over 7.5×10^{12} tonnes representing 2×10^{23} joules or 55.9×10^{15} kWh (thermal). At the bottom of the scale, the oil shale reserves appear to be equivalent to 0.32×10^{15} kWh. This represents only 0.5 per cent of world reserves of fossil fuel. A more complete list is shown in Table 1.1. It must be

Table 1.1 Probable World fossil fuel Reserves

Coal	7.5×10^{12}	tonnes	2×10^{23}	joules
Oil from wells	2×10^{12}	barrels	1.2×10^{22}	joules
Natural gas	10^{16}	cubic feet	1×10^{22}	joules
Oil from tar sands	3×10^{11}	barrels	1.8×10^{21}	joules
Shale oil	2×10^{11}	barrels	1.2×10^{21}	joules
Peat	3×10^{11}	tonnes	7×10^{22}	joules

emphasized that these figures are only estimates; more deposits may be discovered and equally some of the known deposits may prove to be unworkable. What these figures represent is a best guess at what our reserves may be.

In the mid-1950's, M. King Hubbert, then a geologist in the oil industry, showed that it was possible to estimate the probable length of time for which an expendable fuel resource could be exploited. His model was based on the two assumptions that the total fuel reserve is fixed, and that the past and pro-

6

bable future production rates are known. From these premises, it is possible to predict that, starting from zero with a new resource, the production rate will initially rise exponentially. Subsequently the rate of expansion slows down progressively as production begins to outstrip discovery until the production rate eventually reaches a maximum. After this point, the resource is being depleted and the rate of production steadily falls to zero once more. If the past and possible future rates of production are known, then estimates can be made of the probable lifetime for the exploitation of a resource. Figure 1.3

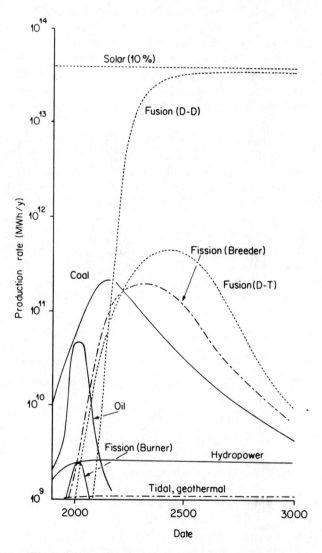

Figure 1.3. Possible fuel consumption levels and projected availability of reserves

shows the curves for the principal fuel reserves based on the assumption that past and present rates of expansion in consumption continue. It is interesting to note that on the curve for oil, we are at the peak at the moment and the actual consumption has followed earlier predictions.

As well as being the largest fossil fuel reserve, coal is the most widely distributed and almost all countries have deposits within their borders. Britain's industrial revolution was based on coal and there are still enormous deposits of coal under the United Kingdom.

Oil reserves are also well known, though the extension of geophysical surveying to the undersea continental shelves has uncovered further fields. Doubtless even more will be discovered by these means. The problem will be to find whether or not extraction will be feasible in every case. The North Sea oil fields are illustrating both the possibilities and the difficulties of deep-water drilling and extraction. As the drilling extends into the North Atlantic with deeper water and worse weather conditions, the difficulties increase dramatically and the practical limits to such operations have yet to be established. A major problem with oil supplies is that, while they are indeed widely distributed throughout the World, more restrictive geological conditions are required for the accumulation of oil and gas than for coal. As a result, deposits are localized. Economics plays a part in determining what is to be considered as a usable reserve. Before the end of 1973 the economically viable oil reserves were dominated by those in the Middle East, but with the price rises imposed at that time the picture changed considerably and other deposits such as the vast Rocky Mountain oil shales became economic. Almost the same amount of oil as in the Middle East suddenly became a usable reserve. It must always be appreciated, however, that the true cost of these reserves is higher in that they are actually more expensive to extract; the cost of Middle East oil is only artificially high, as the cost of extraction is much less than the price paid.

Natural gas is the least contaminated of the fossil fuel supplies. Unfortunately it is also the least common. Because of its cleanliness and its ease of handling it is highly in demand for industrial consumption, power generation and domestic use. In the United States this has been true to such an extent that reserves are being depleted and natural gas has to be imported in large quantities to meet demand.

A further aspect of the energy crisis is that, just when the scarcity of oil is being fully realized, there has been awakened a realization of, and concern for, the protection of threatened plants, animals and areas of scenic beauty from the effects of industrialization and urban development. As a direct consequence, work on the Alaska pipeline was stopped, oil drilling on the American continental shelf was restricted because of fears regarding the effects of oil leaks on marine life, severe criticisms were put in the way of the nuclear power programme and the efficiencies of motor vehicle engines were decreased by the effects of legislation to reduce air pollution. This is in no way a criticism of these events, merely an observation that at the time when we acknowledged that our fuel supplies were more vulnerable than previously thought, we also began to appreciate the effects of man's destructive influence and to take steps to counter them which had the effect of hindering our location of new fuel sources and our development of those already known.

Many governments are placing their short-term and medium-term hopes of maintaining electricity supplies on nuclear energy sources. The United Kingdom had an early lead in the production of electricity from nuclear power, and by 1973 had produced a total of 240×10^6 MWh from nuclear plant. The United States was second in the list with 209×10^6 MWh. Any hopes that nuclear power will be an important contributor to energy reserves in the long run will only be fulfilled if total fuel reserves are extended considerably through its use. Estimates of consumption made by the United States Atomic Energy Commission, the European Nuclear Energy Agency and the International Atomic Energy Agency projected that for the decade 1970–1980 about 400,000 tonnes of uranium oxide would be needed to meet the projected growth rate in nuclear power production. Against this, world reserves of uranium costing less than $10 per pound to extract in 1970 are estimated at 800,000 tonnes. Other reserves are, of course, available but the uranium content is lower and extraction costs are higher. It can be predicted, therefore, that the exploitable lifetime of uranium fission fuels is not significantly greater than that of oil—about 25 or 30 years—if conventional reactors are used for electricity production. One of the problems is that uranium-235, the fission fuel itself, forms only 0·7 per cent of naturally occurring uranium. The rest is a non-fissile isotope, uranium-238, which contributes little to the fission reaction but nevertheless has to be mined and so adds to the cost of uranium fuel. However, it is possible to convert uranium-238 into a useful fissile material, plutonium-239, in a reactor, and by suitable reactor design it can be arranged that for each fissile nucleus consumed more than one new fissile nucleus is produced from the previously useless uranium-238. This is the principle behind the breeder reactor. Consequently, traditional and breeder reactors in a balanced ratio will enable all of the World's supply of uranium to be used as fissile fuel instead of only 0·7 per cent of this amount. In addition, low-grade ores become usable because it is the uranium-238 that is of interest and the energy return per pound of fuel has been increased 140-fold. The exploitable reserves become vast. For example, under most of Tennessee, Kentucky, Ohio, Indiana and Illinois lies, at mineable depths, the Chatanooga black shale. This contains uranium at a content of about 50–60 gm tonne^{-1} in a layer about five metres thick. Taking the density of the rock as 2·2 tonnes m^{-3} leads to the conclusion that about 2,000 km^2 of the shale has an energy content equivalent to the entire United States reserves of fossil fuels. This is about 2 per cent of the area of Tennessee, and a very small fraction of the area under which the shale lies. As a result, the breeder reactor makes available reserves of fuel several orders of magnitude greater than the total of all fossil fuels combined.

The nuclear power programme is subject to two major criticisms. The safety of nuclear reactors has never been fully established, but it must be pointed out that neither has their danger, and that the nuclear power industry has an exceptionally good safety record. Nevertheless, critics insist that the seriousness of the possible accident which could result from a nuclear reactor failure is so great that the risk must not be taken. The other aspect is radioactive wastes. Some of the radioactive materials produced by nuclear reactors are extremely long lived and must, therefore, be stored for considerable lengths of time: 500 years is one of the figures commonly suggested. Thus, if nuclear power were to have been available in the past, we would now be considering the possibility of releasing some of the waste collected and stored before Columbus discovered America. The critics suggest that this is an intolerable burden to place upon our descendents. If mankind were to descend into another period of Dark Ages, the mystique that would develop around the sacred storage farms would resemble that of many science fiction stories.

The other really long-term hope that is advanced for ensuring our supplies of energy is nuclear fusion. Fusion—the joining together of very small nuclei with a consequent release of energy—has had a somewhat mixed history. The first proof that it was possible to release energy in this way came with the hydrogen bomb, in which a conventional nuclear fission explosion is used to drive light nuclei together with sufficient force that they coalesce and release large quantities of energy. Apart from nuclear explosions, the best way of forcing the nuclei together seems to be to heat them inside a container; hence the name thermonuclear power. Unfortunately it is not yet a reality. It has not been possible to produce the high temperatures needed for a long enough time in a gas of sufficient density to make a thermonuclear power plant feasible. Nonetheless, it is still theoretically possible and work is now being done on the eventual design of fusion power plants once the experiments succeed. Most of the problems seem to be understood if not solved, and it seems that fusion may be possible within about ten years. It is expected that a further 20 years will be needed after that stage before the first power plant could be built. The most favourable reaction for fusion is one involving deuterium and tritium—two heavy isotopes of hydrogen. This involves the lowest temperatures and also has a high probability of proceeding once those temperatures have been reached. Tritium occurs naturally in only very small amounts and must, therefore, be manufactured from lithium-6. The world reserves of lithium-6 are thought to be equivalent to a fusion energy of 7×10^{23} joules. This is comparable with the world reserves of fossil fuels. If the more difficult conditions were attained which would allow the deuterium–deuterium reaction to proceed, the world energy crisis would be solved for ever—at least for as long as man could possibly exist. Deuterium is plentiful in the sea, and the resulting energy supply would be sufficient for many million years.

From this we can see that either fission at its best or fusion at its worst would lead to a rather more than doubling of total fuel reserves, giving us something over 1,000 years to find an alternative.

These then are our reserves of fuel and represent what we could call our *energy capital*. Once this is used up, there is no more. There is, however, another side of the coin, what might be termed a *recurrent* energy supply. This comes primarily from two sources; the hot interior of the earth itself, and the incoming radiation from the Sun. There is also a contribution from the Moon. By far the greatest of these is the solar input. The rate at which energy reaches the Earth from the Sun is 173×10^{12} kW. Of this, some 52×10^{12} kW is reflected directly back into space by the atmosphere, leaving a net 120×10^{12} kW for absorption by the Earth and its atmosphere and for conversion to other forms of energy. Since the two functions that are essential for maintaining a food supply, continuance of evaporation and precipitation of water, and photosynthesis, only require 40.4×10^{12} kW, this leaves about 80×10^{12} kW to be disposed of to our advantage if we can develop ways of doing so. This is obviously a simplistic argument but it does provide an upper figure for the potential energy supply from solar radiation. If this figure is compared with man's present power consumption of about 3×10^9 kW, we see that much more comes in than we actually consume. This is the basis of the argument of the solar energy conversion enthusiasts. The critics point out that the cost of using solar energy and converting it to electricity is extremely high, but this seems to be the only hard criticism that has been levied against them.

Solar energy can be regarded as power from heaven; what about power from...? Well, not inappropriately, geothermal power is much smaller at about 0.32×10^{12} kW. Nonetheless, there is a considerable store of energy to be tapped. This has traditionally been done in selected areas such as Iceland and Northern Italy where conditions are favourable for its exploitation. However, interest is spreading to include other areas of the Earth and also to proposals for drilling to the hot rocks below us just as we drill for oil today, followed by pumping of water into the drillhole to be heated and extracted again. In this way heat could be tapped directly from inside the Earth's core and the figure quoted above greatly increased.

Two other power sources, wind and water, have historical importance, and are continually raised as possibilities today. These are in fact manifestations of solar power, and any power derived from them must be subtracted from the total solar power available. They therefore do not add anything to our supply from this source. It may well be, however, that their use on domestic and small industrial scales will yet prove to be one of the best ways of using solar energy efficiently, but this remains to be seen.

The other recurrent power source is lunar power—power from the tides. The total available from this source is not large, and can be realized in only a few places at the moment. Further development in slow-speed turbines will be necessary before the tides can be tapped effectively.

These are our fuel resources. How we choose to use them, or even if we choose to use them, is a social and political decision that must be made. In view of the undoubted fact that the oil and gas supplies are extremely limited and may last for only a few decades if our existing pattern of use continues, the decisions must be made soon.

Further Reading

Committee on Resources and Man, *Resources and Man*, National Academy of Sciences and National Research Council, W. H. Freeman and Co., San Francisco, 1969.
Energy and Power, A Scientific American Book, W. H. Freeman and Co., San Francisco, 1971.
Fisher, J. C., *Energy Crises in Perspective*, John Wiley and Sons, New York, 1974.
Inglis, K. A. D. (Ed.), *Energy, from Surplus to Scarcity?* Applied Science Publishers Ltd., Barking, Essex 1974.
Meadows, D. H., D. L. Meadows, J. Randers and W. W. Behrens III, *Limits to Growth*, Signet, New American Library, New York. 1972.
Readings from Scientific American, *Chemistry in the Environment*, W. H. Freeman and Co., San Francisco, 1973.
Wilson, Carroll L., (Director), *Man's Impact on the Global Environment*, M.I.T. Press, Cambridge, Mass., 1970.

2 Solar Energy and Climate

2.1 The Source of Solar Energy

All living things are dependent on the Sun for a continuous supply of energy to fuel their biochemical reactions. Green plants utilize sunlight directly, storing it in the form of carbohydrate, and it is this stored energy which animals obtain when they eat vegetation. Apart from supplying the nutritional needs of living organisms, the radiation received on Earth from the Sun also ensures that the planetary environment remains suitably hospitable for life as we know it. Even more awesome than the multifarious uses to which the Sun's energy is put is the sheer magnitude of the total energy received. By measuring the solar flux incident at the top of the Earth's atmosphere, we can calculate that the total power received from the Sun is as large as 10^{14} kW. If we assume that the Sun emits radiation isotropically and know the Earth–Sun distance, we can estimate that the Sun is radiating energy at the enormous rate of 10^{23} kW. Given such a magnitude of power output, it is not surprising that a convincing explanation for the source of solar energy was not put forward until the advent of nuclear physics.

Matter in the universe is mainly in the form of hydrogen and helium, most of which is distributed throughout interstellar and intergalactic space. Within our own galaxy the average density of intersteller matter is estimated to be only 1 atom cm^{-3}. Stars such as our Sun are thought to form when random fluctuations lead to a build-up of matter in a certain region of space. When this happens the mutual gravitational attraction between the particles of matter causes condensations of gas and dust to occur, and it is these condensations that eventually lead to new star clusters. We can see examples of this today in, say, the Omega Centauri cluster (Figure 2.1). We would expect this process to continue until all the matter in the star has been condensed into an extremely small volume, but this is not the case. Gravitational potential energy is being lost during the collapsing process, some of it radiating away, the rest going into heating the interior of the star. By the time the star has collapsed to such an extent that its density has reached 10^5 kg m^{-3} in the core, the temperature in these inner regions is approaching 10^7 K and other processes come into play to counteract the gravitational collapse.

Under these conditions the hydrogen atoms are stripped of their electrons to expose the single proton nuclei. We call this extreme state of matter a plasma. Normally protons repel one another by their electrostatic interaction, but,

Figure 2.1. The Great Globular Cluster Omega Centauri. (Photograph taken by E. M. Lindsay at Boyden Observatory, South Africa. Reproduced by permission of E. J. Opik, Armagh Observatory, Northern Ireland)

under such immoderate conditions, they can have sufficient kinetic energy to overcome their mutual Coulomb repulsion and collide. Moreover, the high density within the plasma ensures that such collisions are quite likely to occur. The conditions, in other words, are ideal for thermonuclear fusion reactions to take place (see Chapter 11). Those protons which do collide are able to coalesce, and by a further sequence of nuclear reactions, known as the *proton-proton (p–p) chain*, helium nuclei are produced:

$$H^1 + H^1 \longrightarrow D^2 + \beta^+ + \bar{\nu}_e$$
$$D^2 + H^1 \longrightarrow He^3 + \gamma$$
$$He^3 + He^3 \longrightarrow 2\,H^1 + He^4$$

the overall result of which is

$$4\,H^1 \longrightarrow He^4$$

The positrons (β^+) annihilate themselves in collisions with the electrons in the plasma, while the neutrinos ($\bar{\nu}_e$) escape and are radiated into space.

As we have said, the net result of these reactions is the production of one helium nucleus from four protons. The binding energy curve for nuclei shown in Figure 8.1 illustrates the exceptional stability of the helium nucleus and we expect that a considerable amount of energy will be released in the process of fusing four protons into one helium nucleus. It is this liberated energy which stokes the Sun's fires and maintains this p–p chain of reactions in which we have the solar counterpart to the fusion reactions which man is now attempting to control here on earth. In forming one helium nucleus by the p–p reactions, a total of 26·3 MeV of energy is released (5×10^{-12} J). Thus the thermonuclear 'burning' of 1 g of hydrogen releases 6×10^{11} J, enough to keep a one kilowatt heating element glowing for nearly 20 years! In our Sun it is estimated that the p–p reactions convert hydrogen into helium at the staggering rate of 6×10^{11} kg per second.

The thermal energy generated in the core of a star in this fashion is radiated outwards and the associated radiation pressure prevents the outer layers of the star from undergoing any further gravitational collapse. It has been estimated that there is enough hydrogen in our Sun to prolong this thermonuclear stage in its development for some five thousand million years.

In addition to hydrogen and helium there are also various heavier elements present in stellar material. Some of these are formed by further thermonuclear reactions involving the helium nuclei produced in the p–p chain. If a star contains carbon and nitrogen it has an alternative pathway for the conversion of hydrogen into helium. This *carbon–nitrogen (CN) cycle*, in which the carbon plays the role of a 'nuclear catalyst', was first proposed as a source of stellar energy by H. A. Bethe in 1939, in recognition of which he was awarded the 1967 Nobel Prize for physics. The reaction sequence yields the same net result as the p–p chain, the conversion of four hydrogen nuclei into one helium nucleus with the release of 26·3 MeV of energy:

$$C^{12} + H^1 \longrightarrow N^{13} + \gamma$$
$$N^{13} \longrightarrow C^{13} + \beta^+ + \bar{\nu}_e$$
$$C^{13} + H^1 \longrightarrow N^{14} + \gamma$$
$$N^{14} + H^1 \longrightarrow O^{15} + \gamma$$
$$O^{15} \longrightarrow N^{15} + \beta^+ + \bar{\nu}_e$$
$$N^{15} + H^1 \longrightarrow C^{12} + He^4$$

The cyclic nature of these results is perhaps best illustrated diagrammatically as in Figure 2.2. In stars containing carbon, the CN cycle can compete with the p–p chain. It is believed, however, that the p–p chain dominates when the temperature of the star's core is less than 2×10^7 K, as is the case in the Sun. Consequently, it is the p–p chain of reactions which provides most of the Sun's energy.

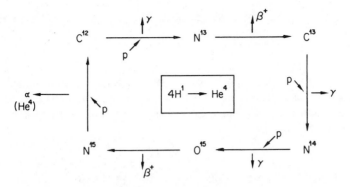

Figure 2.2. The carbon-nitrogen cycle

2.2 The Sun as a Black Body

We now turn our attention from the sources of the Sun's energy to examining the radiation which finally leaves its surface and begins the long journey to Earth. This radiation consists of an almost continuous range of wavelengths, from less than 1 nm (10^{-9} m) to many hundreds of metres. The entire spectrum of electromagnetic radiation is, therefore, present, from X-rays, through the ultra-violet, visible and infra-red regions to low-energy microwaves. Nevertheless, the wavelengths between 250 and 3000 nm carry some 98 per cent of the total emitted energy. Within this range we can effectively treat the Sun as a 'heat radiator', that is to say, a body which emits radiant energy by virtue of its temperature, and compare it with a perfect 'black body'.

Using classical thermodynamics it can be shown that the energy density $\rho(T)$ of the radiation within an enclosure whose walls are opaque is only dependent on the temperature T of the walls according to the equation

$$\rho(T) = sT^4$$

where s is a constant. We can define $\rho_\lambda(T)$ as the energy density of radiation in the wavelength range $\lambda \to \lambda + \partial\lambda$. Now, by Kirchoff's law, the ratio between the emissive and absorptive powers of any body at a given wavelength depends only on the temperature of the body, and not on its particular characteristics; if this were not the case radiative equilibrium could not exist within a cavity containing several types of bodies. (Emissive power refers to the rate of emission of radiant energy and absorptive power to the fraction absorbed of the radiant energy falling upon a body.) A *black body* is defined as one which is a perfect absorber of radiation, whatever the wavelength. If the black body is to remain in equilibrium at the temperature T, it must emit energy at the same rate as it is received. Consequently, the radiation emitted by such a body is a function of temperature only and has the same spectral distribution as the radiation in an enclosure at the uniform temperature T. Both are termed *black-body radiation*. It is a simple matter to show that the power P emitted from unit surface area of a black body is related to the energy density in an enclosure by

$$P = \tfrac{1}{4}c\rho(T)$$

c being the velocity of propogation of electromagnetic radiation, the velocity of light. This leads us to the *Stefan–Boltzman law*, namely that the power emitted per unit area of a black body is proportional to the fourth power of the absolute temperature:

$$P = \sigma\, T^4$$

Stefan's constant, σ, has the value $5\cdot67 \times 10^{-8}\ \mathrm{W\ m^{-2}\ K^{-4}}$.

Let us consider a non-black body at temperature T placed inside an enclosure at uniform temperature T_E. In equilibrium the power absorbed by the body must match the power it emits. Thus, if the body absorbs a fraction ε_λ of the radiation of wavelength λ incident upon it, then it must also emit the same fraction of the radiation that a black body at temperature T would emit. This fraction, ε_λ, is known as the *emissivity* of the surface at wavelength λ. If we assume that we can use an average value ε for the emissivity throughout the wavelength range, then the total power emitted by a body at temperature T may be written as

$$P = \varepsilon\, \sigma\, T^4$$

The distribution P_λ of the power emitted by a black body as a function of wavelength is well known and can be described by the formula due to Planck. (Originally, of course, the study of black-body radiation was instrumental in establishing the concepts of quantum physics.) Thus

$$P = \frac{(2\,\pi\,hc^2/\lambda^5)}{\exp(hc/\lambda\,kT) - 1}$$

In addition to this and the Stefan–Boltzman law, there is a third radiation law, derivable from Planck's formula, and known as the *Wien displacement law*. This states that the product of the absolute temperature and the wavelength, λ_{max}, at which the maximum power output occurs, is a constant.

$$\lambda_{max}\, T = 2\cdot88 \times 10^{-3}\ \mathrm{m\ K}$$

In principle, we can determine the temperature of a black body using any of these laws. Figure 2.3 illustrates the spectral distribution of black-body radiation at various temperatures.

If we attempt to use these radiation laws to estimate the temperature of the Sun we come across a number of difficulties. Firstly, there is a temperature variation across the disc of the Sun as observed from Earth, ranging from nearly 6800 K at its centre to less than 5600 K at its extremity (Figure 2.4). Secondly, solar radiation is emitted simultaneously from layers at varying depths and temperatures and the physical properties of these layers are not

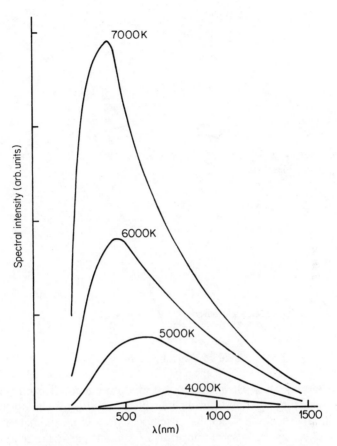

Figure 2.3. Spectral distribution of black-body radiation at various temperatures

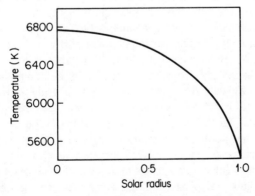

Figure 2.4. Temperature variation across the solar disc as viewed from Earth

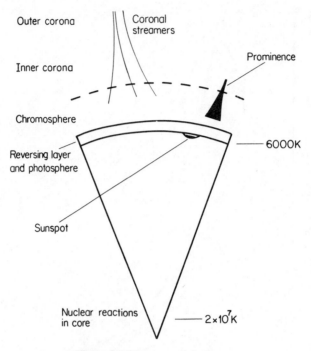

Figure 2.5 Structure of the Sun

the same (Figure 2.5). Most of the Sun's radiation emanates from the photosphere, the reversing layer and the chromosphere. Of these the major radiation source is the photosphere which has a temperature of approximately 6000 K. However, the photosphere is not uniformly bright, one particular manifestation of this being the sunspots which appear darker than the surrounding areas because they are at a lower temperature (about 4500 K). Immediately above the photosphere lies the reversing layer, so called because it has a slightly lower temperature (5300 K) with the result that it absorbs some of the light emitted by the photosphere. This is the origin of the absorption lines in the solar spectrum. The red-coloured chromosphere is an extension of the reversing layer visible as an irregular disc during eclipses of the Sun. This also is of non-uniform brightness due to the spectacular solar flares and prominences which extend for large distances beyond the extremity of the Sun's disc.

For simplicity we can treat the Sun as a uniformly radiating body and compare its spectral curve (as recorded from above the atmosphere by rocket-borne instruments) with black-body distributions. Figure 2.6 depicts such distributions for temperatures of 5000, 6000 and 7000 K together with the smoothed solar curve neglecting the absorption lines in the solar spectrum. It is clear that the solar curve approximates well to the black-body curve for a temperature of 6000 K. The main discrepancies are the occurrence of the intensity maximum of 470 instead of 480 nm, the slight cut-off in the ultra-violet and the slightly lower intensity in the infra-red region of the spectrum.

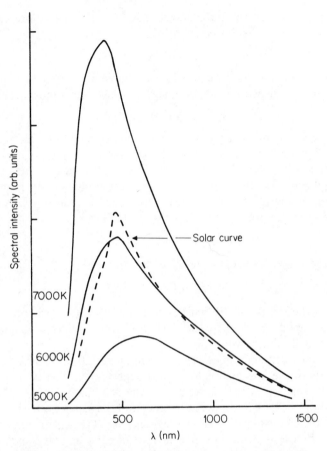

Figure 2.6. Comparison of the solar spectrum (above the atmosphere) with black-body spectra

2.3 The Solar Constant

The *solar constant S* is defined as the quantity of energy per unit time passing through unit area at right angles to the direction of the solar beam, measured just outside the Earth's atmosphere. Throughout this section only effects outside the atmosphere will be discussed. By treating the Sun as a black body of radius R_s and temperature T_s we can calculate S. From the Stefan–Boltzman law the total power emitted by the Sun is $4\pi\sigma R_s^2 . T_s^4$. By the time this radiation has travelled a distance R outwards from the Sun, it has spread out over an area of $4\pi R^2$ so that the power transmitted through unit area is given by

$$S = \sigma \frac{R_s^2}{R^2} T_s^4$$

From astronomical measurements the radius of the Sun may be taken to be

7×10^5 km and the Earth–Sun distance R is 1.5×10^8 km. Consequently, this leads to a value for the solar constant of

$$S = 1600 \text{ W m}^{-2}$$

This is in reasonable agreement with the currently accepted experimental value obtained by measurements from rockets above the atmosphere:

$$S = 1360 \text{ W m}^{-2}$$

The discrepancy is due to approximating the Sun by a disc of a certain diameter radiating at a uniform temperature. In the literature the solar constant is often quoted as 2 ly min^{-1} (langleys per minute). The langley, defined as 1 cal cm^{-2}, is named after the 19th century scientist Langley who pioneered the study of solar radiation and its absorption in the Earth's atmosphere.

It follows that the total power emitted by the Sun is

$$4\pi R^2 S = 3.85 \times 10^{23} \text{ kW}$$

of which the Earth intercepts energy at the rate of

$$\pi R_\text{E}^2 S = 1.8 \times 10^{14} \text{ kW}$$

where R_E is the radius of the Earth. By comparison the present total world power consumption is close to 10^{10} kW, some four orders of magnitude less than the input of solar power.

The solar constant undergoes a slight seasonal variation. The Earth describes an ellipse about the Sun as one focus, although the eccentricity e of the orbit is extremely small—0.0167. Thus if a is the semimajor axis of the Earth's orbit, the shortest distance between the Sun and the Earth is given by $a(1 - e)$, or 1.47×10^8 km, while the largest distance is $a(1 + e)$, or 1.52×10^8 km. The Earth is closest to the Sun (perihelion) in early January and is furthest from the sun (aphelion) in early July. Figure 2.7 illustrates that during the

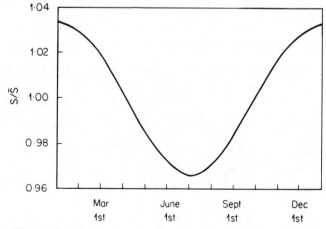

Figure 2.7. Fluctuations in the solar constant, S, caused by the eccentricity of the Earth's orbit around the Sun

course of the year there is an overall variation of nearly 7 per cent in the intensity of solar radiation received at the Earth.

This effect is almost completely swamped, however, by the changes in the pattern of radiation received over the globe as a result of the inclination of the Earth's axis to the plane of its orbit. If this angle were exactly 90°, then the Sun would be permanently overhead at the Equator, and the radiation received per unit surface area, the *insolation on a horizontal plane*, would vary across the globe as cos θ, where θ is the angle of latitude. However, the Earth's axis is actually inclined at an angle of approximately $66\frac{1}{2}°$ to the plane of its orbit which means that the 'effective equator' roams from latitude $23\frac{1}{2}°$N at the summer solstice (June 21st) to latitude $23\frac{1}{2}°$S at the winter solstice (December 21st).

We can describe the fluctuation in the incident solar radiation as a function of latitude, time of day and time of year with the aid of some spherical trigonometry. Consider the Globe as shown in Figure 2.8 with the Sun at some angle of declination δ which varies between $\pm 23°27'$ during the year. The diagram shows a point of observation O at latitude ϕ. The line OZ represents the local vertical at O, so the angle z is the zenith angle of the Sun. It follows that the rate at which solar radiation is passing through unit area parallel to the Earth's surface at O is given by $S \cos z$, where S is the solar constant. (We could also include a factor R_0^2/R^2 to allow for the variation in the Sun–Earth distance R about its mean value R_0, but we shall omit this for the sake of clarity.) The zenith angle of the Sun can be related to three other angles: the latitude ϕ, the solar declination δ and the 'hour angle' h which is the angle through which the Earth must rotate in order to bring the meridian of O directly underneath the Sun. In the diagram the plane of CZ and CS intersects the Earth's surface along a great circle, the arc OP being equivalent to the zenith angle z. By

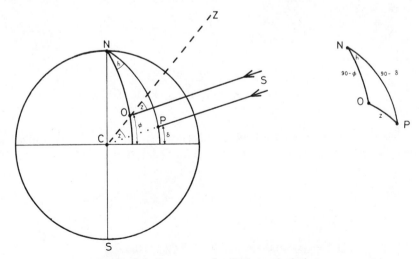

Figure 2.8. Geometry of radiation calculations

solving the spherical triangle NOP we can relate z to h, ϕ and δ. Thus, by a standard result,

$$\cos z = \sin \phi \sin \delta + \cos \phi \cos \delta \cos h \qquad (2.1)$$

The solar declination depends only on the time of year and not on the position of the point of observation, varying, as we have said, from $23°27'$ on June 21st to $-23°27'$ on December 21st. The hour angle h is zero at solar noon and will vary by $+15°$ ($360°/24$) for every hour away from noon. Thus at 15·00 hours the value of h is $+45°$.

Using equation (2.1) we can derive results for some interesting special cases. For example, at the poles $\phi = 90°$ and we find that $\cos z = \sin \delta$. Thus, z and δ are complementary, and the angle of elevation of the Sun equals the declination angle. During the polar day, the Sun merely circles around the horizon, always below $23\frac{1}{2}°$. For six months of the year each pole is in darkness, 'sunrise' or 'sunset' occurring at the spring or autumn equinoxes (March 21st and September 22nd), when the declination angle is zero. In a similar way, we can obtain an expression for the length of the day at different latitudes. At sunrise or sunset z is $90°$ and h may be written as $D/2$, where D is the angular length of the day. Therefore,

$$\cos (D/2) = -\tan \phi \tan \delta$$

Thus at the Equator D equals 12 hours throughout the year, but for other latitudes this is only true at the time of the equinoxes. The latitude above which there is either continuous day or night can be found by putting D equal to zero in this equation, so that the latitude of the polar night or day is the complement of the declination angle.

If we integrate equation (2.1) from sunrise to sunset, we can derive an expression for the total daily insolation I_d on a horizontal surface at the top of the atmosphere. Thus

$$I_d = \int_{\text{daylight}} S \cos z \, dt$$

$$= S \int_{\text{daylight}} (\sin \phi \sin \delta + \cos \phi \cos \delta \cos h) \, dt$$

We can change the variable of integration to h by introducing the angular velocity of the Earth about its axis; $\omega = dh/dt = 2\pi \, \text{rad day}^{-1}$. Therefore,

$$I_d = S \int_{-D/2}^{D/2} (\sin \phi \sin \delta + \cos \phi \cos \delta \cos h) \, dh/\omega$$

$$= \frac{S}{\omega} (D \sin \phi \sin \delta + 2 \cos \phi \cos \delta \sin D/2)$$

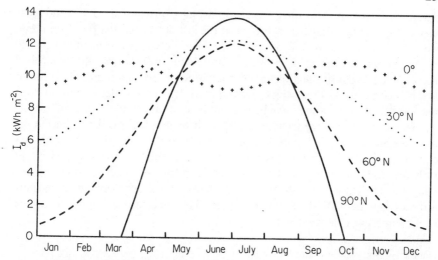

Figure 2.9. Daily insolation, I_d, throughout the year at various latitudes

with D expressed in radian measure. It is convenient to use the units kWh for energy, in which case our final result for the daily insolation on a horizontal plane is

$$I_d = 5 \cdot 2(D \sin \phi \sin \delta + 2 \cos \phi \cos \delta \sin D/2) \, \text{kWh m}^{-2}$$

Using this function it is possible to map out the variation of I_d at different latitudes during the year: this is illustrated in Figure 2.9. In fact the distribution is slightly asymmetric between northern and southern hemispheres owing to the greater proximity of the Sun in January. The maximum daily insolation, $13 \cdot 8$ kWh m^{-2}, occurs at the South Pole on December 21st.

2.4 The Atmosphere

In order to examine the solar radiation reaching the Earth's surface, we must now consider the effect of the atmosphere. First and foremost, the atmosphere provides us with a protective shield against harmful radiation, highly-energetic cosmic-ray particles and meteorites. An appreciation of its size can be gained from the fact that there is nearly one ton of atmosphere directly above each square foot of the Earth's surface.

From Figure 2.6 it is obvious that a sizable proportion of the radiation emitted by a body at 6000 K lies in the ultra-violet region of the spectrum. The energy carried by a photon of ultra-violet light (4 eV or more) is larger than typical chemical bond energies and can, therefore, bring about the destruction of the molecules which are essential to life processes. It is therefore fortunate, or indeed a necessary prerequisite for life as we know it, that our atmosphere has developed in such a way as to shield us from this radiation. We believe that the primordial atmosphere of the earth contained no free oxygen and was extremely transparent to ultra-violet radiation. Consequently, life could exist only in the sea. Once microorganisms in the sea developed the photosynthetic process (Chapter 3), oxygen was released into the environment and, because the metabolic processes of primitive organisms were primarily anaerobic, the oxygen content of the atmosphere increased. As oxygen molecules diffused upwards they were themselves decomposed by ultra-violet radiation into free oxygen atoms, some of which combined to form molecules of ozone, O_3. This gas has the property of strongly absorbing ultra-violet radiation so, once established in the higher reaches of the Earth's atmosphere, it could act as a stratospheric filter for this radiation. Now photosynthetic plants could grow on land and influence the development of the atmosphere to its present composition. Table 2.1 lists the constituents of the atmosphere at various altitudes together with their concentrations. The concentration of water vapour is, of course, extremely variable.

Table 2.1 Composition of the atmosphere

Average concentrations by volume throughout atmosphere for *dry air:*

Oxygen	21 per cent	Argon	0·9 per cent	Methane	1·5 ppm
Nitrogen	78 per cent	Neon	18 ppm	Carbon Monoxide	0·2 ppm
Carbon dioxide	320 ppm	Helium	5 ppm		
		Krypton	1 ppm		
		Xenon	1·1 ppm		

Moist air has a variable water vapour content:

Up to 4 ppm by mass near the surface	A constant proportion of 3 ppm by mass in the stratosphere (above 10 km)

Ozone is hardly present in the troposphere (below 10 km) but reaches a maximum concentration of 10 ppm by volume at a height of 25 km

The atmosphere has a second essential role to play in maintaining the Earth's surface in a condition that is hospitable to life. To illustrate this we shall calculate the *effective temperature* T_e of the Earth's surface on the basis of a radiation balance existing between the incoming solar energy and the energy lost, assuming that the Earth radiates as a black body at a temperature of T_e. The amount of solar radiation intercepted by the Earth is $\pi R_E^2 S$ and to allow for the *albedo* or reflectivity of the earth, denoted by α, we must include a factor of $(1 - \alpha)$, giving us an incoming flux of $\pi R_E^2 (1 - \alpha)S$. In equilibrium this must equal the outgoing flux $4\pi R_E^2 \sigma T_e^4$. Therefore,

$$4\pi R_E^2 \sigma T_e^4 = \pi R_E^2 S(1 - \alpha)$$

$$T_e = \left[\frac{(1 - \alpha)S}{4\sigma} \right]^{1/4}$$

Taking the average value of the planetary albedo to be 0·35 we arrive at a value for T_e of 253 K or $-20°C$. This is patently incorrect for we know that the average temperature of the Earth's surface is much higher than this (approximately 13°C). The reason for this error is that we have calculated the effective temperature for the composite Earth–atmosphere system. If the Earth can be treated as a black body whose temperature is just below 300 K, then the radiation which it emits will be mainly infra-red in character. This is just the region of the spectrum in which molecules such as CO_2 and H_2O are strongly absorbing (a point to which we shall return in due course). The radiation which they absorb is partly used to warm the gases in the atmosphere, but the rest is reradiated in all directions so that some of it is directed back towards the Earth's surface. Consequently, the surface remains warmer than we have predicted. For obvious reasons this is termed the *green-house effect*. Thus, in addition to protecting the surface of the Earth from harmful ultra-violet radiation, the atmosphere also acts as an insulating blanket for our planet.

Needless to say, we can never get something for nothing and, in return for the atmosphere's protective role, we must be tolerant of its behaviour as an absorber of visible light. We have already alluded to the absorption of light by gases such as carbon dioxide, water vapour and ozone, and we must now consider the question of the interaction between electromagnetic radiation and the constituents of the atmosphere in more detail. Of necessity this can only be a brief treatment but for further details of the theory of scattering and absorption of light the reader is referred to the texts on this subject listed in the bibliography. We begin with a preliminary discussion of the extinction of the solar beam, either by scattering or by absorption, as it passes through the atmosphere. In particular, we shall illustrate the treatment of the different path lengths through the atmosphere for various angles of elevation of the Sun.

Consider a beam of light of wavelength λ travelling through some medium such as our atmosphere. The intensity I_λ will vary with distance x according to

$$\frac{dI_\lambda}{dx} = -I_\lambda \rho e_\lambda$$

where ρ is the density of the medium and e_λ is the extinction coefficient per unit mass at wavelength λ. This equation may be integrated to yield

$$I_\lambda = I_\lambda^0 \exp\left\{ -\int_0^x \rho e_\lambda dx \right\}$$

where I_λ^0 is the original intensity of the beam and the integration is carried out along the ray path. (Of course, in our atmosphere, the material responsible for this intensity reduction varies in composition with both height and time.) For light travelling vertically through the atmosphere, the above equation becomes

$$I_\lambda = I_\lambda^0 \exp(-e_\lambda^T)$$

with e_λ^T, the total extinction coefficient in the vertical direction, defined as

$$e_\lambda^T = \int e_\lambda \rho dh$$

For a more general path we define the relative air mass as

$$m_r = \frac{\int e_\lambda \rho \, ds}{\int e_\lambda \rho \, dh}$$

s being measured along the path, and I_λ becomes

$$I_\lambda = I_\lambda^0 \exp(-e_\lambda^T m_r)$$

Alternatively, this equation may be written in terms of the transmittivity, t_λ, of the atmosphere

$$\frac{I_\lambda}{I_\lambda^0} = (t_\lambda)^{m_r}$$

The calculation of m_r, the relative air mass, is by no means straightforward unless we assume some simplified model for the structure of the atmosphere. Thus for an atmosphere in which there are horizontal strata of constant composition, the secant of the zenith angle for the ray path in each stratum is equal to ds/dh and is, therefore, a measure of m_r. For the simplest case, a homogeneous atmosphere of constant density, the height H of the atmosphere may be defined by $p_0 = \rho_0 g H$, where p_0 and ρ_0 are the atmospheric pressure and density at the surface of the Earth. This leads to a value of 8430m for H at 15°C. If R is the radius of the Earth, then a simple calculation shows that

$$m_r = \{(R/H)^2 \cos^2 z + 2R/H + 1\}^{1/2} - (R/H) \cos z \qquad (2.2)$$

Nevertheless, atmospheric refraction and the decrease in density with height must be allowed for, but the calculation becomes exceedingly difficult. It is interesting though that experimental measurements from balloons lead to similar values of m_r to those given by equation (2.2). The variation of m_r with zenith angle, on the basis of equation (2.2), is shown in Figure 2.10. These results do not, of course, apply to materials such as water vapour or ozone which are distributed in certain layers of the atmosphere. In the case of such materials, equation (2.2) can be applied provided that z is taken as being the zenith angle at the altitude of the appropriate layer, and H is interpreted as being its thickness.

As we have said, the two principal mechanisms by which the intensity of the solar beam may be reduced are scattering and absorption. The extinction coefficient may be written as the sum of two components, the scattering coefficient σ_λ and the absorption coefficient α_λ. Scattering occurs when small particles in the atmosphere redirect the solar beam. If we assume that the particles are small (say, an order of magnitude smaller than the wavelength of the light concerned) then the theory of Rayleigh scattering is applicable. Light is a form of electromagnetic radiation and as a light wave passes a small

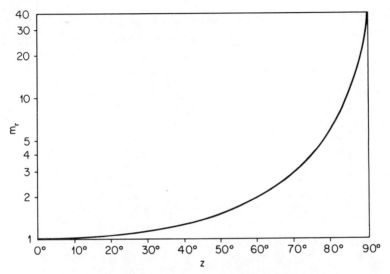

Figure 2.10. Relative air mass, m_r, as a function of zenith angle, z

particle the oscillating electric field induces in the particle (or atom or molecule) an oscillating dipole moment p. The dipole, oscillating with the angular frequency ω equal to that of the light wave, radiates power P according to

$$P = \frac{2\,\omega^4 p^2}{3\,\varepsilon_0 c^3} \propto 1/\lambda^4$$

This power must be derived from the original beam so the amount of radiation scattered by the particles is inversely proportional to the fourth power of the wavelength. This is not entirely correct; a more detailed analysis includes the refractive index n_λ (which varies slightly with wavelength) in the expression for the scattering coefficient per unit length:

$$\sigma'_\lambda = 32\,\pi^3 (n_\lambda - 1)^2 / 3\ N\lambda^4$$

Here N is the number of scattering centres per unit volume. Figure 2.11 shows the variation of σ'_λ with wavelength, together with the variation of t_λ, the fraction of the energy in a vertical beam which penetrates a Rayleigh atmosphere.

This theory is indeed consistent with our everyday observations of the sunlight we receive on Earth. Light scattered from the solar beam is almost entirely responsible for the skylight that we observe. In a pure Rayleigh atmosphere in which only air molecules are responsible for the scattering effects, the sky appears to have a blue colouration as light in the blue region of the spectrum is more easily scattered than the red. As the path length of the Sun's rays through the atmosphere increases, a greater fraction of the blue component of solar radiation is scattered by the larger number of molecules along the ray path. Consequently, the Sun appears to become redder while the sky appears more blue. The effect is most marked, of course, when the Sun is low in the sky,

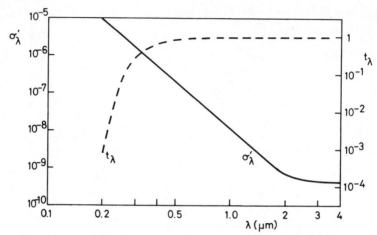

Figure 2.11. Wavelength dependence of the scattering coefficient per unit length, σ'_λ, and the transmittivity, t_λ for a Rayleigh atmosphere

as at sunset. In the presence of large particles, however, light of all colours is scattered to a greater extent with the result that the sky appears less blue and will eventually appear to be white when a sufficiently large number of large particles is present, as in a cloud of water droplets or ice particles.

Whereas scattering takes place at all wavelengths, absorption is a somewhat selective process. The molecules in the atmosphere which are mainly responsible for the absorption of solar radiation are H_2O, CO_2, N_2, O_2 and O_3, while free atoms of oxygen and nitrogen also have a part to play. The visible region is relatively free from absorption bands whereas substantial absorption occurs in the ultra-violet, due to O_2, N_2, O_3, N, O, and in the infra-red, due to H_2O, CO_2. To see why this should be so we must briefly consider the processes by which atoms and molecules absorb electromagnetic radiation.

Quantum theory asserts that electrons in atoms can occupy only certain well-defined energy levels. If a photon has an energy exactly equal to the separation of two such levels it can excite an electron from the lower to the upper level and become absorbed. After some time the excited electron will return to its original level, or to other unoccupied levels, radiating in a random direction the energy which it loses so that the net result is a depletion of the intensity of the original light beam from which the photon was absorbed. When two or more atoms are brought together to form a molecule, the individual atomic levels interact to form a more complex energy level pattern for the molecule as a whole. Nevertheless, the principles of photon absorption remain the same. Absorption due to molecules such as O_2 and N_2 occurs in the ultra-violet and, as the absorption coefficients associated with electronic transitions are usually large, much of the ultra-violet solar radiation is absorbed in the upper layers of the atmosphere. Some of the absorbed energy goes into heating these regions while some goes into the formation of ionized layers.

The lowest ionized layer in the atmosphere, the D-layer, lies between 60 and 90 km high. It is believed that this is formed by photochemical reactions involving nitric oxide, NO. Ionization of the NO molecule occurs when radiation of energy in the range 10–12 eV is incident upon it:

$$NO + hv \longrightarrow NO^+ + e$$

The E-layer, lying above the D-layer at a height of 100 km, is formed by X-rays of about 40 Å wavelength which originate in the solar corona. The ozone layer in the upper atmosphere is also formed by photochemical reactions. The set of equations

$$O_2 + hv \longrightarrow O + O$$

$$O_2 + O + M \longrightarrow O_3 + M$$

$$O_3 + hv \longrightarrow O_2 + O$$

$$O_3 + O \longrightarrow 2\,O_2$$

summarizes the steps involved in the creation and destruction of ozone molecules (the symbol M represents a third atom or molecule whose presence is necessary in order to conserve energy and momentum). The atoms or ions formed in these processes can absorb radiation at even shorter wavelengths than their parent molecules so that the net result of all these absorption processes is to ensure that practically all solar radiation of wavelengths less than 300 nm is prevented from reaching the Earth's surface. Figure 2.12 shows the spectrum of solar radiation in the ultra-violet as observed from the ground.

In addition to electronic energy levels, molecules also have energy levels

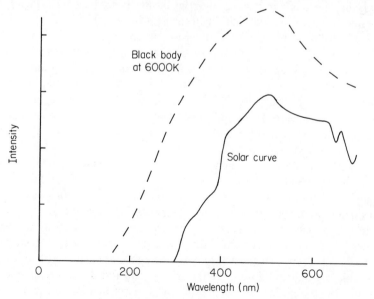

Figure 2.12. Visible and ultra-violet solar spectrum at ground level

associated with the vibrational and rotational motion of their nuclei. (Because of the disparity between the nuclear and electronic masses, it is reasonable to treat the motion of the nuclei and the electrons separately.) For a molecule with moment of inertia I, quantum mechanics predicts that the permitted rotational levels are given by the expression $\hbar^2 j(j+1)/2I$, where j is an integer quantum number and $\hbar$ represents Planck's constant divided by 2π. Theory also predicts that transitions between two such energy levels are only possible if the j values of the two levels differ by unity. Thus the molecule may absorb photons whose energies satisfy

$$E = (\hbar^2/I)(j+1)$$

giving rise to an absorption spectrum made up of a number of lines at equally spaced energies. For a typical diatomic molecule with an internuclear distance of one or two Angstroms, I is of the order of 10^{-47} kg m^2 and, substituting this into the above equation, we find that such a molecule can absorb photons with energies of about 5×10^{-2} eV, corresponding to wavelengths of 25 μm. Although only an estimate, this does suggest that rotational spectra should only be observed in the far infra-red.

A vibrating molecule can be treated as a harmonic oscillator with some natural frequency ω_0. (Of course, a given molecule may have a number of such natural frequencies.) The permitted energy levels of such an oscillator are given by $(n + \frac{1}{2})\hbar\omega_0$, where n is an integer quantum number, and the selection rules are such that transitions can only occur between levels whose n values differ by unity. For each ω_0 then, the absorption spectrum of a molecule consists of a single line at energy $\hbar\omega_0$. The magnitude of the natural frequency depends on the nuclear masses and the bond strengths, and typically these are such that the absorption lines lie in the near infra-red.

Typical absorption spectra for the gaseous constituents of the atmosphere are shown in Figure 2.13 together with the complete solar spectrum in the visible and infra-red regions as observed from ground level. The major absorbers of solar radiation are seen to be carbon dioxide and water vapour. Carbon dioxide has strong absorption bands centred at 2·7, 4·3 and 15 μm, as well as some weak absorption at 2 μm. Water vapour is strongly absorbing in the vicinity of 1·4, 2·0, 2·7 and 6·3 μm. Similarly there is almost total absorption of the solar beam beyond 20 μm due to the rotational bands of water. The vibrational bands of molecular oxygen absorb strongly at 0·76 μm while the ozone layer in the stratosphere absorbs at 9·6 μm. In addition, a careful inspection of Figure 2.13 reveals the slight absorption due to the minor constituents of the atmosphere: carbon monoxide, methane and nitrous oxide.

Figure 2.13 also shows clearly the presence of the so-called 'atmospheric windows' in the solar spectrum. Thus there is a region of high transparency between 7 and 13 μm, interrupted only by the ozone absorption band at 9·6 μm. It is in this region, therefore, that the many infra-red devices available today (for example, the guidance systems of heat-seeking missiles or, less violently, night photography systems) must operate if they are to be effective.

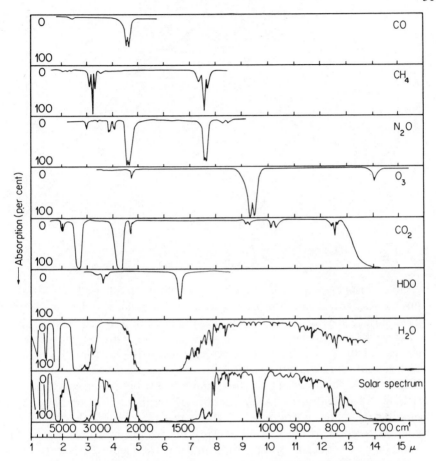

Figure 2.13. Visible and infra-red spectra of atmospheric gases and of the sun from ground level. (J. T. Houghton and S. D. Smith, *Infra-red Physics*, 1966. Reproduced by permission of the Clarendon Press, Oxford)

We now turn from this somewhat qualitative description of atmospheric effects and examine the overall radiation balance in the Earth–atmosphere system, with the obvious implications that this has for the climate of our planet.

2.5 The Radiation Balance in the Atmosphere

The total incident radiation may be represented as the sum of a number of components as follows.

$$Q_T = (Q_d + Q_s)(1 - A) + (Q_d + Q_s)A + Q_r + Q_a \qquad (2.3)$$

In this equation Q_d and Q_s are the direct and scattered radiation reaching the Earth's surface, Q_r and Q_a are the amounts of radiation reflected and absorbed

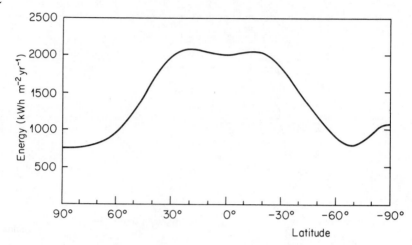

Figure 2.14. Latitudinal variation of the average annual solar radiation received at the ground

by the atmosphere, and A is the albedo or reflectivity of the Earth's surface. Thus the first term in equation (2.3) represents the amount of radiation absorbed by the surface and the second term gives the amount reflected by the surface. In Figure 2.9 we saw the variation in solar insolation with latitude, neglecting the effect of the atmosphere. Now we can present the data for the radiation which actually does reach the surface.

Figure 2.14 illustrates the average annual distribution of solar radiation reaching the ground as a function of latitude. Similarly, Figure 2.15 shows a 'contour map' of the World in which the 'contours' represent the annual solar radiation received at the ground. There are a number of comments which should be made concerning these diagrams. Firstly, Figure 2.14 clearly shows that equatorial regions receive the most energy during the year, as we should expect. But it also shows the secondary maximum at the South Pole which is a result of the extremely transparent Antarctic atmosphere. Again, in Figure 2.15, we see that maximum values of the annual radiation budget are to be found in the principal desert areas of the World. Thus the Sahara Desert of North Africa, the deserts of Pakistan and North America, and perhaps the central regions of Australia are all encircled by the 200 contour. This means that up to three-quarters of the incident solar energy actually penetrates the atmosphere in these regions. At the other end of the scale, less than half this amount is available in Arctic regions.

To digress for a moment, it is obvious that the variations in solar radiation received are of great importance for climate and hence for agriculture (as will be discussed in more detail in the following chapter). They also influence energy consumption across the Globe, a point of paramount importance for the theme of this book and yet so obvious that it need hardly be made. During the winter months, European and North American countries have to consume a large portion of their fuel stocks simply to keep warm. (For example, it has been

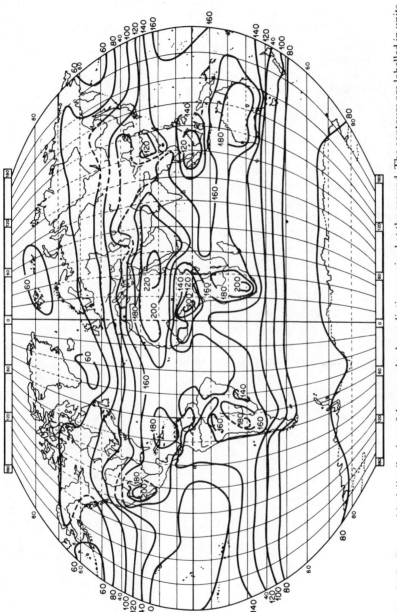

Figure 2.15. Geographical distribution of the annual solar radiation received at the ground. The contours are labelled in units of kly yr^{-1}, or kcal cm^{-2} yr^{-1}. These can be converted to kWh m^{-2} yr^{-1} by multiplying by 11.7 (W. D. Sellers, *Physical Climatology*, 1965. Reproduced by permission of the Chicago University Press Copyright © 1965 The University of Chicago Press)

estimated that in the United Kingdom some 40 per cent of electrical power production is used for heating during the peak demand period.) By contrast, many of the so-called under-developed countries are in much warmer climes and can, therefore, concentrate a high proportion of their energy resources on the production of power for industrial and agricultural use.

What are the relative magnitudes of the various terms in equation (2.3)? Let us consider the yearly figures averaged over the Globe. The term Q_r, representing the radiation reflected by the atmosphere, accounts for approximately 30 per cent of the incident radiation. This figure is made up of about 5 per cent due to air molecules, dust and water vapour in the atmosphere, and some 25 per cent due to reflection by clouds. The reflectivity of clouds is usually 50 per cent but may be as large as 90 per cent in the case of the tall cumulus clouds common in summer. Approximately 20 per cent of the Sun's energy is absorbed by clouds and atmospheric constituents—the term Q_a. This leaves a total of 50 per cent which actually reaches the Earth's surface, 20 per cent having been scattered downwards by the atmosphere to reach the ground as diffuse shortwave radiation (Q_s), and 30 per cent reaching the surface directly (Q_d).

Of the radiation which reaches the Earth's surface, some is absorbed and some is reflected. The reflected radiation (unlike the re-emitted infra-red radiation from the Earth) is mostly lost to space as it is mainly in the visible and not easily absorbed by the atmosphere. Some 5 per cent of the incident radiation is lost in this way so, together with the 30 per cent from Q_r, this leads to an overall planetary albedo of 35 per cent. The breakdown of solar radiation into these various components is illustrated schematically in Figure 2.16.

The reflectivity of the Earth's surface is extremely variable. Thus for a snow or

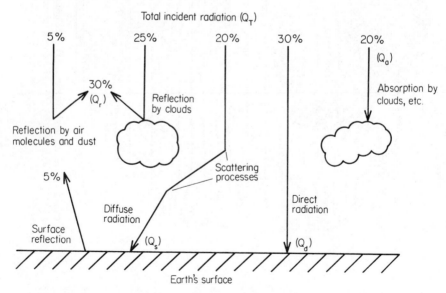

Figure 2.16. Breakdown of solar radiation in the Earth–atmosphere system

ice-covered surface the reflectivity can be as high as 50 per cent. This drops to 25 per cent for dry sand and to 15 per cent for wet sand, forest or grassland. (These are only average reflectivities for wavelengths of less than 4 μ.) The reflectivity of a water surface is surprisingly low, as we can show using Fresnel's formula for normal incidence:

$$R = \frac{(n-1)^2}{(n+1)^2} \times 100 \text{ per cent}$$

As the refractive index n for water is about 1·33, this leads to a reflectivity of only 2 per cent. However, this can be increased threefold if the water is agitated, and can be as high as 17 per cent for the reflection of diffuse radiation.

The surface of the Earth is, of course, radiating energy by virtue of its temperature. As this is close to 300 K the radiation is predominantly in the infra-red, as we have seen before. Of course, the emissivity of the surface will vary from place to place but, on average, one square metre of the Earth's surface will radiate 600 kWh in infra-red energy each year. In the same time it will receive 1400 kWh of energy, which means that there is a net positive radiation balance for the Earth's surface of 800 kWh yr^{-1} m^{-2}. The radiation balance of the complete Earth–atmosphere system must average to zero through the year, so that for each square metre of ground area there must be a net negative radiation balance for the atmosphere of 800 kWh yr^{-1}.

The atmosphere acts as a heat sink at all latitudes whereas the surface acts as a heat source except near the Poles. Thus we find a positive radiation balance for the whole Earth–atmosphere system near to the Equator (less than about 40° latitude) and a negative balance near the Poles. To counteract this imbalance, energy must, therefore, be transferred from the Equator to the Poles. Overall then, there must be a two-way heat transfer: from the surface to the atmosphere in most regions and from low to high latitudes. There are two ways in which this can be achieved; by conduction or convection, that is by the transfer of sensible heat; or by the transfer of latent heat.

The transfer of heat from the surface to the atmosphere is accomplished in a number of ways. Firstly, some of the infra-red radiation emitted by the surface is absorbed by the water vapour and carbon dioxide in the atmosphere which is, therefore, heated. Secondly, there is straight-forward convective cooling of the surface. Most important of all, however, is the transfer of latent heat. The Sun's energy causes water to evaporate from the surfaces of rivers, lakes and seas. This water vapour is carried high into the atmosphere before condensing and releasing its latent heat of vaporization.

The transfer of heat from low to high latitudes works on essentially the same principles. Because the Earth's surface is warmer at the Equator than the atmosphere above it and also warmer than at the Poles, convection currents are set up in which warm air rises in equatorial regions and descends in polar regions. Thus convection currents not only transfer heat from surface to air but also from the Tropics to the Poles. Latent heat is also transferred in this way: the air rising from the Equator is very moist and gives out its latent heat of vaporization en route to the Poles. The circulation is completed by a flow of air from the Poles to the equator along the surface of the Earth.

In addition to these mechanisms, heat can also be transferred by means of ocean currents. It is generally true that such currents (e.g. the Gulf Stream) transport warm water from the band of ocean between latitudes $\pm 20°$ to higher latitudes where the most significant heat losses occur between 50 and 70°. As much as 20 per cent of the meridional heat transfer may be due to this mechanism.

2.6 Climatic Effects of Power Production

The climate of the Earth has been changing continuously over a very long period of time without man's help. The evidence of previous Ice Ages, when the average temperature of the Earth is thought to have been 8 or 9°C less than is now the case, is enough to remind us of that. Moreover, meteorologists are of the opinion that there have been severe variations in the climate of the Earth since the last Ice Age. It would appear that there have been four climatic epochs during which the climate has differed substantially from that pertaining to the present day. During the post-glacial warm period of c. 5000–3000 BC vegetation belts were displaced polewards and the equatorial monsoon belt was widened. This was followed by the cold climatic epoch of the early Iron Age culminating at about 900–450 BC. The growth of peat bogs during this cold epoch indicates that there was a sharp increase in rainfall. Thus peat-bog sections from Ireland or Northern Europe exhibit a colour change from the black lower peat to the lighter upper layers. The early Middle Ages, c. 1000–1200 AD, marked the appearance of a secondary climatic optimum, which was in turn followed by the 'Little Ice Age' of c. 1430–1850 AD. On a much smaller scale we know that the average temperature of the Earth rose by about 0·4°C between 1900 and 1945, since when it has decreased somewhat. Recent analysis of the data obtained from weather satellites suggests that this trend is well established and that the snow and ice cover of the Earth increased by more than 10 per cent during the years 1967–1972.

Fluctuations in climate can be identified with a number of factors, such as alterations in the output of radiation by the Sun or in the Earth's orbit. Apart from these potential external factors we can expect climatic change to be related to variations in the composition of the atmosphere, in the reflectivity of clouds and of the Earth's surface, and in the balance between water in the ice caps, oceans and atmosphere, not to mention the effect of any changes induced in the patterns of circulation in the atmosphere or oceans. It is obvious that before the extent of man's influence on climate can be estimated, the detailed interaction and balance between the various natural influences must be understood.

When such large vacillations in weather conditions as we have described can occur without man's influence, it is reasonable to ask if it is really necessary to worry about any future part he may play in climatic change. But it may well be that man's activities could cause instabilities to occur which could lead to very large perturbations in the radiation balance. For example, a slight rise in the average temperature of the Earth caused by man's power production and consumption would lead to an increased water vapour content in the atmosphere, thus enhancing the greenhouse effect and increasing the temperature still further. Similarly, a reduction in temperature brought about by air pollution would lead to an extension of the polar ice caps and, as ice reflects more incident radiation than water, this would increase the planetary albedo and lead to a further decrease in temperature. That such positive feedback mechanisms could occur means that it is prudent to examine the effect that man's power production is having on the radiation balance in the Earth–atmosphere system and to attempt to predict the future consequences of his actions.

Many different pollutants are produced during power production. The combustion of fossil fuels releases carbon dioxide, water vapour and smaller quantities of other gases such as sulphur dioxide and the oxides of nitrogen. Of these, carbon dioxide and water vapour are the most important in their influence on climate. In addition, combustion, and especially incomplete combustion, leads to the emission of dust particles into the atmosphere. Nuclear reactors present the problem of dealing with the highly radioactive waste material which they produce but this is not our immediate concern. (This will be dealt with in Chapter 10.) However, in common with fossil fuel power stations, they release huge quantities of waste heat, and this may have an effect on local and global climate. Indeed, all of the energy consumed by man is eventually returned to the environment as heat.

Although carbon dioxide has only a concentration of 320 ppm by volume in the atmosphere, it has, as we have seen, an important role to play in regulating the Earth's surface temperature. That climatic change might result from a variation in this carbon dioxide concentration was recognized as long ago as 1863 by the British physicist Tyndall. Nevertheless, it was not until the International Geophysical Year of 1958 that the first systematic observations of this quantity were made. It would appear that the concentration of carbon dioxide is increasing annually by some 0·2 per cent (0·7 ppm in 320) and that the concentration at the North Pole remains greater than that at the South Pole by approximately 1 ppm, which is equivalent to an 18 month time lag between the concentrations at these locations. The mean variation of carbon dioxide concentration since 1958 is shown in Figure 2.17.

It is not necessarily true, of course, that this increase in carbon dioxide concentration is a direct result of fossil-fuel combustion. It has been estimated (United Nations, 1956) that up to 1950 the burning of fossil fuels produced a quantity of carbon dioxide equivalent to 10 per cent of that present in the atmosphere in 1950. (The 1950 data were obtained by extrapolating the results of Figure 2.17.) Analysis of tree-rings and other biological specimens of known age indicates a 1–2 per cent decrease in C^{14} concentration over the same period. C^{14}, a radioactive isotope of carbon with a half-life of 6000 years, is

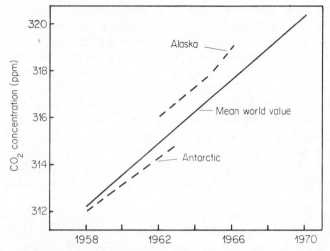

Figure 2.17. Atmospheric concentration of carbon dioxide. (Reprinted from SCEP report, *Man's Impact on the Global Environment*, 1970, by permission of the M.I.T. Press, Cambridge, Massachusetts. Copyright © 1970. The Massachusetts Institute of Technology Press)

produced in the atmosphere when C^{12} atoms are bombarded by cosmic rays. Consequently, some C^{14} is taken up by plants but will be virtually non-existent in fossil fuels which have been deprived of contact with the atmosphere for so long a time. Therefore, burning fossil fuels dilutes the C^{14} in the atmosphere and the reduction in C^{14} concentration, the Suess effect, is indeed consistent with fossil-fuel consumption (Suess, 1965).

Data for carbon dioxide production and accumulation in the years 1950–1960 are given in Table 2.2. It is known that the concentration is increasing by 0·7 ppm each year, which corresponds to a 5500 Mt increase during 1960. Some 10,800 Mt were introduced into the atmosphere that year, so it would appear that approximately 50 per cent of the carbon dioxide produced is actually being retained by the atmosphere. The rest is presumably taken up by the biosphere and ocean reservoirs. Living matter will respond to an increase in carbon dioxide concentration by increasing its rate of photosynthesis, unless limited by nutrients (see Chapter 3), and will grow faster. Similarly, ocean water can take up carbon dioxide in the form of carbonates or simply as dissolved gas. Very little is known about the rate at which these reservoirs can accept carbon dioxide or indeed about their total capacity. At present we can only assume that the ocean and biosphere will continue to account for half of the carbon dioxide which is released into the atmosphere.

A model based upon this balance between the reservoirs for carbon dioxide and using the projected growth rates for fuel consumption suggests that the atmospheric concentration of carbon dioxide could have increased from 320 to 380 ppm by the year 2000 (Table 2.2). More dramatically, if we then assume

Table 2.2 Carbon dioxide in the atmosphere

Year	Amount added from fossil fuel (Mt yr^{-1})	Cumulative amount over previous decade (Mt)	Concentration by volume (ppm)	Total Amount (10^6 Mt)
1950	6,700	52,200	306	2·39
1960	10,800	82,400	313	2·44
1970	15,400	126,500	321	2·50
1980	22,800	185,000	334	2·61
1990	32,200	268,000	353	2·75
2000	45,500	378,000	379	2·95

Existing and potential concentrations

(Reference SCEP Report 1970. Reproduced by permission of the M.I.T. Press.)

that all the known reserves of fossil fuel were to be burned, it would be possible for man to increase the carbon dioxide concentration by a factor of four or more by the next century. Needless to say, the implications of this must be investigated.

To understand the implications of such changes it is necessary to develop reasonable mathematical models for describing global climate. The physics of the problem is well understood. To describe the atmosphere the pressure, temperature, density, air velocity and moisture content must be specified at every point. These quantities can be related by four fundamental physical laws: Newton's Second Law, the First Law of Thermodynamics, the Gas Equation of State and the Continuity Law for Mass. To these, restraints and boundary conditions must be added, such as the incident flux of solar radiation. This is fine in theory but it must be realized that the atmosphere is capable of change over a few millimetres or over many kilometres, within a few seconds or within years, and that it contains an infinite number of particles whose motion must be described. It is by no means surprising, then, that the current models are so crude.

Nevertheless, some simple calculations have been attempted relating the change in surface temperature to variations in the carbon dioxide concentration. As long ago as 1896 Arrhenius estimated that a rise in surface temperature of 9°C would result from a threefold increase in carbon dioxide concentration. (It is interesting to note that in order to estimate the importance of carbon dioxide absorption, he used data obtained by Tyndall for the absorption of moonlight passing through the atmosphere at various angles of elevation!) Modern calculations are indeed in rough agreement with his prediction. Manabe (1970) assumed a fixed degree of relative humidity for his model and found that doubling the carbon dioxide concentration led to an increase in the surface temperature of 2·4°C. On the basis of the data in Table 2.2, this implies an increase of 0·5°C by the year 2000.

Of course, at the same time as he is injecting large quantities of carbon dioxide into the atmosphere, man is also using up atmospheric oxygen to burn fossil fuels. However, no noticeable change in the concentration of oxygen has been recorded this century and it seems that there is a very stable equilibrium between the photosynthetic production of oxygen (Chapter 3) and its consumption by living creatures.

Balanced against this trend towards warmer conditions is the effect of increasing atmospheric turbidity as a consequence of the emission of dust particles in burning fuel and in other industrial processes. Natural dust emitters—volcanoes—do, of course, contribute to the turbidity from time to time in a most spectacular fashion, but man's influence cannot be discounted. Particles in the atmosphere affect our weather in two ways: by blocking out the Sun's rays and hence cooling the Earth, and by providing nuclei on which water vapour can condense and form clouds. One famous example of increased rainfall from atmospheric pollution is at La Porte, U.S.A., sited 30 miles downwind from the steel-works in South Chicago which emit huge quantities of smoke. Meteorological records for the early 1960's show that La Porte had nearly half as many hail and thunderstorms again as surrounding districts.

Fortunately the atmosphere is very efficiently cleansed of these particles by snow and rainfall. By using radioactive tracer techniques to follow the movement of atmospheric dust, it is found that particles in the troposphere, the lower atmosphere, usually reside there for not more than two weeks before being washed down to the surface. On the other hand, if particulate emissions do reach as high as the stratosphere they can remain there for many months, or even years, and their influence on global climate is by no means understood. Large concentrations of dust in stratospheric regions would contribute to the heating of those regions by absorption of some solar radiation, and would, therefore, deprive the lower atmosphere and the Earth's surface of much of its heat input.

The hyperbolic cooling towers at power stations emit large quantities of steam and water vapour into the atmosphere. In stable weather conditions it is possible for these vapours to become trapped under an inversion layer and thus generate dense fog. Of course, power stations generate water vapour not only during the cooling process, but also in the burning of hydrocarbons; this is also true of domestic heating systems, factories and motor vehicles. During cold winter conditions the atmosphere may not be able to absorb all this water vapour, in which case foggy weather results.

Waste heat from power stations, factories and so on is as much a contaminant as are gases and particles. A single motor vehicle, capable of producing 40 bhp (approximately 30 kW), is only about 25 per cent efficient and is, therefore, emitting some 90 kW of waste heat. (It must be remembered that the 30 kW of useful power also finds its way eventually into the environment as heat). There are two ways in which heat enters the environment, either directly as sensible heat or as latent heat, for example, of water evaporated from cooling towers. In the latter case, water vapour is transported high into the atmosphere before condensing and releasing its latent heat. Meteorologists know that heat sources such as large urban areas can affect local climate, but the question remains as to whether or not they could influence global climate. In a recent study (1970) of the 4000 square mile area of the Los Angeles basin, Lees has calculated that this area generates thermal energy at the rate of 5 per cent of the solar power received at the ground. It is estimated that in North America $2 \cdot 2 \times 10^6$ MW of waste heat are produced per annum, and that this could rise to $7 \cdot 5 \times 10^6$ MW by the year 2000. Corresponding world figures are shown in Table 2.3. Large as these figures may seem, when distributed over the entire globe they are but insignificant amounts.

Table 2.3 Waste heat production

	Assumed yearly increase (per cent)	1970 (10^6 MW)	1980 (10^6 MW)	2000 (10^6 MW)
World	5·7	5·5	9·6	31·8
North America	4	2·2	3·4	7·5
Central America	6	0·12	0·2	0·68
South America	6	0·09	0·16	0·5
West Europe	4	1·08	1·6	3·5
West Asia	10	0·05	0·13	0·81
Far East	10	0·44	1·1	7·2
Australasia	8	0·07	0·145	0·64
Africa	6	0·1	0·18	0·57
East Europe	7	1·4	2·7	10·4

(Reference United Nations, Statistical Papers Series J, No. 14, *World Energy Supplies 1966–1969*, United Nations New York 1971. Reproduced by permission of the United Nations.)

In this section we have only been able to sketch an outline of the many ways in which power production may influence the global climate. However, this should be enough to emphasize the importance of the task now facing the atmospheric and environmental scientists in attempting to predict the eventual climatic trends which will result from the many conflicting influences exerted by man.

2.7 Summary

Apart from describing the basic energy flows on which the Earth depends, this chapter lays some important groundwork for the rest of this book. The analysis of the radiation incident upon the surface of the Earth and its spectral composition will be of importance in the discussion of photosynthesis (Chapter 3). The geographical distribution of this radiation has immense implications for agriculture and general considerations of growth rates. Similarly, it is of considerable importance in deciding whether the application of solar power is feasible in a given locality (Chapter 12).

References

Arrhenius, S., *Phil. Mag.*, **41,** 237 (1896).
Bethe, H. A., *Phys. Rev.*, **55,** 434 (1939).
Lees, L., in SCEP report (see below) (1970).
Manabe, S., in *Global Effects of Environmental Pollution*, S. F. Singer, (Ed.), D. Reidel, Holland, 1970.
SCEP, 'Man's Impact on the Global Environment', a report by the Study of Critical Environmental Problems work group, M.I.T. Press, Cambridge, Mass. 1970.
Suess, H. E., *J. Geophys. Res.*, **70,** (1965).
Tyndall, *Phil. Mag.*, **25** (1863).
United Nations, 'World energy requirements', *Proc. Int. Conf. on Peaceful Uses of Atomic Energy*, Vol. 1, UN Department of Economic and Social Affairs, New York, 1956.

Further Reading

Bates, D. R., (Ed.), *The Planet Earth*, Pergaman, Oxford, 1964.
Lipson, S. G. and H., *Optical Physics*, Cambridge University Press, Cambridge, 1969.
Robinson, N., (Ed.), *Solar Radiation*, Elsevier, Amsterdam, 1966.
SCEP Report, *Man's Impact on the Global Environment*, M.I.T. Press, Cambridge, Mass., 1970.
Sellers, W. D., *Physical Climatology*, University of Chicago Press, Chicago, 1965.

3 Photosynthesis

Were it not for photosynthesis, the trapping and storing of the Sun's energy by green plants, life as we know it on this planet, and consequently the fossil fuels which man uses, could not have existed. The energy obtained by burning these fuels is essentially the solar energy that was stored in their parent plants millions of years ago. In later chapters we shall examine how deposits of coal and oil were formed from the fossil remains of these ancient plants but for the moment we shall concentrate on the photosynthetic process itself. Our account will of necessity be brief and, save for the barest details, the experimental basis of the currently accepted theories will be absent.

At first glance photosynthesis is a deceptively simple process. Plants grow by using light energy to convert carbon dioxide and water into organic matter with the evolution of oxygen. The elements of water account for three-fifths of the growth while most of the remainder is constituted from carbon dioxide which, as we have seen, is present in our atmosphere only to the extent of 320 ppm. Of course, many photochemical reactions are known, but the important point about photosynthesis is the fixing of light energy into stable organic products. It is worth noting that the light energy is used to convert materials of low chemical potential energy (H_2O and CO_2 are already fully oxidized) into organic matter of high chemical potential energy.

That photosynthesis only takes place in the green parts of plants has been known since the 18th century. This was first reported by Ingen-Housz in the first major publication on the subject of photosynthesis (1779). Nearly 60 years were to elapse before it was shown that the green colour of plants was confined to intracellular organelles known as *chloroplasts*. Meanwhile, in 1782, Senebier had demonstrated that the presence of carbon dioxide was necessary before the photosynthetic reaction could take place. The first proof that carbohydrates were the end product of photosynthesis was provided by Sachs in 1862, who, after exposing one half of a leaf to light while keeping the other half dark, exposed the whole leaf to iodine vapour. The illuminated half changed colour to dark violet indicating the presence of starch iodine complex. The energy stored in these carbohydrates represents the total energy stock from which plants and animals extract their requirements. It was indeed noted by Ingen-Housz that photosynthesis in plants is superimposed on the reverse process of respiration, which is the slow conversion of organic matter to carbon dioxide and water with release of energy.

3.1 Evolution of the Photosynthetic Process

In Chapter 2 we mentioned the role of photosynthesis in the development of the Earth's atmosphere to its present composition. That organisms utilizing this process could have developed in the primordial atmosphere is not immediately obvious, so, in order to see how photosynthesis evolved, we must look at the evolution of life itself. What follows is but a cursory account of the presently accepted theory of the origin of life; in any case, this is still somewhat conjectural.

As the Universe consists mainly of hydrogen and helium, it is reasonable to suppose that the Earth's atmosphere was originally made up of these two gases. However, because of their extreme lightness and consequent rapid diffusion, molecules of these gases were able to escape from the gravitational pull of the planet over a period of time. Next the Earth developed a secondary atmosphere of a strongly reducing nature. Any water present was dissociated by the Sun's ultra-violet radiation into hydrogen and oxygen, the latter reacting with reducing substances to form, for example, metal oxides. Carbon dioxide was formed by volcanic activity or during meteorite impact. Meteorites could generate enough heat on impact to cause the iron and water which they contained to be converted into iron oxide and hydrogen; the iron oxide could oxidize free carbon to carbon dioxide and carbon monoxide. Carbon dioxide, once formed, did not persist in the atmosphere but reacted with the free hydrogen present to form methane and other hydrocarbons. Similarly hydrocarbons, ammonia and hydrogen sulphide were produced by the reactions between water and metal carbides, nitrides and sulphides.

Thus the secondary atmosphere was essentially composed of methane, ammonia, hydrogen and water with smaller amounts of hydrogen sulphide, higher hydrocarbons, nitrogen and carbon monoxide. Carbon dioxide was not present (having been reduced to non-volatile carbonates), nor was oxygen in any significant quantity. It is from these components that life must have evolved. The theories of Haldane and Oparin suggest that the formation of organic molecules from inorganic matter was quite probable in the environment just described. When organic molecules of sufficient complexity were produced it was then possible for a living system to evolve.

This theory received considerable support from the experiments of Miller (1953, 1955). He subjected mixtures of hydrogen, water, methane and ammonia to an electrical spark discharge (simulated lighting) for a prolonged period of time. Analysis of the mixture then revealed the presence of a number of biologically important molecules (for example, the amino acids glycine, alanine and sarcosine) together with carbon dioxide, carbon monoxide, nitrogen and various polymers. Further work has demonstrated that this abiogenic synthesis occurs in reducing atmospheres of widely varying compositions and that many complex organic compounds can be produced. It is also significant that the synthesis can be carried out using ultra-violet light to which the secondary atmosphere described above is transparent.

According to Haldane and Oparin, the first organisms were heterotrophic rather than photosynthetic, that is they obtained their energy by anaerobic conversion of available organic compounds. Without other processes which could produce organic material this early life form would probably have failed to survive, and this may indeed have happened a number of times before pigments such as the porphyrins appeared. The magnesium–porphyrin complex chlorophyll is capable of promoting a photochemical reaction driven by visible light and in which there are no back reactions. This means that an accumulation of energy in stable compounds can be achieved. As we shall see later, this is the essence of photosynthesis. Once living organisms containing these pigments had developed in the sea, where they were protected from the Sun's ultra-violet radiation, the oxygen they evolved led in time to a protective layer of ozone in the atmosphere (see Section 2.4). Now similar organisms could develop on land.

The time scale of this evolutionary process can be estimated from geological evidence. We believe that the Earth's atmosphere contained no free oxygen until the Pre-Cambrian era, about 10^9 years ago. Fossils similar to the bacteria and algae of today have been dated back to 1.6×10^9 years and it appears that oxygen production in photosynthesis developed about 10^9 years ago. Higher organisms such as corals and sponges appeared only just before the Cambrian era, 6×10^8 years ago.

3.2 Structural Aspects of Photosynthesis

The structure of a typical chloroplast is depicted schematically in Figure 3.1. This shows the chloroplast surrounded by a unit membrane and containing an inner membranous structure. The pigment in the chloroplasts is concentrated in regions called *grana*. The spinach chloroplast, for example, has some 50 grana, each about 500 Å in diameter. In cross section (also Figure 3.1) the grana are seen to be stacks of flattened closed sacs or thylakoids, some of which extend from granum to granum. In some lower plants, however, these grana do not exist.

The chloroplasts are thought to be the reaction centres for the photosynthetic process. Thus it is found that when the plant is illuminated, oxygen gas is evolved only from the chloroplasts and not from the surrounding colourless regions, as was first proved by Engelmann in 1883 with the aid of bacteria which tend to migrate towards concentrations of oxygen. Similarly, carbo-hydrates are found to accumulate in the chloroplasts. The most convincing evidence for this viewpoint is undoubtedly the work of Trebst, Tsujimoto and Arnon who succeeded, in 1958, in preparing isolated chloroplasts which could photosynthesize completely.

The most important photosynthetic pigments are the *chlorophylls* which are present in all photosynthetic cells (with the exception of photosynthetic bacteria which contain the related pigment *bacteriochlorophyll*). These chloro-phylls are the phytol esters of dihydroporphyrin carboxylic acids. The structures

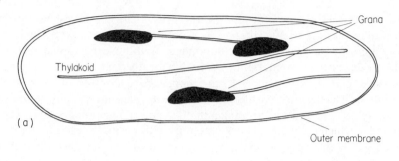

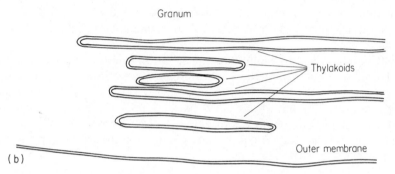

Figure 3.1. (a) Schematic representation of a typical chloroplast; (b) cross-sectional view illustrates the formation of grana

of chlorophylls a and b, first elucidated by Fischer in 1940, are shown in Figure 3.2. The porphyrin structure, a nearly flat arrangement of four pyrrole rings about a magnesium atom, is similar to that found in haemoglobin, but with a magnesium atom instead of an iron atom at the centre of the ring. Attached to this ring structure is a long hydrocarbon side-chain of phytol which can rotate relative to the porphyrin ring. The phytol residue is soluble in lipids so it is presumably embedded in the phospholipid of the membrane structure. Chlorophylls a and b are the predominant pigments in all green plants and algae.

Also present in all photosynthetic cells are a number of *carotenoids*, similar to the pigment in carrot roots. Carotenoids are usually orange or yellow but this colour is often masked by chlorophyll. There are two major groups of carotenoids: the *carotenes* which are hydrocarbons and the *carotenols* (or *xanthophylls*) which are alcohols or ketones. The common carotenes are all stereoisomers with the common molecular formula $C_{40}H_{56}$. The structure of β-carotene is shown in Figure 3.2. The oxygen in the carotenols is contained in hydroxyl, carbonyl or carboxyl groups attached to the ring groups of β-carotene.

Chlorophylls absorb light in two regions of the spectrum, the blue–violet and the red. The blue absorption band, the Soret band, occurs in all porphyrin

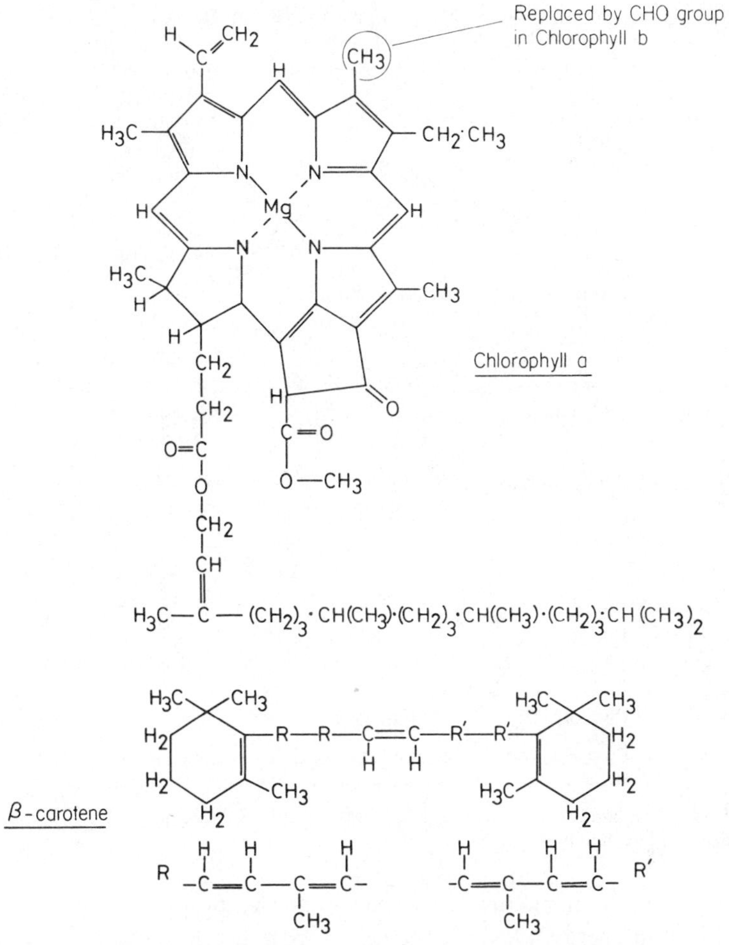

Figure 3.2. Chemical structure of the photosynthetic pigments chlorophyll a, chlorophyll b and β-carotene

derivatives but the absorption in the red is a property of chlorophylls only. Carotenoids, on the other hand, only absorb light in the blue region of the spectrum. Absorption peaks for various common pigments are listed in Table 3.1. There is considerable evidence for the existence of several types of chlorophyll a in plants and this appears to hold generally for all pigments. However, if extraction of these pigments is attempted, only one type can be isolated outside the cell.

3.3 Chemical Aspects of Photosynthesis

There are essentially three phases to the basic energy cycle of life: assimilation and storage of solar energy, extraction of energy from the store and utilization

Table 3.1 Absorption maxima of the photosynthetic pigments

	Pigment	Occurrence	Absorption maxima in organic solvents (nm)	Absorption maxima in cells (nm)
Chlorophylls	Chl a	All photosynthetic plants	420, 660	435, 675
	Chl b	Higher plants green algae	453, 643	480, 650
	BChl a	Bacteria	365, 605, 770	800, 850, 890
Carotenoids	α-Carotene	Major carotene of red algae	420, 440, 470 (hexane)	—[a]
	β-Carotene	Major carotene of most plants	425, 450, 480 (hexane)	—[a]
	Luteol	Green leaves, green and red algae	425, 445, 475 (ethanol)	—[a]
	Fucoxanthol	Brown algae	425, 450, 475 (ethanol)	—[a]
Phycobilins	Phycoerythyrin	Red algae		490, 546, 576
	Phycocyanin	Blue-green algae		618

[a] The absorption bands of carotenoids in cells are difficult to locate due to the strong overlap of the chlorophyll blue bands.

of this energy. Solar energy is trapped in the chloroplasts and is stored as chemical bond energy by the reduction of carbon dioxide to large molecules such as carbohydrates, lipids and proteins. The energy is taken up by the many new chemical bonds that must be formed in these large molecules. Plant and animal cells can extract the energy stored in fats, carbohydrates and proteins by the process of respiration but they cannot do so simply by releasing the stored energy as heat. Energy transfer by thermal means, although the most common physical mechanism, is not practical in biological systems which are usually isothermal. Therefore, energy transfer in cells takes place by the exchange of chemical bonds, or electrons, together with their energy, between neighbouring molecules. In order to transfer energy from the large molecules produced by photosynthesis to a particular region of the cell where energy is required for some purpose, the energy is first stored again in smaller, more manageable molecules which can easily be sent off to the region in question.

Adenosine triphosphate, ATP, is the most ubiquitous of these mobile, high energy content molecules. As depicted in Figure 3.3 it is highly negatively charged when in a neutral solution (pH 7). Energy is stored in forming this

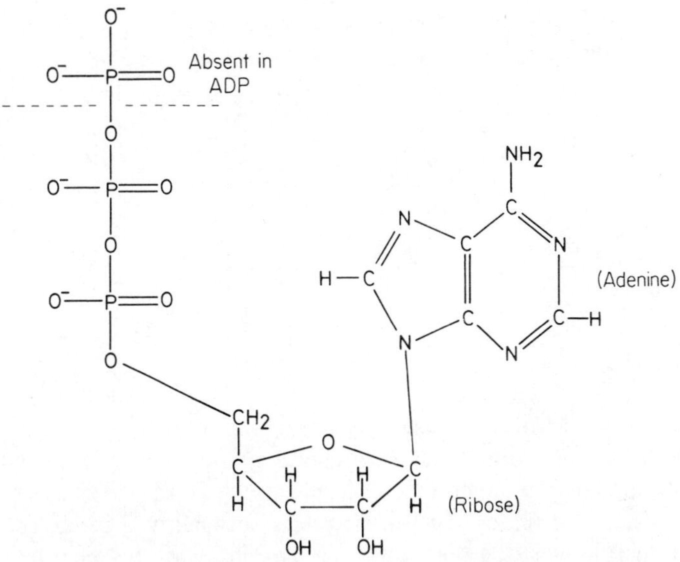

49

Figure 3.3. Chemical structure of adenosine triphosphate, ATP. In neutral solutions ATP is highly charged as illustrated in the diagram

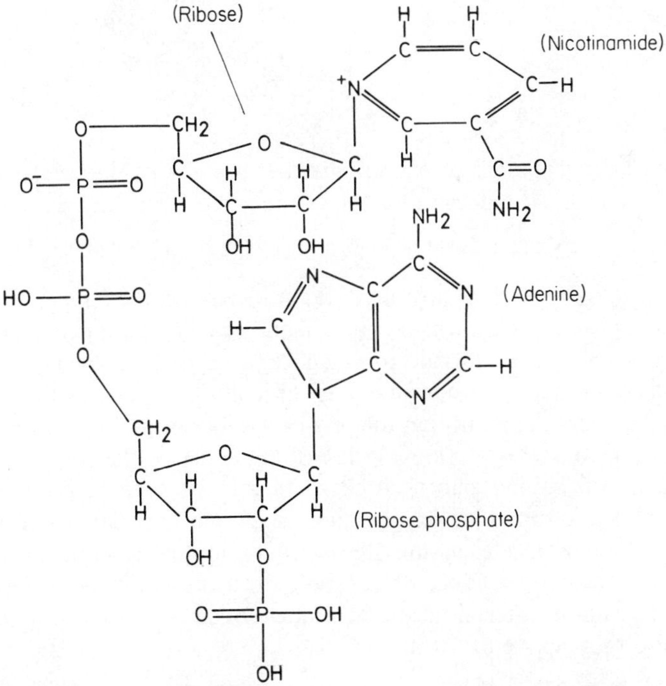

Figure 3.4. Chemical structure of nicotinamide adenine dinucleotide phosphate, NADP

high concentration of charge in the pyrophosphate bonds during the production of ATP from ADP, the diphosphate. Enzymatic breakdown of the unstable terminal link in ATP releases the stored energy once again. Thus the hydrolysis of ATP into ADP and phosphoric acid yields 0·3 eV per molecule (8 kcals per mole):

$$H_2O + R-O-\overset{\overset{\displaystyle O}{\|}}{\underset{\underset{\displaystyle OH}{|}}{P}}-OH \quad \rightleftharpoons \quad R-OH + HO-\overset{\overset{\displaystyle O}{\|}}{\underset{\underset{\displaystyle OH}{|}}{P}}-OH + 0{\cdot}3 \text{ eV}$$

$$\text{(ATP)} \qquad\qquad \text{(ADP)}$$

Another molecule of immense importance in photosynthesis is NADP, nicotinamide adenine dinucleotide phosphate, shown in Figure 3.4. There is a definite similarity between this molecule and ATP in that one of the phosphate groups can be removed with the release of energy. The molecule plays an important role in redox reactions, readily converting between its oxidized form, $NADP^+$, and the reduced form NADPH:

$$\text{(NADP}^+) \qquad + \ 2e \ + \ H^+ \ \rightleftharpoons \qquad \text{(NADPH)}$$

We have already alluded to the apparent simplicity of the photosynthetic reaction. Thus the basic equation for the process may be written as

$$CO_2 + H_2O + nh\nu \longrightarrow (CH_2O) + O_2 \tag{3.1}$$

where (CH_2O) represents a carbohydrate molecule (for example $C_6H_{12}O_6$ glucose) and n gives a measure of the overall efficiency of the process. However, the conversion of carbon dioxide to sugar requires some 5 eV per molecule and, as a photon of red light has an energy of just under 2 eV, and one of blue light about 3 eV, it is apparent that the process cannot be a direct one.

We shall conclude this section by describing some of the important clues to the mechanism behind photosynthesis. In all photosynthetic reactions it is found that the number of oxygen molecules evolved equals the number of carbon dioxide molecules assimilated so it is tempting to conclude that the carbon dioxide is merely reduced to carbohydrate while giving up its oxygen content. This simple interpretation was questioned when the reaction was studied using radioactive O^{18} tracers:

$$2\,H_2O^{18} + CO_2 \longrightarrow H_2O \ + (CH_2O) \ + O_2^{18}$$

$$2\,H_2O \ + CO_2^{18} \longrightarrow H_2O^{18} + (CH_2O^{18}) + O_2$$

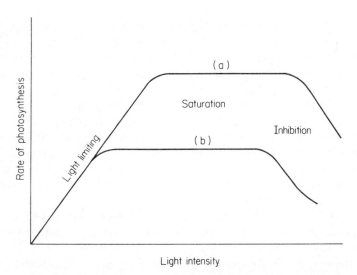

Figure 3.5. Photosynthetic rate as a function of light intensity. Curve (a) corresponds to a higher concentration of available carbon dioxide than does curve (b)

Thus we can deduce that the water molecules are split and that the evolved oxygen is derived from the water.

Considerable insight into the mechanism can also be achieved from studies of reaction kinetics. For low light intensities the reaction rate only increases slightly with increasing temperature, as might be expected for a photochemical reaction. For high light intensities, on the other hand, the rate is a rapidly changing function of temperature, in common with most chemical reactions which double their rate every $10°C$ or so. It would seem, therefore, that there are in fact two reactions involved, a photochemical reaction and a 'dark' reaction. At low light intensities the overall rate is effectively limited by the photochemical reaction. Figure 3.5 illustrates the reaction rate as a function of light intensity for different concentrations of carbon dioxide. At low intensities the rate is limited by the amount of light available but at higher intensities saturation occurs, implying that the dark reactions are proceeding as quickly as possible but are unable to keep pace with the light supply. The higher carbon dioxide concentration raises the saturation plateau by increasing the rate of the dark reaction. The inhibitory effect at very high intensities represents damage caused to the plant cells by the high light flux.

The presence of two reactions can also be inferred from experiments involving the use of flashing lights. It was first demonstrated by Warburg in 1919 (using the unicellular algae *chlorella*) that for a given amount of light the photosynthetic yield is actually greater when the illumination is intermittent. This has a simple interpretation in terms of the dual-reaction theory: the rate limiting dark reactions catch up with the light reactions during the 'off' periods thus giving a greater overall rate.

Warburg and Negelein in 1922 were the first to measure the number n in equation (3.1). They found that only four quanta of visible light, of any wavelength, were necessary to reduce one molecule of carbon dioxide. As the reduction of carbon dioxide requires some 5 eV per molecule and as four quanta of red light carry about 7·5 eV, an efficiency of 70 per cent is implied for the process. This appears to be somewhat high and, indeed, there has since been considerable debate as to the correct value. The measurement is exceedingly difficult as it must necessarily be carried out at very low light intensities when the corrections for respiration are large and unpredictable. It is now generally accepted though that at least eight quanta are needed per molecule of carbon dioxide. This implies that at least two quanta are used for the transfer of each hydrogen atom from water to carbon dioxide.

Figure 3.6 shows the absorption spectrum of chlorophyll together with its action spectrum in situ. The action spectrum (maximum quantum yield as a function of wavelength) is much flatter than the absorption curve. This is, of course, attributable to the presence of other pigments besides chlorophyll (Section 3.2). The action spectrum does show a dip in the blue, the region in which carotenoids are the major light absorbers. This suggests that carotenoids are not as effective as other pigments in photosynthesis. Moreover, the action spectrum also drops off at long wavelengths. This is particularly noticeable in red algae where the 'red drop' occurs in the centre of the main absorption band of chlorophyll a, at 650 nm. Emerson has shown that the red drop can be removed by simultaneous illumination with shorter wavelength light (see, for example, Emerson (1958)). Thus oxygen productivity from chlorella using combined beams of 650 and 700 nm light is greater than the sum of the producti-

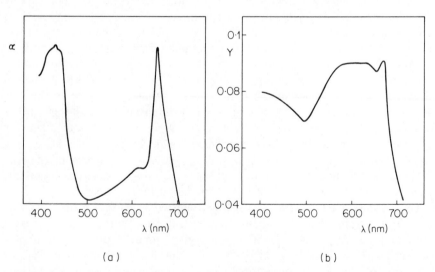

Figure 3.6. (a) Absorption spectrum of chlorophyll. (b) Action spectrum—quantum yield, Y, as a function of wavelength, λ, of chlorophyll. (R. Emerson and C. M. Lewis, *Am. J. Bot.*, **30,** 165, 1943)

vities for the separate beams. This enhancement effect was to lead to the idea of two pigment systems driving two separate light reactions, a model to which we shall return in Section 3.4. The enhancement occurs even if the far-red and shorter wavelength lights are incident as alternating flashes several seconds apart.

The flashing light experiments described earlier also led to another important aspect of photosynthesis coming to light. In 1932 Emerson and Arnold showed that not more than one molecule of carbon dioxide per 2500 molecules of chlorophyll is reduced by a single intense flash. Assuming a quantum efficiency of eight, this implies the existence of a *photosynthetic unit* comprising 300 chlorophyll molecules. Granular units of approximately the correct size for accommodating this number of molecules have been observed in electron micrographs of chloroplasts and have been dubbed *quantasomes*.

Perhaps the most important step forward in our understanding of photosynthesis was made by van Niel (1962) with his comparative studies of photosynthetic bacteria and green plants. Bacteria that contain the pigment bacteriochlorophyll can grow using light energy but without evolving oxygen. In addition, they can only assimilate carbon dioxide in the presence of a suitable reducing agent such as hydrogen sulphide or certain organic compounds. The typical reaction in this case can be written as

$$CO_2 + 2\,H_2A \longrightarrow (CH_2O) + H_2O + 2\,A$$

where H_2A represents the reducing agent. By analogy with this process, van Niel proposed that green-plant photosynthesis could be described by

$$CO_2 + 2\,H_2O \longrightarrow (CH_2O) + H_2O + O_2$$

instead of the simpler equation (3.1). Thus he suggested that there are two parallel processes in photosynthesis, oxidation of water and reduction of carbon dioxide. From his studies of other photosynthetic bacteria he concluded that neither of these processes is exclusive to green-plant photosynthesis. The step that is unique is the photochemical reaction involving chlorophyll that forms separate oxidizing and reducing agents to drive the remainder of the process.

3.4 Mechanism of Photosynthesis

We are now in a position to describe the currently accepted models for the photosynthetic mechanism. For convenience, we can divide the basic process into four parts:

(1) the light reaction leading to the formation of oxidizing and reducing agents,

(2) oxidation of water to oxygen,

(3) carbon dioxide intake and its reduction to carbohydrates,

(4) interaction of oxidants and reductants to provide stored energy in the form of ATP.

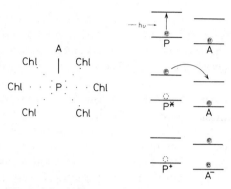

Figure 3.7. A photosynthetic reaction centre.
P represents a specially sensitized chlorophyll
molecule and A a suitable electron acceptor

The most fundamental part of the photosynthetic process is the photo-chemical reaction. The various biochemical changes are initiated at photo-chemical reaction centres, first identified by Duysens in 1952. He showed that when photosynthetic bacteria are illuminated, there is a slight decrease in the intensity of their long wavelength absorption band. The absorption rapidly returns to normal in the dark. Oxidation of a small portion of the total bacterio-chlorophyll is responsible for this reversible bleaching, and it is this fraction which is specially sensitized as a reaction centre. Most of the pigment acts merely as a light gatherer. It is thought that all reaction centres are of the form shown in Figure 3.7. The sensitizing pigment P is a specialized chlorophyll (or bacteriochlorophyll) close to a suitable electron acceptor, A. When exposed to light, P becomes oxidized and A reduced.

Preparations enriched in reaction centres have been obtained using suspensions of broken cells and their absorption spectra have been studied. The light induced oxidation of the pigments is now particularly noticeable. Thus in the bacterium *rhodopseudomonas spheroides* the photochemically active pigment is P870, 870 nm being the wavelength of maximum absorption. Analogous pigments are found in all photosynthetic organisms. In green plants and algae the main pigment is P700, probably chlorophyll a. Beyond this the nature of a pigment such as P700 is uncertain: chemically we can only say that P700 is a chlorophyll molecule in an environment suitable for photochemical reactions.

It is equally true that the nature of the acceptor, A, is not yet understood. As Figure 3.7 shows, excitation and electron transfer in a reaction centre leads to the primary photochemical products P^+ and A^-. Before the reaction can take place again these must return to their original states P and A. The direct back reaction is wasteful so the return is via some more circuitous route. The recovery rate of the complete system has been measured by observing photo-synthesis in rapidly flashing lights. It is found that the dark reactions can keep pace with 25 reactions per second in the reaction centres.

In reality the specialized reaction centres are surrounded by additional chlorophyll molecules which provide a common light gathering system. This system, which has been evolved to make full use of the available light, may be linked, of course, with the photosynthetic unit described earlier and containing 300 chlorophyll molecules. Thus the photosynthetic unit comprises the photo-chemical reaction centre together with the associated light gathering pigment. Within this unit, energy absorbed by one chlorophyll molecule can easily be transferred to the reaction centre, either by electron migration or by excitonic means. Thus the chlorophyll molecule's delocalized system of alternate single and double bonds facilitates electron transfer over the dimensions of a single molecule and also between neighbouring molecules whose π-orbitals overlap. Excitonic transfer involves the migration of the excitation energy throughout the molecular assembly. Such a mechanism is well known in solids. The excited electron and the associated 'hole' left behind in the ground state migrate together through the molecules carrying their excitation energy with them.

At this point we must return to the comment made in the previous section concerning the existence of two photochemical systems in photosynthesis. There are three major pieces of evidence which support this view. Firstly, it should be possible, if there is a single photochemical reaction, to obtain equal rates of photosynthesis for different wavelengths of light simply by adjusting the relative intensities of the two light sources. (Only the absorption coefficients and efficiency of energy transfer should vary with wavelength.) This can indeed be done for sustained illumination but switching from one light source to another leads to 'chromatic transients' in the rate of photosynthesis (Blinks effect). Secondly, there is the red-drop effect (Section 3.3) and corresponding enhancement. Thirdly, light absorbed by phycobolin pigments was found to be exceedingly efficient for photosynthesis. Studies of the energy transfer between phycobolin and chlorophyll molecules suggested that there are two forms of chlorophyll: an active form which can receive energy from phycobolin and an inactive form which cannot. Nevertheless, the inactive form becomes active when 'enhanced' by shorter wavelength light.

The concept of two light reactions was reinforced by investigations of *cytochromes* in plants. A cytochrome (Cyt) is an iron-containing tetrapyrrole molecule, *heme*, covalently bonded to a low molecular weight protein. (The heme structure is similar to that of chlorophyll.) The cytochromes are effectively electron-transport enzymes, converting between their oxidized and reduced states as they transport electrons from substance to substance. They have a key role to play in respiration but, as shown by Hill in 1954, they are also found in all photosynthetic cells. The connection with photosynthesis was clarified by Duysens in 1961 when he demonstrated the effect of light of different wavelengths on the redox state of Cyt f. In darkness Cyt f is in a reduced state but far-red light (685 nm) oxidizes it. However, if shorter wavelength light is used together with the red light, the Cyt f partly returns to its reduced form. These different wavelengths are, as we have seen, associated with the different pigment systems so it is reasonable to suppose that the cytochromes act as electron carriers between the two photochemical systems.

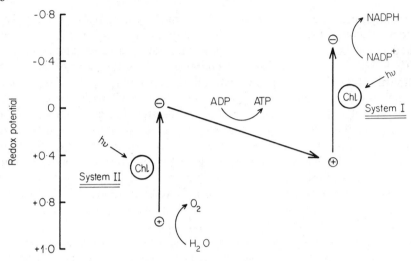

Figure 3.8. The 'z scheme' for photosynthesis

These observations were to provide considerable confirmation for the so-called *z scheme* for photosynthesis illustrated in Figure 3.8. The vertical scale in this diagram is a measure of the redox potential. The more negative the oxidation potential is, the higher is the electron energy and the stronger is the reducing power. The overall separation between the strong oxidant made by system II and the strong reductant made by system I is 1·4 eV which could be provided by a single quantum of red light but instead two photochemical steps are used. The electron transfer from the top of system II to the bottom of system I releases sufficient energy (0·5 eV) to produce one molecule of ATP from ADP. Thus two quanta, one trapped by each photochemical system, can transfer one electron through the whole 'z' while making a molecule of ATP. This means that eight quanta can transfer four electrons through the 'z' to form two molecules of NADPH and four molecules of ATP, sufficient for the needs of the Calvin cycle for carbon assimilation (see below) and with an ATP molecule to spare. In reality, though, phosphorylation is less efficient than this but the scheme still agrees with the known quantum efficiency of about eight.

The red-drop and enhancement effects can now be understood in terms of this z scheme. Far-red light only actuates system I but short wavelength light is capable of driving both systems. Enhancement occurs when the additional short wavelength light causes system II to catch up with and run at the same rate as system I, thus leading to maximum efficiency. Evidence to support this z scheme has been obtained by a number of techniques. For example, chemical isolation of the two systems can be achieved with the aid of various poisons that prevent the flow of electrons between them. Similarly, some degree of physical isolation has been obtained via centrifugation. Mutant algae have also been developed which lack certain cytochromes with the result that one of the photochemical systems is inoperative.

The incorporation of carbon dioxide into photosynthetic organisms and its conversion into carbohydrate with the aid of a suitable reductant (such as NADPH) provided by the photochemical reactions is perhaps the best understood part of the photosynthetic mechanism. This is in the most part due to the use of radioactive C^{14} as a tracer element, a technique ingeniously utilized by Calvin and his collaborators in the 1950's. For his work in elucidating the path of carbon in photosynthesis Calvin was awarded the 1961 Nobel Prize for Chemistry. The original technique involved the use of suspensions of the unicellular alga chlorella. These were allowed to reach a steady state of photosynthetic activity in the presence of ordinary carbon dioxide before being permitted a short period of photosynthesis in the presence of labelled carbon dioxide. Chromatographic techniques were then employed to ascertain the reaction products.

Many such reaction products were identified, but, by reducing the period of exposure, it was established that an increasing proportion of C^{14} was found in molecules of phosphoglyceric acid (PGA—see equation (3.2)). Thus, PGA would appear to be the first stable product during carbon assimilation. It was also found that the C^{14} atoms appeared in the carboxyl group of the PGA, implying that this had been produced by a carboxylation reaction. Further studies confirmed that this was the case and that the primary carbon dioxide acceptor was ribulose diphosphate (RDP), which is then converted into PGA as follows:

$$
\begin{array}{l}
\text{H}_2\text{C}-\text{O}\cdot\cdot\text{P} \\
|\\
\text{H C}-\text{OH} \\
|\\
\text{H C}-\text{OH} \quad \xrightarrow{\text{CO}_2} \\
|\\
\text{C}=\text{O} \\
|\\
\text{H}_2\text{C}-\text{O}\cdot\cdot\text{P} \\
\\
\text{(RDP)}
\end{array}
\qquad
\begin{array}{l}
\text{H}_2\text{C}-\text{O}\cdot\cdot\text{P} \\
|\\
\text{H C}-\text{OH} \\
|\\
\text{C}=\text{O} \quad \longrightarrow \\
|\\
\text{HOO C}-\text{C}-\text{OH} \\
|\\
\text{H}_2\text{C}-\text{O}\cdot\cdot\text{P}
\end{array}
\qquad
2
\begin{array}{l}
\text{H}_2\text{C}-\text{O}\cdot\cdot\text{P} \\
|\\
\text{H C}-\text{OH} \\
|\\
\text{COOH} \\
\\
\text{(PGA)}
\end{array}
\qquad (3.2)
$$

The next step in the reaction sequence involves ATP and NADPH, both produced from the initial photochemical reactions in the chloroplasts. PGA is first phosphorylated by ATP in the presence of a kinase enzyme to give diphosphoglyceric acid (DPGA) which is then reduced by NADPH to triose phosphate (TP).

$$
\begin{array}{l}
\text{H}_2\text{C}-\text{O}\cdot\cdot\text{P} \\
|\\
\text{H C}-\text{OH} \quad \xrightarrow{\text{ATP}} \\
|\\
\text{COOH} \\
\\
\text{(PGA)}
\end{array}
\qquad
\begin{array}{l}
\text{H}_2\text{C}-\text{O}\cdot\cdot\text{P} \\
|\\
\text{H C}-\text{OH} \quad \xrightarrow[\text{H}^+]{\text{NADP}} \\
|\\
\text{O}=\text{C}-\text{O}\cdot\cdot\text{P} \\
\\
\text{(DPGA)}
\end{array}
\qquad
\begin{array}{l}
\text{H}_2\text{C}-\text{O}\cdot\cdot\text{P} \\
|\\
\text{H C}-\text{OH} \quad + \quad \text{NADP}^+ + \text{P}\cdot\cdot\text{OH}\\
|\\
\text{H C}=\text{O} \\
\\
\text{(TP)}
\end{array}
$$

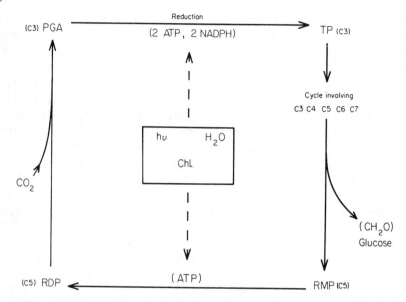

Figure 3.9. The Calvin cycle for the assimilation of carbon in photosynthesis

The triose phosphate then enters into a complex cycle involving compounds with between three and seven carbon atoms per molecule and, eventually, glucose and ribulose monophosphate (RMP) are formed. The overall cycle is completed when RMP is converted into RDP with the aid of ATP.

Figure 3.9 is an extremely simplified version of the Calvin–Benson cycle for the path of carbon in photosynthesis. As indicated, the cycle is fuelled by the light energy which is converted in the chloroplasts into high energy phosphate, ATP, and reducing power, NADPH. The cycle uses two molecules of NADPH and three molecules of ATP for each molecule of carbon dioxide fixed. It should be noted that the incorporation of carbon dioxide into the cycle does not require light energy but rather that it does depend on the presence of ribulose diphosphate. The Calvin–Benson cycle is beautifully simple in that it involves only one reduction process and one carboxylation, but this can be no guarantee of its uniqueness and there are indeed many reasons for supposing that other, more complex cycles do exist in nature. Thus, tracer studies have established the existence of many substances among the reaction products (e.g. malic acid, glycolic acid and various amino acids) which do not appear in the Calvin–Benson cycle.

The initial acceptor in the Calvin–Benson cycle is ribulose diphosphate but a cycle based on phospho-enol-pyruvate (PEP) has been discovered in a number of plants (for example, sugar cane, maize and crabgrass). PEP reacts with carbon dioxide to form malic acid and aspartic acid, both four-carbon compounds. (As PGA, the initial product in the Calvin–Benson cycle, is a three-carbon compound, we can, therefore, distinguish the two systems as C3 or C4.) However, unlike their counterpart in the Calvin–Benson cycle, these molecules cannot be converted into carbohydrate directly. Instead they are broken down enzymatically to yield a three-carbon compound, pyruvic acid, with the release of carbon dioxide. The pyruvic acid is converted by the action of ATP back into PEP, thus closing the cycle. The released carbon dioxide is now assimilated by the plant's Calvin–Benson cycle and leads to carbohydrate formation.

The reason for this apparently unnecessary complication to the C3 cycle may be understood by examining the reactivities of PEP and RDP with carbon dioxide. PEP is much more reactive with carbon dioxide than is RDP and the latter's activity is actually inhibited by high oxygen concentrations. Consequently, under conditions of low carbon dioxide concentration and high oxygen concentration, the C4 pathway is actually more efficient than the C3 pathway given sufficient light intensity. In a hot arid environment with the leaf stomata closed to prevent water loss (Section 3.5) C4 plants can continue to photosynthesize without a supply of carbon dioxide through the stomata. The C4 pathway can utilize the low concentrations of carbon dioxide in the leaf spaces and can effectively transport the carbon dioxide into the inner parts of the cells where it is used in the Calvin–Benson cycle. It is indeed found that C4 plants are better photosynthesizers than C3 plants when there is intense solar radiation and a low humidity.

By contrast with this development in our understanding of the metabolic pathways involved in carbohydrate formation, the mechanism of oxygen evolution is virtually a mystery, and not much more is known about the assimilation of oxygen in respiration! It is known that the mechanism of oxygen production is enzymatic and that the enzyme system requires the presence of manganese. It may be the case that the Mn^{2+} ions, complexed with enzyme proteins, are oxidized photochemically to Mn^{3+} ions which are then capable of oxidizing water molecules in dark reactions.

To complete this account of the photosynthetic mechanism, we now examine the light-induced production of ATP, *photophosphorylation*. This phenomenon was first observed in illuminated chloroplasts by Arnon in 1954. He distinguished between two types of photophosphorylation: cyclic and non-cyclic (Figure 3.10). Non-cyclic photophosphorylation is associated with the transfer of an electron through the z scheme, as we have already noted. Cyclic photophosphorylation is associated with a back reaction in photochemical system I. Such a back reaction is always likely to occur inasmuch as the whole aim of the photosynthetic process is to drive electrons against the potential gradient. Cyclic photophosphorylation saves some of the energy waste in these back reactions by coupling them to ATP production.

Electron transport in photosynthesis is accompanied by the movement of ions, noticeably hydrogen ions, across the thylakoid membranes with a consequent change in the electric potential across the membrane. It is found that by controlling either this ion flow or the potential change the production of ATP can be stopped. Thus it seems that photophosphorylation and electron transport are coupled. Early theories on the coupling mechanism have concentrated

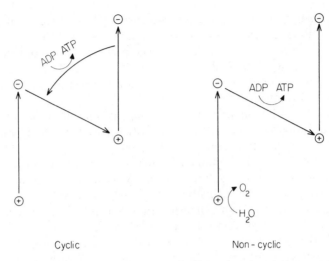

Cyclic   Non - cyclic

Figure 3.10. Cyclic and non-cyclic photophosphorylation

on chemical coupling. Without entering into details, the link is provided by some intermediate whose chemistry involves both redox and hydration reactions, the former to couple with the photochemical redox reactions and the latter to couple with the formation of the anhydride bond when ATP is produced from ADP and phosphate. Unfortunately, the chemical intermediate has never been identified or isolated.

More recently the chemiosmotic theory of coupling has suggested that ATP synthesis occurs when the redox reactions produce a hydrogen ion concentration gradient across the thylakoid membranes. ATP is formed at the expense of the free energy of this ionic gradient. It has indeed been demonstrated that ATP formation can occur using this mechanism with an artificially maintained hydrogen ion gradient. Once again, this theory cannot be said to be established until it is shown that the mechanism is as efficient as is the reality of photosynthesis and respiration.

There are still many questions to be answered concerning the mechanism of photosynthesis. The detailed mechanism for energy transfer within the photosynthetic units has still to be clarified, as has the detail for the enzyme pathways in carbon assimilation. Similarly, the primary photochemical reactions are still very unclear, as are the mechanisms for oxygen evolution and ATP synthesis. Nevertheless, many significant advances in our understanding of photosynthesis have been made in recent decades, and this rate of progress appears to be continuing unabated.

3.5 Productivity of Plants

Photosynthesis has two essential roles to play on this planet. Firstly, it maintains the balance in the atmosphere between oxygen and carbon dioxide.

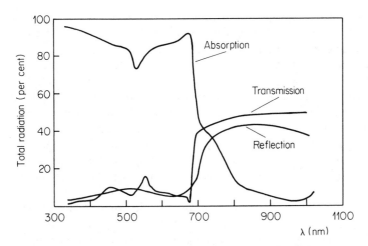

Figure 3.11. Optical properties of green leaves. (W. Tranquillini, in *Encyclopaedia of Plant Physiology*, Vol. 5, 1960. Reproduced by permission of Springer-Verlag New York Inc.)

Secondly, it supplies the total food requirements for all organisms, whether photosynthetic or not. In this section we shall briefly examine this second role of photosynthesis and look at the productivity of plants, both terrestrial and aquatic. To begin, we must first consider the supply of the various raw materials needed by plants for photosynthesis.

The first of these raw materials is sunlight. In Chapter 2 we discussed the spectral composition of solar radiation reaching the Earth's surface. It turns out that only 25 per cent of this radiation is capable of activating the photosynthetic process and, as plants only photosynthesize at maximum rate during part of the day, a mere fraction of this 25 per cent is actually used. Figure 3.11 shows typical curves for the absorption, transmission and reflection of light by green leaves as a function of wavelength. The low absorption in the infra-red is not wasteful as this light cannot be used for photosynthesis and would only succeed in heating the leaf if it were absorbed.

Plants can control the amount of light they absorb simply by orientating their leaves. Thus the higher leaves in plants are usually nearly vertical to reduce the amount of sunlight intercepted (full sunlight can actually have an inhibitory effect) while the lower leaves are horizontal. In addition there can be a movement of chloroplasts within the plant cells. Chloroplasts will gather near the illuminated surface of a leaf and will orientate their surfaces to compensate for the intensity of the light. It is worth noting that the saturation region of Figure 3.5 is not reached in a complete plant as the shade leaves are always in the limiting intensity region no matter how bright the sunlight.

Water is the second raw material. It is taken in by the plant's root structure along with the minerals vital to the photosynthetic process. However, for photosynthesis to function, a wide variation in its availability can be tolerated because the relevant tissues already contain more than 80 per cent water.

The third raw material is carbon dioxide. In aquatic environments this exists in solution either as free carbon dioxide or as carbonic acid. The degree of dissociation of the carbonic acid into bicarbonate or carbonate ions depends on the pH of the water. Between pH 6 and 10 the acid is partially dissociated into bicarbonate ions. At a pH greater than ten carbonate ions are the dominant species. All aquatic plants use free carbon dioxide and carbonic acid as their sources of carbon, and some also use bicarbonate ions. There are in addition many algae which can photosynthesize in water with a pH as high as 11, implying that they can use carbonate ions effectively. In all aquatic conditions the diffusion rates of the various ions is more of a limiting factor than their concentrations.

Terrestrial plants obtain their carbon dioxide from the atmosphere through pores in their leaves called *stomata*. When fully open these stomata account for one per cent of the surface area of the leaves and the rate at which carbon dioxide is taken in is similar to that which would be obtained if gaseous exchange could take place over the entire surface. (This can be proved using elementary kinetic theory.) The carbon dioxide diffuses in the air-space between the plant cells and dissolves in the water of the cell walls. Once dissolved it then has to diffuse to the chloroplasts in the aqueous phase. This part of the journey, although only a few microns long, provides the stiffest resistance to the passage of the carbon dioxide.

The stomata do not only control the intake of carbon dioxide, they also permit water loss from the leaves. The stomata remain open only in the light and if the concentration of carbon dioxide is less than 300 ppm in the vicinity of the leaf. They close in the dark and if the supply of water is low. Thus, they respond in such a way as to aid photosynthesis and reduce the loss of water by the plant. For example, it may be the case that a particular lower leaf is not receiving enough light for photosynthesis, in which case the stomata will stay closed to limit water loss.

The productivity of land plants can be estimated fairly easily by measuring the harvested dry weight of vegetation and making suitable corrections for the root structure and losses due to grazing and leaf fall. Aquatic productivity is by no means as simple to predict. The rate of photosynthesis by photoplankton is not simply related to the existing concentration of the species because of the high growth rate and the rapid removal by predatory zooplankton. Consequently a direct determination of the rate must be attempted. This can be done by monitoring the rate of oxygen production but the method is very inaccurate except for rich waters. A more accurate method, devised by Nielsen (1962), is the measurement of the rate of assimilation of C^{14} supplied as bicarbonate ions.

The results of such measurements show that the productivity of the oceans is much less than that of land per unit area. There is a number of reasons for this. Terrestrial plants have a leaf structure that avoids extremely high incident intensities and permits good penetration to the lower leaves. In contrast, surface plankton are exposed to inhibitory intensities on bright days and much of the available light is absorbed by the water and suspended particles of the surface layer. It has been estimated that the maximum amount of chlorophyll above one square metre of land in a tropical rain forest is an order of magnitude greater than that below a similar surface area of water.

Table 3.2 gives the estimated total yields for the various types of vegetation on the Earth's surface. The data for terrestrial productivity, although somewhat dated, are in good agreement with the most recent estimates (Vallentyne, 1966).

Table 3.2 Photosynthetic yield on Earth

	Area (10^6 km^2)	Yield $(10^9$ tonnes carbon per annum)
Forests	44	11
Grassland	31	1
Farmland	27	4
Deserts	47	0·2
Total land[a]	149	16·3
Ocean[b]	361	22
Total yield	510	38
Gross yield (15 per cent correction for respiration)		44

[a] Schroeder, *Naturwiss.*, **7**, 976 (1919).
[b] A. Steeman Nielsen, 1963, *The Seas*, M. N. Hill, (Ed.), Vol. 2, pp. 129–164, Wiley Interscience, New York.

The data for the oceans must be treated with some caution as there are considerable geographical variations in productivity due to fluctuations in the supply of minerals. Applying a correction of 15 per cent to allow for respiration, the total gross yield of photosynthesis is seen to be approximately $4·5 \times 10^{10}$ tonnes of organic carbon per annum. This is equivalent to the production of $1·3 \times 10^{11}$ tonnes of oxygen and the fixation of 2×10^{11} tonnes of carbon dioxide. By comparison, man releases an estimated $1·5 \times 10^{10}$ tonnes of carbon dioxide into the atmosphere each year by the combustion of fossil fuels as we saw in Chapter 2.

Using the above figures and estimating the total visible radiation incident on the Earth's surface (Chapter 2), it is possible to calculate the average efficiency for photosynthesis. This turns out to be 0·25 per cent, the figure for land plants being slightly higher at 0·4 per cent. Even a well-cultivated crop such as maize is only about 1 per cent efficient and, when it is realized that only 25 per cent of its total dry mass can be consumed, the overall efficiency of food production is only a meagre one-quarter per cent. An order of magnitude increase in this efficiency is well within the limits set by the photosynthetic mechanism so it is imperative that man should examine the possibility of cultivating more productive crops.

The principal reason for the low yields from crops is that most of them are annuals, which means that for some time after sowing there are no leaves capable of absorbing sunlight. Thus peak photosynthetic efficiency is only achieved for a relatively short period during the growing season. This could be improved by reorganizing agricultre on a continuous basis in such a way as to maintain a leaf canopy throughout the year. Harvesting would then be an all the year round process involving the collection of the older leaves which could then be treated to yield protein.

An alternative possibility is the mass culture of unicellular algae, for example chlorella, which, in addition to being very efficient for photosynthesis, have a high protein content. Indeed, more than half the dry weight of chlorella is protein. Prototype algae farms are already in operation. The algae are cultured in large ponds through which carbon dioxide is continuously pumped. A daily yield of 5 g m^{-2} of carbon has already been obtained. Although this is less than half the daily yield of maize or sugar cane, the protein production of these chlorella ponds is much higher than that of conventional crops. Not surprisingly, the major problem with the technique is its extremely high cost. Perhaps this can be overcome by linking the culture of algae with the disposal of sewage. Rapid growth of algae and bacteria takes place in shallow sewage ponds exposed to sunlight. Bacterial decomposition of sewage produces carbon dioxide which stimulates photosynthesis by the algae. Their oxygen evolution in turn stimulates the bacterial decomposition. While it may appear impossible to produce food for human consumption in this way, there is some evidence that algae do produce antibiotics which kill pathogenic bacteria.

Perhaps a less repugnant and cheaper approach to the problem is to use our extensive knowledge of genetics and cross-breeding of strains to produce a crop with a high photosynthetic efficiency. In Section 3.4, plants which utilize the four-carbon pathway for photosynthesis were described. These have a more efficient photosynthetic apparatus and can grow in hostile conditions of extreme heat and low humidity. (It is true that three-carbon plants can also grow in these conditions but they cannot photosynthesize during the hottest hours of the day.) If suitable crops could be bred incorporating these characteristics, the productivity of poor agricultural land would be greatly enhanced.

References

Arnon, D. I., M. B. Allen and F. R. Whatley, *Nature*, **174**, 394 (1954).
Calvin, M., *Science*, **135**, 879 (1962).
Duysens, L. N. M., Ph. D. Thesis, University of Utrecht, 1952.
Duysens, L. N. M., J. Amesz and B. M. Kamp, *Nature*, **190**, 510 (1961).
Emerson, R., *Ann. Rev. Plant Physiol.*, **9**, 1 (1958).

Emerson, R. and W. Arnold, *J. Gen. Physiol.*, **15**, 391 (1932).
Engelmann, Th. W., *Archiv. ges. Physiol.*, **30**, 95 (1883).
Fischer, H., *Naturwiss*, **28**, 401 (1940).
Haldane, J. B. S., 'Origin of life' in *Rationalist Annual*, 1929.
Hill, R., *Nature*, **174**, 501 (1954).
Ingen-Housz, J., 'Experiments upon vegetables, discovering their great power of purifying the common air in the sun-shine and of injuring it in the shade and at night', London, 1779.
Miller, S. L., *Science*, **117**, 528 (1953).
Miller, S. L., *J. Am. Chem. Soc.*, **77**, 2351 (1955).
van Niel, C. B., *Ann. Rev. Plant Physiol.*, **9**, 1–24 (1962).
Nielsen, A. S., *The Seas*, Vol. 2, M. N. Hill, Ed., pp. 129–64, Wiley Interscience, New York, 1962.
Oparin, A. I., *The Origin of Life*, 3rd ed., Academic Press, New York, 1957.
Sachs, J., *Bot. Z.*, **20**, 365 (1862).
Senebier, *Memoires Physico-Chimiques*, Gènève, 1782.
Trebst, A. V., H. Y. Tsujimoto and D. I. Arnon, *Nature*, **182**, 351 (1958).
Vallentyne, J. R., *Primary Productivity*, **309** (1966).
Warburg, O. and E. Negelein, *Z. Physik. Chem.*, **102**, 235 (1922).

Further Reading

Calvin, M. and J. A. Bassham, *The Path of Carbon in Photosynthesis*, Prentice-Hall, Englewood Cliffs, N. J., 1957.
Clayton, R. K., *Light and Living Matter*, Vol. 2, McGraw-Hill, New York, 1971.
Fogg, G. E., *Photosynthesis*, English Universities Press, London 1968.
Rabinowitch, E., *Photosynthesis and Related Processes*, Vol. 1 (1945); Vol. 2, Part 1 (1951); Vol. 2, Part 2 (1956); Vol. 3 (1973), Wiley Interscience, New York.
Rabinowitch, E. and Govindjee, *Photosynthesis*, Wiley, New York, 1969.

4 Fossil Fuels

Man is heavily dependent on fossil fuels, and it is the rapidly increasing demand for them, together with a realization that the world's resources are finite, that has led to the preoccupation with energy and fuel which is so prevalent today. The events which precipitated the 'Energy Crisis' of 1973–1974 were of a political nature, and public interest in fuel supplies will probably last no longer than it does over any other so-called crisis. Nevertheless, in exposing our dependence on an exhaustible fuel supply, the short-term crisis has focused attention on one of the most pressing long-term problems which man has ever faced.

It is undoubtedly true that the World's fossil fuels are finite, in the sense that the rate of accumulation is so slow as to be unimportant on a human time-scale. However, the known useful reserves have a surprising ability to expand just in time to save the World from collapse. It is interesting to note, for example, that at the very time when the news media were full of forecasts that the World's oil supplies would run out by 1995 unless the oil companies started drilling in the 'impossible' depth of 200 metres of water, the oil companies were quietly doing just that, and that six months after the height of public concern the reserves of oil in the North Sea area were publicly stated to be of the same order of magnitude as those of the Arab oil states. Various cynical reasons have been advanced for this sudden change, but the most likely explanation is the simple economic one that if the price of a commodity rises, it becomes worthwhile to put more effort into finding it, coupled with the technological explanation that when the difficult techniques of drilling in shallow water have been mastered, the even more difficult problems presented by deeper water seem less frightening. But considerations such as these will be dealt with in later chapters. For the present we shall concern ourselves with the nature of fossil fuels, and with the ways in which they occur in nature.

The energy in fossil fuels comes originally from the Sun's radiation. As was described in Chapter 3, the process of photosynthesis enables plants to store, in the form of carbohydrates, a small fraction of the energy which they receive from the Sun, and in doing so they liberate oxygen. When the plant dies, it decays, and most of the carbohydrate is oxidized, releasing energy. Under some conditions, the decay may be incomplete, and some of the energy is retained. It is this energy which is tapped when fossil fuels are burnt. Coal, oil and peat are alike in that they are the decayed remains of vegetation, but the nature of the vegetation, and of the conditions under which the vegetation grew and the decay occurred, are different for the different fuels. The conditions of formation of oil are fundamentally different from those of solid fuels, and we shall consider the two separately.

4.1 Formation of Coal and Peat

Coal and peat are thought to be the partially-decayed remains of vegetation which grew on wet land such as swamps, bogs and marshes. When the vegetation died, it fell into the swampy environment and was transformed under anaerobic conditions into peat. The microbiological process involved in the generation of peat have been the subject of much research, and although the details are not fully understood, it is thought to proceed approximately as follows:

(1) Bacterial decomposition of water-soluble constituents and of cellulose, perhaps represented very approximately by the equation

$$2\,C_6H_{10}O_5 \longrightarrow C_8H_{10}O_5 + 2\,CO_2 + 2\,CH_4 + H_2O$$
$$\text{cellulose} \qquad\qquad \text{humified}$$
$$\text{residue}$$

Decomposition rates are frequently very low in comparison with other eco-systems at comparable temperatures, especially if the ground is waterlogged and acidic, and the rate of decomposition diminishes with increasing depth, partly because of the anaerobic conditions in which relatively few micro-organisms can function (Clymo, 1965).

(2) Accumulation of compounds which resist decomposition, such as lignin, resins and waxes, and of the dead bacteria. These will eventually form the principal constituents of peat and lignite (brown coal).

This process is referred to rather loosely as *humification*, and the extent to which the process has occurred is measured in a semiquantitative manner by the humification number or H number, which is determined by colorimetry. During humification, much of the water is retained in the peat in a colloidal form, and this, as we shall see in Chapter 5, presents problems when peat is used as a fuel.

The transformation of peat into coal proceeds by a sequence of geological processes. First of all, the peat deposit becomes covered with sand and silt, thus bringing to an end the biological processes of peat formation. Over the course of time, the thickness of sediments increases by deposition of further sand and silt, and the peat is subjected to rising pressure. Water and volatile components are expelled, and the remaining material becomes relatively impoverished in oxygen and richer in carbon. Hydrogen ceases to be combined with oxygen as water, and instead becomes attached to carbon, forming hydro-carbons. This process, which may take millions of years, eventually transforms the spongy, fibrous peat into hard, brittle coal.

The series of conditions described above, namely a swamp subsequently

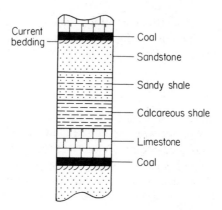

Figure 4.1. Sequence of strata in a cyclothem

covered by an increasing thickness of sediment, is most likely to occur in a river delta or estuary, and it is thought, therefore, that coal deposits were laid down under such conditions. Geological examination of coal measures reveals that very often they have accumulated in a cyclic sequence (known as a *cyclothem*) consisting of, in ascending order, limestone, calcareous shale, sandy shale, sandstone and coal, as shown schematically in Figure 4.1. It is suggested that the sandstone accumulated under clear-water marine conditions, the shale under deltaic conditions in the outer reaches of the delta and the coal under swampy conditions at the higher end of the delta. Such a sequence would arise naturally if the sea invaded the land gradually, for example as a consequence of the land subsiding. Further evidence of this mechanism is provided by markings on the sandstone layer, immediately beneath the coal seam, which are known as current-bedding (Figure 4.1) and which are produced by the action of currents in shallow water on the depositing sand. In most coal measures there are several such cyclothems, yielding a series of coal seams at different depths.

The origin of cyclothems is a matter of considerable interest to geologists. The fact that they occur worldwide, wherever late-Palaeozoic coal measures occur and in coal measures of other ages as well, indicates either that they are caused by worldwide cyclic alternations of conditions which affected all the coal-forming areas together, or that they arise from an inherent cyclic process which is characteristic of the mechanism of coal formation, but in which each coal-forming area behaved independently of the others (see, for example, Verhoogen and coworkers, 1970).

The first group of theories assumes worldwide or other large-scale cyclic variations of conditions. One theory proposes a sequence of tectonic oscillations (oscillatory movements of the Earth's crust), each oscillation consisting of a long, gradual subsidence followed by a short, rapid uplift. Other tectonic hypotheses propose variations in the rate of subsidence, with constant sedimentation, or variations in the rate of sedimentation, due to uplift of the land at the source of the river, while the rate of subsidence at the estuary was constant. In all these cases, when sedimentation exceeded subsidence, the shoreline would advance and the conditions of land would prevail, and when the rate of subsidence exceeded sedimentation, the shoreline would recede and marine conditions would prevail. Oscillations of a different sort, but still on a large scale, are proposed by theories involving changes of climate. One such theory attributes cyclothems to the advance and retreat of large glaciers. Another assumes that variations in aridity in the source regions produced rapid erosion of the land up-river and the consequent rapid deposition of sediment at the delta gave rise to a swamp which was subsequently inundated owing to subsidence.

Theories such as those just described, which involve changes affecting large areas of the Earth, clearly imply a correlation between the cyclothems in all the coal measures over a wide area, or even over the whole World. It has proved very difficult to find evidence of such a correlation, and consequently there are serious doubts about the validity of these theories. There is a special objection to the glacial theory, in that the duration of the cyclic sedimentation period appears to be much longer than the period over which glaciation is known to have existed.

An entirely different set of theories assumes that periodic variations in sedimentation are an inherent characteristic of deltaic conditions. Such a situation might arise if the rate of compaction of the sediments varied, as is assumed by one theory of this type. The deltaic sediment is imagined to build up to sea level, whereupon it is covered by a coal swamp, yielding a deposit of peat. As sediment continues to build up, the pressure on the peat deposit increases steadily until some sort of mechanical yield point is reached at which the peat starts to compact rapidly. The swamp is then flooded by the sea, and the cycle repeats. To say the least, this theory implies unusual mechanical properties for the compacting layers. Another theory of this type proposes a mechanism involving changes in the positions of distributories (the small watercourses in a delta). If a swamp grows in a delta in fresh water supplied by a distributory, and the distributory then shifts, salt water invades the swamp, killing the freshwater vegetation. With continued slow subsidence, deposition of shale and limestone would occur under salt-water conditions, until the advancing movement of the distributaries brought the deposition of sand and clay to sea level or above. The cycle would then repeat. A similar theory of this type involves a change in the nature of swamp vegetation with time. After deposition of sand and clay up to or above sea level, the initial swamp consists of dense tree-like vegetation which keeps out sea water. As the peat layer thickens under the trees, and essential mineral nutrients are rendered unavailable, the vegetation deteriorates, and eventually it is unable to keep out the sea. The resulting marine inundation kills the swamp vegetation. With continued subsidence and compaction of peat into coal, the water depth increases, and the resulting open-sea conditions lead to deposition of limestone and marine shale. This situation continues until the deltaic distributaries return, carving channels in the marine sediments and depositing silt, sand and clay. The cycle then repeats. These last two theories appear to coincide rather closely with present-day experience in, for example, mangrove swamps.

It is impossible at present to establish which, if any, of the foregoing theories is correct, and it may even be that different cyclothems have their origins in different mechanisms. Moreover, this is not the only area in which there are doubts about the mechanisms of coal formation, for there are doubts also about the conditions under which the vegetation grew and subsequently decayed. For example, it has been widely believed that the only conditions under which accumulation of organic matter as fossil fuel can occur are those typified by the mangrove swamps which today occur only in tropical and sub-tropical regions of the Earth. Such hot swampy conditions provide both a luxurious growth of vegetation and the necessary wet anaerobic environment for decay into fossil fuels. In view of the obvious fact that the bulk of the World's coal supplies come from regions which today do not have such a climate, it becomes necessary to postulate changes in climate on a very pronounced scale between the time when the deposits of organic material were laid down (around 300 million years ago in the case of the coal deposits of the British Isles) and the present day. Various theories have been proposed, including suggestions of major changes in global climate on a par with those of the Ice Ages, and even of large shifts in the axis of rotation of the Earth, but a more favoured view nowadays involves the drift of continents about the surface of the Earth (see, for example, Verhoogen and coworkers 1970). Such a theory may indeed be close to the truth. However, there is evidence that accumulation of coal-forming organic material need not require a tropical climate, but rather that any climate from temperate to tropical may suffice (Meyerhoff, 1970; Meyerhoff and Teichert, 1971). If this is the case, it is possible that at least some of the World's deposits of coal can be accounted for without postulating enormous variations in climate.

It is further widely believed that fully anaerobic conditions are necessary for formation of peat deposits from decaying vegetation. However, some evidence has been advanced (Kurbatov, 1963) to suggest that aerobic conditions are adequate and even advantageous for peat formation.

The above warnings indicate the need for caution in any discussion of the origins of fossil-fuel deposits. It is evident that while the general mechanism may be understood, the details are by no means certain.

4.2 The Classification of Coal

The physical and chemical nature of coal is highly variable, depending on such factors as the nature of the original vegetation, the geological conditions under which it was transformed, and the age of the deposit. The latter is probably the most important. During the formation of coal from the original vegetation, the percentage of carbon increases with time, as will be shown later in this chapter, so in general, younger deposits are lower in carbon content per unit mass, older deposits are higher, and the oldest deposits are often almost pure carbon. It is important to note, however, that this is a generalization.

Coal is classified according to *rank*, which is a concept based largely but not entirely on carbon content. Although the types of coal merge into each other in a continuous manner, it is possible to recognize five distinct classes.

Peat

This is the lowest rank of the classification. It is characterized by well-preserved remains of the vegetation from which it was formed. It is soft and spongy, and in its natural state has an exceptionally high water content, often 92–94 per cent (Green, 1963). Even after drying the water content is very high, and it is often burnt while still containing as much as 50 per cent water (see Chapter 5). Because peat represents an early stage of decomposition, the carbon content is low (about 50 per cent of the dry weight is a typical figure) and there is a high proportion of volatile matter, including water and hydrocarbons. Consequently it burns with a long flame. It is used as a fuel only in relatively few parts of the World, where its principal uses are as a domestic heating fuel and for electricity generation (see Chapter 5).

Lignite or Brown Coal

This occupies the next rank. It is a partially-transformed type of peat, and still contains recognizable woody material from the lignin of the original plant tissues. Chemically it is not much different from peat, and it still contains a large percentage of water and volatiles, so its burning characteristics are similar. Mechanically it is soft but not noticeably spongy. Lignite is used extensively in Germany and Central Europe, and also to a small extent in America, as an industrial fuel and for electricity generation. It has also been an important source of chemical feedstock, and, in the form of briquettes, it is used as a domestic fuel.

Bituminous Coal

Next is bituminous coal, so called because when distilled to make coal-gas and coke, one of the by-products, coal tar, is rather like bitumen in appearance. Bituminous coal is hard, black and often banded (see later). It contains recognizable vegetable material but only in a fossilized form. It comprises about 30–40 per cent of volatile matter, but is low in water. It burns with a smoky yellow flame, and has a high calorific value (energy content per unit mass), considerably higher than lignite or peat. Domestic coal is bituminous, and for this reason it is the most familiar of all the classes. It is also of great importance as an industrial fuel and for electricity generation. Its volatile content is mainly hydrocarbons, and for this reason it can be used for the manufacture of town-gas supplies; indeed, until quite recently, almost all the United

Kingdom's town-gas supplies were made in this way. The residues from gasification are coke, which is a useful smokeless fuel, and coal tar, which is an important source of chemicals for such purposes as pharmaceuticals and artificial fibres. Certain grades of bituminous coal produce a very hard, strong coke which is used as a reducing agent by the steel industry. Manufactured smokeless fuels are made from bituminous coal by controlled distillation.

Anthracite

The highest rank of coal is anthracite, which is over 90 per cent carbon and often has a very low volatile content. It is hard, brittle and stony in appearance. Because of the absence of volatiles it burns with a short blue smokeless flame. Its calorific value is high, but that of some bituminous coals can be higher because they have a greater hydrogen content. It has little or no tendency to form a coke. Anthracite is used only as a fuel, and because of its high cost it is used mainly in the domestic sector, where it is regarded as a 'premium-grade' fuel with natural smokeless properties.

Cannel Coal

This is a special category of coal, characterized by a very high percentage of hydrocarbon volatiles. It can be regarded as a special type of bituminous coal, and is usually marketed in that category rather than on its own. It is particularly notable for the ease with which it can be distilled to yield oils of a petroleum type.

The foregoing classification is rather inexact, especially as the ranks merge imperceptably. To improve the precision of classification, the American Society for Testing Materials had adopted a system based on fixed (i.e. non-volatile) carbon content and calorific value, and this is reproduced in Table 4.1.

As well as chemical and energy considerations, the size of pieces in which the coal is sold is of considerable importance to the user. For this reason it is unusual to despatch coal directly from the mine to the consumer; instead it is first sieved into several different size ranges. Domestic coal is also subjected to a cleaning operation, usually by washing with water, to remove rock and other incombustible dirt.

As was noted earlier, the lower-rank coals tend to be younger than those of higher rank. This is partly because the conversion of peat to coal is slow, and partly because younger deposits are generally not buried to such great depths as older deposits and therefore have not experienced such great pressures. Table 4.2 lists the ages of some important coal deposits together with an indication of their rank.

4.3 The Composition of Coal

Although coal is complex and highly variable, it is possible to generalize to some extent about both its physical and its chemical nature.

Table 4.1 Classification of coals by rank

Class	Group	Fixed carbon limits, per cent (dry, mineral-matter-free basis)		Volatile matter limits, per cent (dry, mineral-matter-free basis)		Calorific value limits, Btu per pound (moist, mineral-matter-free basis)		Agglomerating character
		Equal or Greater Than	Less Than	Greater Than	Equal or Less Than	Equal or Greater Than	Less Than	
I. Anthracitic	1. Meta-anthracite	98	—	—	2	—	—	Non agglomerating
	2. Anthracite	92	98	2	8	—	—	
	3. Semianthracite	86	92	8	14	—	—	
II. Bituminous	1. Low volatile bituminous coal	78	86	14	22	—	—	Commonly agglomerating
	2. Medium volatile bituminous coal	69	78	22	31	—	—	
	3. High volatile A bituminous coal	—	69	31	—	14,000	—	
	4. High volatile B bituminous coal	—	—	—	—	13,000	14,000	
	5. High volatile C bituminous coal	—	—	—	—	11,500	13,000	
		—	—	—	—	10,500	11,500	Agglomerating
III. Subbituminous	1. Subbituminous A coal	—	—	—	—	10,500	11,500	Non agglomerating
	2. Subbituminous B coal	—	—	—	—	9,500	10,500	
	3. Subbituminous C coal	—	—	—	—	8,300	9,500	
IV. Lignitic	1. Lignite A	—	—	—	—	6,300	8,300	
	2. Lignite B	—	—	—	—	—	6,300	

(Reproduced by permission of the American Society for Testing Materials.)

Table 4.2 Age of some coal deposits in relation to their rank

Geological period (in millions of years)		Occurrence of coal
Recent		
Tertiary 60	Lignite	Hungary, Russia, U.S.A. (West), New Zealand, England (West), Germany, Spitsbergen
Cretaceous 120	Lignite	Germany, Hungary, U.S.A. (West), Alaska, Japan
Jurassic 145	Lignite	Alaska, Japan, Siberia, Mexico, Scotland (North), England (North)
Triassic 170	Bituminous	Germany (South), U.S.A., China, Japan
Permian 210	Bituminous	France (Central), Germany (East), South Africa, Australia, India, China, South America
Carboniferous 285	Bituminous	England, Scotland, Wales, France, Belgium, Germany, U.S.A. (East)
Devonian 325	Bituminous	Russia, Bear Island
Silurian 350 Ordovician 410 Cambrian 500 Pre-Cambrian 2000	No coal recorded	

(Modified from P. J. Adams (1960). Based on Crown Copyright Geological Survey diagram. Reproduced by permission of the Controller, Her Majesty's Stationery Office.)

Bituminous coal usually shows a well-defined banded structure, with alternating bright and dull bands (see, for example, Holmes, 1965). The material in the brightest bands, known as *vitrain* from its glassy appearance, is derived from wood and bark. Other bands are finely laminated and consist of shreds of vitrain in a fine-grained matrix; this is known as *clarain*. The dull-grey bands are made of a material known as *durain*, which is derived from non-woody vegetation, including spores. A fourth material, *fusain*, is extremely flaky and friable, and gives rise to coal-dust; this is also derived, like vitrain, from wood and bark.

Lignite and anthracite consist of the same basic constituents as bituminous coals, but the proportions and the degree of alteration from the original vegetation are different. Anthracite is usually unbanded. Cannel coal is mainly durain and contains very large amounts of algal remains and spore cases. There is a high non-combustible ash content, which arises from muddy sediment in the original swamp.

The chemistry of coal can be summarized approximately by describing it as a hydrocarbon material deficient in hydrogen. This is utilized in certain industrial processes in which hydrogen is added to coal to produce synthetic petroleum. However, such a description is an oversimplification, and a rather fuller account is in order.

As we have noted previously, the non-volatile carbon content increases with increasing rank, and the volatile content decreases. This is illustrated clearly in Table 4.3, which lists analyses of coals of different ranks, and compares them with analyses of peat and wood. Table 4.3 also illustrates some other useful generalizations:

(1) The moisture content decreases with increasing rank, owing to the gradual transformation from partly-decayed, spongy, waterlogged peat to hard, stony anthracite.

(2) The oxygen content decreases with increasing rank, for the same basic reason as the decrease in water content.

(3) The hydrogen content decreases slightly with increasing rank. This is because of the reduction in the volatile fraction, which is emitted largely as hydrocarbons and is thus rich in hydrogen. However, the decrease is very slight, and much less than might be expected in view of the great decrease in volatiles. The explanation of this is not straightforward, but it can be explained approximately in terms of a transfer of hydrogen from water molecules in low-rank coal to hydrocarbon molecules in high-rank coal, with the elimination of oxygen, possibly as carbon dioxide.

(4) The sulphur content is rather variable and cannot be correlated with rank. A high sulphur content is undesirable, because it burns to sulphur dioxide during combustion and is liberated into the atmosphere, where it is one of the more unpleasant pollutants. Moreover, excessive sulphur content can make the coal liable to spontaneous combustion in storage, and in extreme cases it can be harmful to the plant in which it is burnt.

(5) The ash content is highly variable. It appears that in some cases the ash is derived from muddy conditions in the original coal swamp; this is particularly true of cannel coals.

As well as these major constituents of coal there are numerous minor ones, which are often present in remarkably large amounts. This arises from the well-known ability of living organisms to concentrate trace elements from their environment; the example of manganese nodules on the sea bed has received occasional publicity, but the concentration of trace elements in coal is just as remarkable. Table 4.4 shows average contents of some elements in

Table 4.3 Some typical analyses of fuels of different rank (% by weight)

	Peat		Lignite		Sub-bituminous		High-volatile bituminous		Low-volatile bituminous		Semi-anthracite		Anthracite	
Moist	91.20	71.14	35.96	31.06	23.93	16.04	14.08	13.99	3.20	9.50	5.19	5.81	3.33	2.70
Ash	0.35	7.41	7.75	7.88	6.07	11.53	10.96	14.32	5.87	28.06	14.01	17.04	9.12	5.83
Volatile	75.11	48.44	56.71	45.32	51.07	43.34	47.48	41.01	20.50	21.17	12.98	16.25	3.73	3.63
F. Carbon	20.91	25.88	43.29	54.68	48.93	56.66	52.52	58.99	79.50	78.83	87.02	83.75	96.27	96.37
Hydrogen	5.23	3.67	4.51	5.04	5.47	5.23	5.35	5.31	4.93	3.82	3.97	4.48	3.09	3.25
Carbon	56.82	45.77	73.61	73.21	72.83	78.83	78.03	79.76	88.45	82.39	89.64	88.91	92.91	93.34
Nitrogen	1.48	1.66	2.15	1.48	1.54	1.60	1.20	1.55	1.39	1.36	0.63	2.16	0.91	1.22
Oxygen	32.26	19.58	17.69	18.66	18.99	14.31	9.73	10.16	3.84	6.75	3.23	3.97	2.41	1.32
Sulphur	0.23	3.64	2.04	1.61	1.17	0.84	5.69	3.22	1.39	5.68	2.53	0.48	0.68	0.87
MJ kg⁻¹	22.5	18.1	29.2	30.0	29.3	31.9	33.3	33.5	36.6	33.0	35.9	35.5	35.5	35.9

Ash-free, dry applies to the Hydrogen, Carbon, Nitrogen, Oxygen and Sulphur rows.

(Modified from Raistrick and Marshall (1938). Reproduced by permission of English Universities Press Ltd.)

Table 4.4 Minor elements in coal ash

Element	Average content in coal ash (g tonne^{-1})	Average content in Earth's crust (g tonne^{-1})	Factor of enrichment
B	600	10	60
Ge	500	1·5	330
As	500	1·8	280
Bi	20	0·2	100
Be	45	2·8	16
Co	300	25	12
Ni	700	75	9
Cd	5	0·2	25
Pb	100	13	8
Ag	2	0·1	20
Sc	60	22	3
Ga	100	15	7
Mo	50	1·5	30
U	400	2·7	150

(From B. Mason (1966). Reproduced by permission of John Wiley and Sons, Inc.)

coal ash, together with the average content in the Earth's crust, so giving an indication of the ability of the coal-forming vegetation to concentrate these elements. It must be noted, however, that such figures are approximate, and the actual concentration of an element in a particular sample of coal ash may well differ from these figures by an order of magnitude or more.

The presence of some of these elements in coal ash has caused some anxiety about atmospheric pollution, especially by such poisons as arsenic, mercury, cadmium and lead, and by the radioactive element uranium. However, as we shall see in Chapter 5, modern coal-fired power stations have very effective ash-arresting equipment, and it seems unlikely that serious pollution can arise from this particular problem. A more positive viewpoint is to emphasize the mineral wealth to be found in coal ash, and to wonder if extraction of such valuable elements as germanium might be economically worthwhile.

The foregoing discussion on the chemistry of coal has concentrated on elemental analysis, which, although interesting and important, does not give us any real insight into the chemical structure of coal. The complexity of coal, together with its variability, make exact chemical description impossible, but reviews, for example by Lessing (1955) and van Krevelen (1961), indicate that some progress has been made. Coal appears to contain a large percentage of aromatic rings, in which the hydrogen atoms can be replaced by various atomic groups such as hydroxyl, short aliphatic side chains and ether linkages. The carbon ring may also be fused with other rings in units ranging from about four fused rings in low-rank coals to between five and ten in bituminous coals, and up to 30 in anthracite. In a given coal the units all belong to the same skeletal species, although they vary in size, shape and properties. About 75 per cent of the carbon in coal exists in such structures. Although any such description is bound to be inexact, it is of considerable importance, for a knowledge of the basic structure of coal is advantageous if it is to be used as a chemical raw material (see Chapter 5) as well as a fuel.

4.4 World Distribution of Peat and Coal

Peat deposits are widely distributed, but only very few countries have really large and important quantities. Table 4.5 shows the percentage of world peat

Table 4.5 World peat resources

Country	Per cent of World's resources
U.S.S.R.	60·8
Finland	9·5
Canada	9·1
U.S.A. (excluding Alaska)	5·0
Germany (G.D.R. and F.R.G.)	3·8
Great Britain and Ireland	3·5
Sweden	3·4
Poland	2·3
Indonesia	0·9
Norway	0·7
Cuba	0·3
Japan	0·2
Denmark	0·08
Italy	0·08
France	0·08
New Zealand	0·056
Hungary	0·04
Netherlands	0·03
Austria	0·03
Argentina	0·03
Rumania	0·03
Yugoslavia	0·01
Czechoslovakia	0·01
Spain	0·004
Other countries	0·02

(From A. S. Olenin (1963). Reproduced by permission of the Controller, H.M.S.O.)

resources in different countries. The absolute amount of peat available is rather uncertain; Moore and Bellamy (1974) have estimated the reserves as $3\cdot3 \times 10^{11}$ dry tonnes, and they estimate that this represents an energy store of about 7×10^{22} joules, while O'Donnell (1974) quotes the reserves as about 2×10^{11} dry tonnes. The discrepancy between these figures is not great, and they can be regarded as reasonably trustworthy, although how much of this peat can be exploited economically as fuel is debatable. Peat is being formed continuously, and in favourable circumstances the rate of accumulation may be as high as three tonnes per hectare per year, (Moore and Bellamy, 1974), but the average rate of accumulation is much lower than this, and it is estimated that the present world rate of exploitation at 9×10^7 tonnes per year is greater than the rate at which the World's supply is being replaced. It is reasonable, therefore, to regard peat as a non-renewable resource, but it is less so than other fossil fuels.

Coal occurs in commercially useful quantities almost everywhere. Table 4.6 shows the location of important deposits. Because coal deposits are so

Table 4.6 World coal resources, in tonnes $\times 10^9$
(G tonnes)

Country	Hard coal	Lignite
Australia	16	95
Brazil	11	—
Canada	61	24
China	1011	—
Colombia	13	—
Czechoslovakia	12	10
Germany (East)	—	30
Germany (West)	70	—
India	106	—
Japan	19	—
Poland	46	15
South Africa	72	—
U.S.S.R.	4121	1406
United Kingdom	16	—
United States of America	1100	406
Yugoslavia	—	27

(Notes: Includes measured reserves and inferred resources in seams of 30 cm or thicker at depths of less than 1200 m (hard coal) or 500 m (lignite). National resources of less than 10 G tonnes are excluded. From United Nations Statistical Year Book (1972). Copyright of the United Nations, 1972. Reproduced by permission.)

widely distributed, it is not usually necessary to import and export large amounts, and moreover, most of the major industrial nations have their own coal-mining industries (for obvious historical reasons—see Chapter 1), which further reduces the need for large-scale international trade in coal. In this way coal is very markedly different from oil, which is one of the most important items in world trade.

It is impossible to estimate accurately the amount of mineable coal in the World, for it depends on the state of geological knowledge, and perhaps more important, on the state of development of mining engineering. The current estimate by the United States Geological Survey (Averitt, 1969) is about 7.6×10^{12} tonnes. In 1928 the same organization quoted a figure of about 3.2×10^{12}

tonnes (Zimmermann, 1951). The difference between these figures arises in part from extensions of geological surveys, but also from a decision to include coal at greater depths, presumably reflecting improved techniques; the earlier figures excluded coal at depths greater than 3000 feet (915 metres), while the later figures include coal at 4000 or even 6000 feet (1220 or 1830 metres). Although the definition of mineable reserves is, therefore, arbitrary, the two estimates are remarkably close to each other in relative terms, so it is reasonable to expect the estimates to be accurate. Accepting the larger figure, this represents a store of energy of about 2×10^{23} joules or about 5×10^{16} kWh.

The processes which form peat and which transform it into coal are still continuing today. Opinion is divided on whether the rate of accumulation of coal-forming material is of the same order as, or much less than it was in, for example, the Carboniferous period. However, there is no doubt that the rate of consumption is vastly greater than the rate of accumulation, so it is a perfectly acceptable approximation to regard coal as a non-renewable resource of energy.

4.5 The Chemistry of Crude Oil

Crude oil is a complex mixture of variable composition, but in general terms it may be described as a mixture of many different hydrocarbons, together with several different classes of non-hydrocarbon compounds. Elemental analyses of some crude oils are shown in Table 4.7, and it is evident that although carbon and hydrogen predominate there are appreciable quantities of sulphur, and a little nitrogen and oxygen. In addition to these major constituents, there are often traces of vanadium and nickel, and small traces of chlorine, arsenic, lead and other elements. It is probably true to say that every crude oil is a highly individual, unique mixture, but despite the variability, it is possible to identify important groups of compounds which are common to most if not all crude oils.

The hydrocarbons can be classified according to the arrangement of the chain of carbon atoms. The majority are paraffins, and may be either normal (straight chain) or *iso* (branched chain), see Figure 4.2. Other important groups are the cycloparaffins and aromatics. Perhaps surprisingly, olefins (e.g. ethylene) are absent from crude oil (Peel, 1970) despite their great importance in the petrochemical industry.

A list of the predominant hydrocarbons in petroleum is shown in Table 4.8, and an example of the percentage distribution of different chain lengths and classes in a typical crude oil is shown in Figure 4.3. Crude oils from different regions have remarkably different percentages of the different classes of hydrocarbons, as Table 4.9 shows. Even within a class, different chain lengths occur in different percentages from one region to another, as shown in Table 4.10. In general, 'young' petroleum tends to predominate in hydrocarbons of high molecular weights, while 'older' petroleum tends to have larger percentages of the lighter hydrocarbons.

Table 4.7 Elemental composition of some crude oils, mineral wax and asphaltic substances

		C	H	S	N	N+0	O
				Percentage composition			
Crude oils	Mecock, W. Va.	83·6	12·9	0·37			3·6
	Humboldt, Kans.	85·6	12·4	0·76			
	Healdton, Okla.	85·0	12·9	0·6			
	Coalinga, Calif.	86·4	11·7	0·7			
	Beaumont, Texas	85·7	11·0			2·61	
Wax	Ozokerite	84–86	14–16	0–1·5	0–0·5		
Natural asphalts	Athabaska Tar	84·4	11·23	2·73	0·04		
	Bermudez Lake	82·88	10·79	5·87	0·75		
	Trinidad Pitch Lake	80–82	10–11	6–8	0·6–0·8		
	Asphalt from Utah limestone	89·9	9	0		1·1	
	Native asphalt	82	11	2	2–2·5		
Asphaltites	Gilsonite	85–86	8·5–10	0·3–0·5	2–2·8		0–2
	Glance pitch	80–85	7–12	2–8		0·2	
Asphaltic pyrobitumens	Wurtzilite	79·5–80	10·5–12·3	4–6	1·8–2·2		
	Albertite	83·4–87·2	8·96–13·2	Tr–1·2	0·42–3·1		1·97–2·22

(From G. D. Hobson (1967). Reproduced by permission of Elsevier Publishing Company.)

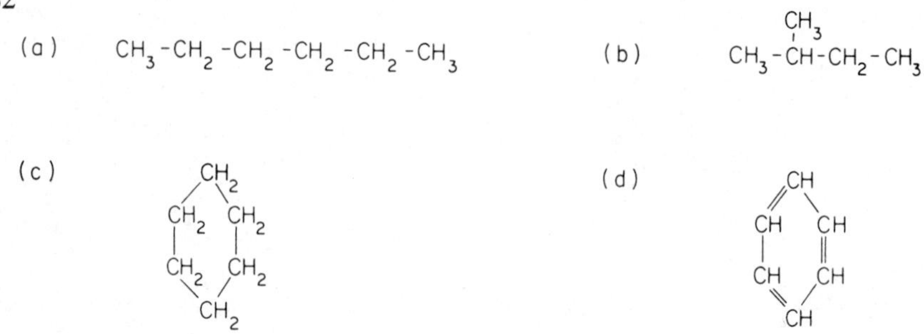

(a) CH3-CH2-CH2-CH2-CH2-CH3
(b) CH3-CH-CH2-CH3
CH3
(c)
CH2
CH2 CH2
CH2 CH2
CH2
(d)
CH
CH CH
CH CH
CH

Figure 4.2. Arrangements of carbon chains in hydrocarbons: (a) normal or straight-chain paraffin (normal hexane); (b) *iso*- or branched-chain paraffin (*iso*-pentane); (c) cycloparaffin (cyclohexane); (d) aromatic (benzene). Note that the bonds in the ring are in fact equivalent, rather than alternately double and single

%
4
3
2
1
0
1
2
3
4
5
5
10
20
30
40
50
N

Figure 4.3. Distribution of hydrocarbons in a light crude oil, by class and number of carbon atoms per molecule.

Curve	*Class*	*Weight (per cent)*
1	Normal paraffins	23·3
2	*iso*-Paraffins	12·8
3	Cycloparaffins	41·0
4	Aromatics	6·4
5	Aromatic cycloparaffins	8·1

(Balance to 100 per cent made up of resins and asphaltenes)

(From Bestougeff (1967). Reproduced by permission of Elsevier Publishing Company)

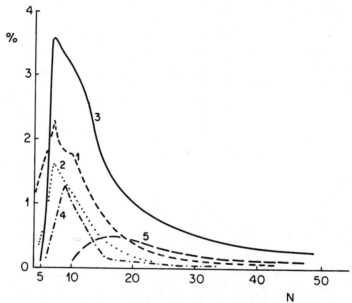

Table 4.8 Predominant constituents of petroleum

Series and hydrocarbon	Carbon atom number	Per cent of crude oil	
		min.	max.
Normal paraffins			
Pentane	C_5	0·2	3·2
Hexane	C_6	0·04	2·6
Heptane	C_7	0·03	2·5
Octane–Decane	C_8–C_{10}	0	1·8–2·0
Undecane–Pentadecane	C_{11}–C_{15}	0	0·8–1·5
Hexadecane and higher	C_{16} and higher	0	< 1·0
Isoparaffins			
2-Methylpentane	C_6	0·2	1·16
3-Methylpentane	C_6	0·06	0·9
2-Methylhexane	C_7	0·03	1·1
3-Methylhexane	C_7	0·02	0·9
2-Methylheptane	C_8	0·03	1·0
3-Methylheptane	C_8	0·02	0·4
2-Methyloctane	C_9	0·02	0·4
3-Methyloctane	C_9	0·01	0·2
2-Methylnonane	C_{10}	—	0·3
3-Methylnonane	C_{10}	—	0·1
4-Methylnonane	C_{10}	—	0·1
Pristane (isoprenoid)	C_{19}	—	1·12
Cycloparaffins			
Methylcyclopentane	C_6	0·11	2·35
Cyclohexane	C_6	0·08	1·4
Methylcyclohexane	C_7	0·25	2·8
1,*trans*-2-dimethylcyclopentane	C_7	0·05	1·2
1,*cis*-3-dimethycyclopentane	C_7	0·04	1·0
1,*cis*-3-dimethylcyclohexane	C_8	—	0·9
1,*cis*-2-dimethylcyclohexane	C_8	—	0·6
1,1,3-trimethylcyclohexane	C_9	—	0·7
Aromatics			
Benzene	C_6	0·01	1·0
Toluene	C_7	0·03	1·8
Ethylbenzene	C_8	0·01	1·6
m-Xylene	C_8	0·02	1·0
1-Methyl-3-ethylbenzene	C_9	—	0·3
1,2,4-Trimethylbenzene	C_9	—	0·6
1,2,3-Trimethylbenzene	C_9	—	0·4
1,2,3,4-Tetramethylbenzene	C_{10}	—	0·3
2-Methylnaphthalene	C_{11}	—	0·3
2,6-Dimethylnaphthalene	C_{12}	—	0·4
Trimethylnaphthalene	C_{13}	—	0·3

(From Bestougeff (1967). Reproduced by permission of Elsevier Publishing Company.)

Table 4.9 Composition of some crude oils in terms of the main hydrocarbon groups, resins and asphaltenes

Type of crude	Sources	Paraffins (%)	Naphthenes (%)	Aromatics (%)	Resins and asphaltenes (%)
Paraffinic		40	48	10	2
Paraffinic–	Oklahoma City	36	45	14	5
naphthenic	East Texas	33	41	17	9
Naphthenic	Emba-Dossor	12	75	10	3
	Balachany	9	66	19	6
	Bibi Eibat	11	60	19	10
Naphthenic–	Santa Fe	20	45	23	12
aromatic	Borneo	15	35	35	15
Mixed asphaltic	Inglewood	8	42	27	23
	Perm	13	15	40	32

(Data extracted from A. N. Sachanen (1950). Reproduced by permission of Oxford University Press.)

Table 4.10 Normal paraffins distribution in light crude oils according to the molecular weight

Fraction of n-paraffins	Content in crude oil (vol. per cent)		
	Iraq (Tertiary)	Ponca City (Ordovician)	Hassi Messaoud (Cambrian)
C_5-C_{10}	9·55	11·5	14·05
$C_{11}-C_{20}$	7·50	10·13	5·07
$C_{21}-C_{30}$	2·10	~2·23	~1·82
C_{31} etc.	~0·6	~0·25	~0·7
Total content in crude oil	19·75	24·11	21·64

(From Bestougeff (1967). Reproduced by permission of Elsevier Publishing Company.)

Clearly, the foregoing account indicates a high degree of complexity. However, it is interesting to note that in view of the enormous number of possible isomers among the hydrocarbons, an even greater complexity might be expected, so that the total number of constituents, although somewhat high, represents only a very small proportion of theoretically possible structures (Bestougeff, 1967).

Among the non-hydrocarbons, the most important from a practical point of view are sulphur and its compounds. Free sulphur is present in some but not all crude oils (Constantinides and Arich, 1967). Common sulphur compounds in light distillates include hydrogen sulphide, and organic sulphur compounds such as mercaptans, monosulphides, disulphides and cyclic

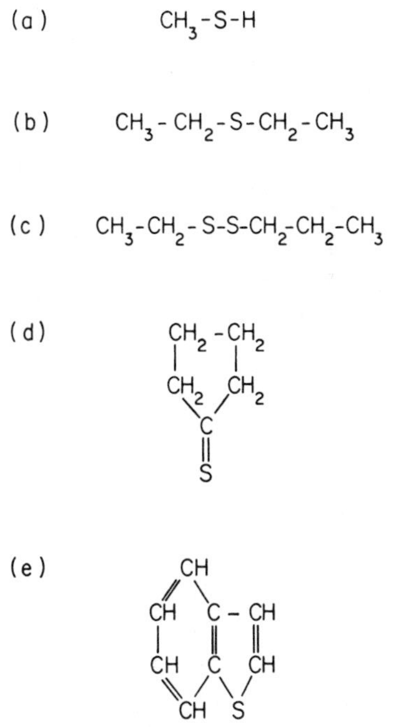

(a) CH_3-S-H

(b) $CH_3-CH_2-S-CH_2-CH_3$

(c) $CH_3-CH_2-S-S-CH_2-CH_2-CH_3$

(d)

$$\begin{array}{c} CH_2-CH_2 \\ | \quad\quad | \\ CH_2 \quad CH_2 \\ \diagdown C \diagup \\ \| \\ S \end{array}$$

Figure 4.4. Sulphur compounds in crude oil: (a) mercaptan; (b) mono-sulphide; (c) disulphide; (d) cyclic sulphide; (e) benzothiophene

(e)

$$\begin{array}{c} CH \\ \diagup\diagdown \\ CH \quad C-CH \\ | \quad \| \quad \| \\ CH \quad C \quad CH \\ \diagdown\diagup\diagdown\diagup \\ CH \quad S \end{array}$$

sulphides (Figure 4.4). The majority of the sulphur content is contained, however, in the heavier fractions, where it occurs mainly as complex ring structures of the benzothiophene type.

Nitrogen generally occurs in complex ring compounds. Oxygen generally occurs as acids, phenols and ketones. Metals are found in the form of oil-soluble organometallic compounds known as porphyrins.

As we have seen earlier, oil is often found in association with natural gas. The principal constituent is always methane, but other light hydrocarbons occur in significant quantities which vary from one source to another, as indicated in Table 4.11.

4.6 Formation of Oil

It is fairly well established that oil is formed from marine plankton which were deposited on the sea bed and decayed under anaerobic conditions. There are several pieces of evidence which contribute to this belief:

(1) Formation of methane during the decay of organic material is very common, and is familiar in such situations as 'Will-o'-the-Wisp' in marshes, and in the powering of plant in sewage works from the methane evolved in the decomposition of the sewage. Methane, as we have seen, is the principal constituent of natural gas.

(2) Emery and Hoggan (1958) found methane, ethane, propane, butane, isobutane, ethylene, hexane, isopentane, cyclobutane, cyclopentane, benzene,

Table 4.11 Constituents of selected natural gases

(per cent by volume)

Area	North Sea	Germany	Holland	France	Algeria	New Zealand	U.S.A.	Kuwait	
Methane	94·4	74·7	81·8	69·3	83·5	46·2	80·9	76·7	57·6
Ethane	3·1	0·1	2·8	3·1	7·0	5·2	6·8	13·2	18·9
Propane	0·5	—	0·38	1·1	2·0	2·0	2·7	5·3	12·6
Butane	0·2	—	0·13	0·6	0·8	0·6	1·1	1·7	5·8
Pentane and higher hydrocarbons	0·2	—	0·12	0·7	0·4	0·1	0·5	0·8	3·0
Nitrogen	1·1	7·2	14·0	0·4	6·1	1·0	7·9	—	—
Carbon dioxide	0·5	18·0	0·77	9·6	0·2	44·9	0·1	2·2	2·1
Hydrogen sulphide	—	—	—	15·2	—	—	—	0·1	—
Calorific value MJ m^{-3}	38·2	29·1	32·8	30·9	42·1	24·2	39·6	46·9	57·6

(From M. H. Lowson (Ed.), *Our Industry—Petroleum*. Reproduced by permission of the British Petroleum Co. Ltd.)

toluene and xylene in samples of marine sediments from the relatively shallow depth of about 3·5 metres into the sediment layer, while Smith (1952) examined samples from greater depths, down to 30 metres, and obtained cycloparaffins, aromatic and asphaltic fractions. Such hydrocarbons, as we have seen, are normally found in crude oil, perhaps with the exception of ethylene.

(3) Porphyrins are an important group of organometallic compounds in the chemical composition of crude oil, and as was discussed in Chapter 3, green plants depend on porphyrins and similar compounds for photosynthesis. This is a powerful argument that green vegetation of some sort was involved in the formation of oil.

(4) Oil always occurs in or close to sedimentary rocks, and these rocks are usually of marine origin.

The detailed mechanism by which the organic material was converted into oil is not known with any certainty, neither is it known which parts of the vegetable matter gave rise to oil. However, it is believed that the following mechanism is approximately correct.

Fine-grained muds rich in planktonic remains were deposited on the sea bed in still water in an anaerobic environment. It is thought that fine-grained mud was preferred to coarser-grained material because examination of sedimentary rocks has shown that generally the finer-grained rocks are richer in organic matter than the coarser-grained. It is thought that still water was needed because the required sedimentation of fine particles and the required anaerobic conditions can be achieved only in relatively still water. Anaerobic conditions are thought to have been necessary, as in the case of coal and peat, to prevent complete decay. Slow decomposition by anaerobic bacteria turned the plankton remains into *sapropel*, an amorphous organic material which is known to form when plankton and similar organisms decay under water (Neishtadt, 1963). The sapropel was buried and subjected to conditions of elevated temperature and pressure, which, although not known in detail, must not have been too severe, since porphyrins are unstable above about 200°C. As was mentioned earlier, older petroleum deposits tend to have greater proportions of the lighter hydrocarbons, and it may be speculated that this is due to the higher temperatures or pressures to which they may have been subjected, perhaps bringing about the cracking of heavier molecules to yield lighter products. However, the nature of the environment of the oil-forming beds may have been of greater importance, as has been suggested by Haeberle (1951).

The sedimentary rocks in which this process occurred are known as the *source rocks*. In order to accumulate significant quantities of oil, it is necessary for the hydrocarbons to *migrate* from the source rocks to *reservoir rocks*, in which the oil and gas can be stored, and from which it can flow at acceptable rates when tapped by man. Clearly the reservoir rock must be appreciably porous and permeable, and this is achieved most readily in coarse-grained sediments, which, it will be recalled, are usually low in organic content and thus cannot in general act as source rocks themselves.

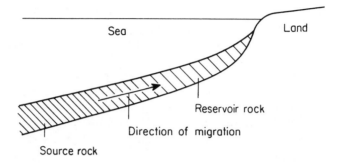

Figure 4.5. Highly simplified sketch of migration of oil from source rocks to reservoir rocks

   The migration of oil from the source rocks to the reservoir rocks is thought to occur in response to the effect of water. Some very simple theories have been proposed, such as that sketched in Figure 4.5, where it is proposed that the oil moves up under the influence of gravity, i.e. buoyancy, because it is less dense than the water evolved during the compaction of the sedimentary rock. Despite the attractive simplicity of such models, they are unsatisfactory because they appear to be unable to account quantitatively for the forces required to move the isolated droplets of oil from the source rock into the reservoir rock. The difficulty is that the mechanical forces acting on an isolated droplet of oil in the pore spaces between the grains of a rock such as sandstone are few; capillary attraction, surface tension, buoyancy and the pressure differences due to the flow of water through the rock pores. Surface tension acts to resist movement, and all the others act to produce movement, but the only one which is great enough to overcome the surface tension is the pressure difference due to flow, and then only if the oil droplet is streaked out in the direction of movement until the pressure difference between the ends of the droplet is sufficiently large. Numerical estimates of the extent of elongation needed to achieve the required pressure imply an elongation of the order of one metre, which at first sight appears excessive, but which might well be achieved during compaction of fine-grained muddy sediments. Having achieved the required degree of elongation, the droplets would move in the direction of water circulation. Baker (1967) has reviewed the proposed mechanisms of migration, and the reader is referred to his review for further information.
   Migration will continue in this way until some barrier intervenes. The reservoir rock is unable to act as such a barrier, for it is coarse-grained, and the restraining effect of surface tension will be less than in the fine-grained source rock. Instead, a very fine-grained rock known as a *cap-rock* is needed to restrain the oil, and if it is to prevent further migration it must in general overlie the reservoir rock. Unless such a cap-rock is present, accumulation is impossible, and the oil will continue to migrate until it reaches the surface as a seep.
   A combination of a reservoir rock overlain by a suitably-shaped cap-rock

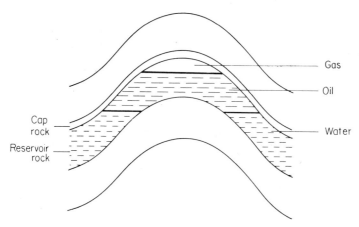

Figure 4.6. Oil trap—anticline. (This figure, and also Figures 4.7 to 4.11, are modified from Fox (1970) and are reproduced by permission of British Petroleum Company Ltd.)

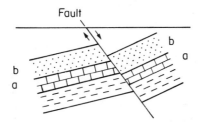

Figure 4.7. Oil trap—fault. Impervious rock b is brought opposite permeable rock a

produces a structure known as an *oil trap*, in which accumulation of oil can occur. There are numerous types of oil trap known to petroleum geologists, and only a few basic types will be discussed here.

The simplest type of trap is an *anticline*, sketched in Figure 4.6, in which the strata are folded gently into a concave-downwards arched shape, generating an arch of impermeable cap-rock below which oil can accumulate. Anticlines can cover vast areas, and consequently they are the most important type of trap in terms of the volume of oil retained in them.

Another type of geological structure which can trap oil is the *fault*, in which tension or compression in the rocks has caused the strata to slip relative to each other (Figure 4.7). Accumulation of oil can occur where an impermeable layer is brought opposite a permeable layer on the other side of the fault, sealing it off.

In another type of trap, an *unconformity*, one series of rocks has been deposited, tilted and eroded away. Further deposition has occurred, followed by

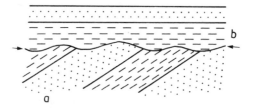

Figure 4.8. Oil trap—tilted discontinuity. An impervious deposit b overlies a permeable rock a

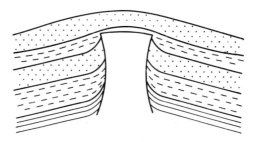

Figure 4.9. Oil trap—salt dome

further tilting, giving rise to a tilted discontinuity (Figure 4.8). Traps occur where a permeable layer in the earlier deposits is sealed by an impermeable layer in the later deposits.

A salt dome (Figure 4.9) can give rise to traps in two ways. Salt is impermeable and may cut off a permeable layer, creating a trap directly, and furthermore, the anticlines over a salt dome can create an anticline trap in the manner described earlier.

Compaction of the sea floor can give rise to traps in a manner similar to that sketched in Figure 4.10, and, finally, marine reefs and sandbars can give rise to *stratigraphic* traps in which the permeability of the rock changes suddenly, producing the situation shown in Figure 4.11.

Having arrived at the trap, secondary migration may occur, in which, because of differences in density, the water, oil and perhaps gas, separate out into layers. The extent to which this happens is influenced by the porosity of the reservoir rock (high porosity tends to encourage separation) and also by the prevailing pressure, which may force the gas to stay in solution in the oil.

4.7 World Distribution of Petroleum

Petroleum appears to be less widely distributed than coal, and the majority of the World's known reserves are located in a relatively small number of

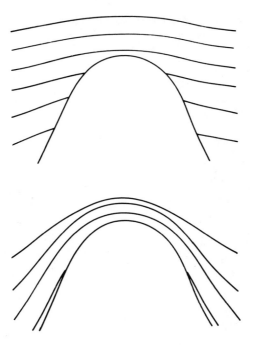

Figure 4.10. Oil trap formed by compaction
of deposits on the sea floor

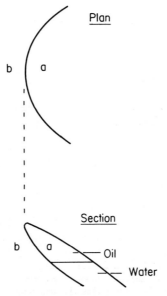

Figure 4.11. Stratigraphic
oil trap

Table 4.12 World oil reserves 1974 in barrels $\times 10^9$

Asia	
Australia	2·3
Brunei	1·6
China	20·0
Indonesia	10·5

Europe	
Norway	5·0
United Kingdom	13·0
U.S.S.R.	80·0

Middle East	
Abu Dhabi	21·5
Dubai	2·5
Iran	60·0
Iraq	31·5
Kuwait	64·0
Neutral Zone	17·5
Oman	5·3
Qatar	6·5
Saudi Arabia	132·0
Sharjah	1·5
Syria	7·1

Africa	
Algeria	7·6
Angola	1·5
Congo	4·9
Egypt	5·1
Libya	25·5
Nigeria	20·0

Western Hemisphere	
Argentina	2·5
Canada	9·4
Colombia	1·4
Ecuador	5·7
Mexico	3·6
Peru	1·1
Trinidad	2·2
United States	34·7
Venezuela	14·0

(Source: *Oil and Gas Journal*, December 31, 1973. Figures for the North Sea (United Kingdom and Norway) have been revised on the basis of figures from British Petroleum Ltd. Reproduced by permission of the Petroleum Publishing Co. and British Petroleum.)

places. Table 4.12 shows the known reserves in 1974, and it is clear that, for example, about 50 per cent of the World's proven reserves are located in the Middle East. However, it should be borne in mind that the rate of discovery of oil at the present time is so rapid that this figure is likely to be outdated very rapidly.

From Table 4.12 it is clear that many of the World's industrial nations, including the United States of America and Europe, have a relatively small share of the currently producing oil reserves, although it must be admitted that the picture will change when Alaska and the North Sea are in full production. For the present, however, the major industrial nations are obliged to import large quantities of oil, and consequently the import and export of petroleum is of great importance in World trade, thus contrasting markedly with coal, where in general the largest consumers are the largest producers.

As in the case of coal, it is impossible to give an exact figure for the amount of extractable oil in the World, particularly as the oil industry is engaged at present in major explorations in offshore areas. A fairly recent figure is 2×10^{12} barrels (Gaskell, 1973), which compares quite well with estimates made in the 1940's of about 6×10^{11} barrels. Since much of the difference arises from changes in recovery techniques, rather than discovery, it is reasonable to accept the recent estimates as fairly accurate. In energy terms, this represents about 10^{22} joules or 3×10^{15} kWh. Gas reserves associated with the oil amount to a further 10^{22} joules, doubling the total. There is an additional resource in oil shales and tar sands, which may well amount to more than the total of liquid oil quoted above, but how much of this can be exploited economically is uncertain. Thus it is evident that the World's resources of oil are appreciably less than those of coal, amounting to about 10 per cent of the energy stored in fossil fuels. Oil, like coal, is probably still being formed, but the rate of consumption is so vastly greater than the rate of accumulation that it is entirely correct to regard the World's oil resources as non-renewable.

4.8 Oil Shales and Tar Sands

So far we have considered only liquid oil resources. In addition there are large resources of hydrocarbon oils in the form of oil shales (mostly in the Green River area of Colorado, Utah and Wyoming, U.S.A.) and tar sands (mostly in the Athabasca area of Alberta, Canada). The Green River oil shales were formed some 50 million years ago by the deposition of silt and organic matter in shallow freshwater lakes, to be transformed by geological processes into a fine-grained marlstone impregnated with a hydrocarbon-rich polymer known as kerogen. The Athabasca tar sands are geologically rather more like normal oil reserves, in that they consist of reservoir 'rock' of sand impregnated with a viscous crude oil.

The extraction of useful oil from these deposits is clearly more difficult than from liquid oil wells, and until the recent rises in the price of oil it was commercially uneconomic. Nowadays, however, it is considered worthy of exploitation on a pilot-plant scale, and the Canadian sands are already in production.

The extent of hydrocarbon resources in these deposits is uncertain; Hubbert (1971) has estimated that the total recoverable oil from all these sources is about 5×10^{11} barrels, which is about a quarter of the current estimate of the world total of liquid oil, while de Nevers (1966) has estimated that the Green River shales alone can yield recoverable oil of 1×10^{12} barrels, equal to half the current figure for liquid oil, with a further 3×10^{11} barrels from the Athabasca tar sands.

References

Adams, P. J., *The Origin and Evolution of Coal*, H. M. S. O., London, 1960.

Averitt, P., U. S. Geological Survey Bulletin 1275, 1969.

Baker, E. G., in Nagy and Colombo, (Eds.), *Fundamental Aspects of Petroleum Geochemistry*, p. 299, Elsevier, Amsterdam, 1967.

Bestougeff, M. A., in Nagy and Colombo, p. 76, 1967.

Clymo, R. S., *Journal of Ecology*, **53**, 747 (1965).

Constantinides, G. and G. Arich, in Nagy and Colombo, p. 109, 1967.

Corbett, P. F., in M. H. Lowson, (Ed.), *Our Industry-Petroleum*, British Petroleum. London, 1970.

de Nevers, *Scientific American*, **214**, (2), 21, (1966).

Emery, K. O. and D. Hoggan, *Bulletin of the American Association of Petroleum Geologists*, **42**, 2174, (1958).

Fox, A. F., in Lowson, 1970.

Gaskell, T. F., 1973, *Energy—Erom Surplus to Scarcity?*, K. A. D. Inglis, (Ed.), Applied Science Publishers, Barking 1974.

Green, W. A. R., 1963, *Transactions of the 2nd International Peat Congress*, Leningrad, R. A. Robertson, (Ed.), p. 291, H. M. S. O., Edinburgh, 1968.

Haeberle, F. R., *Bulletin of the American Association of Petroleum Geologists*, **35**, 2238 (1951).

Hobson, G. D., in Nagy and Colombo, p. 6, 1967.

Holmes, A., *Principles of Physical Geology*, Thomas Nelson and Sons Ltd., London, 1965.

Hubbert, M. K., *Scientific American*, **225**, (9), 31, (1971).

Kurbatov, I. M., *Transactions of the 2nd International Peat Congress*, Leningrad, p. 133, 1963.

Lessing, L. P., *Scientific American*, **193**, (1), 58, (1955).

Mason, B., *Principles of Geochemistry*, Wiley, New York, 1966.

Meyerhoff, A. A., *J. Geology*, **78**, 1 (1970).

Meyerhoff, A. A. and C. Teichert, *J. Geology*, **79**, 285 (1971).

Moore, P. D. and Bellamy, D. J., *Peatlands*, Elek Science, London, 1974.

Neishtadt, M. I., *Transactions of the 2nd International Peat Congress*, Leningrad, p. 223, 1963.

O'Donnell, S., *New Scientist*, **63**, (904), 18 (1974).

Olenin, A. S., *Transactions of the 2nd International Peat Congress*, Leningrad 1963, p. 1, H. M. S. O., Edinburgh, 1968.

Peel, D. H., in Lowson, *Our Industry—Petroleum*, p. 158, British Petroleum, London, 1970.

Raistrick, A., C. E. and Marshall, *The Nature and Origin of Coal and Coal Seams*, English Universities Press, London, 1938.

Sachanen, A. N., in A. E. Dunstan, A. W. Nash, B. T. Brooks and H. Tizard, (Eds.), *The Science of Petroleum*, Oxford University Press, 1950.

Smith, P. V., *Bulletin of the American Association of Petroleum Geologists*, **36,** 411, (1952).

United Nations, *Statistical Yearbooks 1972 and 1973*, New York, 1973 and 1974.

van Krevelen, D. W., *Coal—Typology, Chemistry, Physics, Constitution*, Elsevier, Amsterdam, 1961.

Verhoogen, J., F. J. Turner, L. E. Weiss, and C. Wahrhaftig, *The Earth: an Introduction to Physical Geology*, Holt, Rinehart and Winston, Inc., New York, 1970.

Zimmermann, E. W., *World Resources and Industries*, (2nd ed.), Harper and Bros., New York, 1951.

Further Reading

Holmes, A., *Principles of Physical Geology*, Thomas Nelson and Sons, Ltd., London, 1965.

Lowson, M. H., (Ed.), *Our Industry—Petroleum*, British Petroleum Co. Ltd., London, 1970.

van Krevelen, D. W., *Coal—Typology, Chemistry, Physics, Constitution*, Elsevier, Amsterdam, 1961.

Verhoogen, J., F. J. Turner, L. E. Weiss and C. Wahrhaftig, *The Earth: an Introduction to Physical Geology*, Holt, Rinehart and Winston, Inc., New York, 1970.

5 The Use of Coal and Peat

5.1 Coal

The history of the industrial world is based on coal. Man has mined and used coal for about 3000 years; the Chinese are said to have used coal in 1000 B.C. Its use as an industrial fuel began in Britain in the 17th and 18th centuries, when the iron industry, faced with an increasing shortage of wood for charcoal, experimented with coal and later with coke as both fuel and reducing agent for blast furnaces. The partnership of coal and iron was an important stimulus to the Industrial Revolution, and machinery made of iron and powered by the coal-fired steam engine (itself made of iron) became the basis of a new way of life. The steam engine not only consumed coal, but also made coal-mining easier, because it simplified the problem of pumping water from deep mines, with the result that yet more coal was produced and consumed.

During the 19th centrury, coal was the sole important source of fuel for industrial process in Britain, and as the United States of America and Western Europe became industrialized, production of coal became important in those countries also. During the 20th century coal became increasingly challenged by oil in the more advanced industrial nations, particularly in the United States of America where native oil resources were available. In these countries the coal industry suffered a decline, not merely in its percentage share of the market for fuel (see Chapter 1), but also in the tonnage produced. Meanwhile, several less industrialized countries, notably China, were rapidly expanding their coal production, and the world total of coal production continued to expand slowly. A set of figures for the more important coal-producing countries is shown in Table 5.1a and figures for lignite are shown in Table 5.1b. It should be borne in mind that these figures are for production rather than consumption, and do not take imports and exports into account. It is evident that, although coal may have declined in relative importance, its position as a fuel is still of very great significance in the industrial world.

The uses to which coal is put have changed considerably over the years. Table 5.2 shows this changing pattern in the United Kingdom for the 50 years from 1923 to 1972. The following trends are apparent:

(1) Electricity generation has become by far the most important use of coal, and the tonnage increased steadily until very recently, despite competition from oil. At the same time, direct use of coal as fuel has declined. This reflects

97

Table 5.1a Coal production (millions of tonnes)

	Australia	Belgium	Canada	China (excl. Taiwan)	Czechoslovakia	France	West Germany	India	Japan	Korea (North and South)	Netherlands	Poland	Spain	South Africa	U.S.S.R.	U.K.	U.S.A.	World total
1950	17	27	15	41	18	51	126	33	38	3	12	78	11	26	185	220	505	1436
1951	18	30	15	51	18	53	136	35	43	—	12	82	11	27	202	226	520	1512
1952	20	30	14	64	20	55	141	37	43	1	13	84	12	28	215	230	458	1498
1953	19	30	13	67	20	53	142	37	47	1	12	89	12	28	224	228	440	1495
1954	20	29	12	80	22	54	146	37	43	2	12	92	12	29	244	228	379	1475
1955	20	30	11	94	22	55	149	39	42	3	12	94	12	32	277	225	442	1597
1956	20	29	11	106	23	55	153	40	47	4	12	95	13	34	304	226	477	1686
1957	20	29	10	124	24	57	150	44	52	5	11	94	14	35	329	227	468	1734
1958	21	27	9	271	26	58	150	46	50	7	12	95	14	37	353	219	389	1762
1959	21	23	8	348	25	58	143	48	47	10	12	99	14	36	365	209	390	1890
1960	23	22	8	420	26	56	143	53	51	12	12	104	14	38	375	197	392	1982
1961	24	22	7		26	52	144	56	54	13	13	107	14	40	377	194	379	1807
1962	25	21	7		27	52	142	61	54	16	12	110	13	41	386	201	396	1852
1963	25	21	8		28	48	143	66	52	18	12	113	13	42	395	199	430	1924
1964	28	21	8		28	53	143	62	51	21	11	117	12	45	409	197	455	1992
1965	32	20	9	327	27	51	135	67	50	25	11	119	13	48	428	190	475	2042
1966	32	17	8	227	27	50	126	68	51	27	10	122	13	48	407	177	493	2047
1967	33	16	8	300	26	48	112	68	47	29	8	124	12	49	414	175	508	1949
1968	38	15	8	325	26	42	112	71	47	31	7	129	12	52	416	167	501	2009
1969	40	13	8		27	41	112	75	45	32	6	135	12	53	426	153	513	2053
1970	44	11	12	360	28	37	111	74	40	36	4	140	11	55	433	145	550	2131
1971	45	11	14	390	29	33	111	70	33	37	4	146	11	59	341	147	503	2124
1972	49	11	16	400	28	30	103	75	28	—	3	151	11	58	451	120	535	2145

(Derived from United Nations Statistical Yearbook 1972, United Nations, New York 1973. Reproduced by permission of the United Nations.)

Table 5.1b Lignite production (millions of tonnes)

	Australia	Bulgaria	Czechoslovakia	East Germany	West Germany	Hungary	Poland	U.S.S.R.	Yugoslavia	World total
1950	7	6	28	137	76	12	5	76	12	381
1951	8	6	30	151	83	14	5	79	11	413
1952	8	7	33	158	84	17	5	86	11	435
1953	8	8	34	173	85	19	6	96	10	465
1954	9	9	38	182	88	19	6	103	13	494
1955	10	10	41	201	90	20	6	115	14	535
1956	11	10	46	206	95	18	6	125	16	565
1957	11	12	51	213	97	19	6	135	17	593
1958	12	13	57	215	94	22	8	143	18	613
1959	13	14	54	215	94	23	9	141	20	617
1960	15	15	58	225	96	24	9	138	21	638
1961	17	17	65	237	97	25	10	134	23	662
1962	17	19	69	247	101	25	11	131	24	684
1963	19	20	73	254	107	27	15	137	26	718
1964	19	24	76	257	111	27	20	145	28	747
1965	21	24	73	251	102	27	23	150	29	742
1966	22	25	74	249	98	26	25	144	28	737
1967	24	27	71	242	97	23	24	142	25	723
1968	23	28	75	247	102	23	27	136	26	738
1969	23	29	79	255	107	22	31	138	26	765
1970	24	29	82	261	108	24	33	145	28	793
1971	23	27	85	263	104	23	35	150	30	804
1972	24	27	86	248	110	22	38	152	30	806

(Derived from United Nations Statistical Yearbook 1972, United Nations, New York 1973. Reproduced by permission of the United Nations.)

the demand for 'convenience' fuels—solid fuel is usually the least convenient source of energy in most applications, while electricity is often the most convenient. This trend can be expected to continue. Over the period covered by the table, the efficiency with which coal is converted into electricity has increased considerably, and the roughly tenfold increase in tonnage of coal represents a much larger increase in electricity consumption. Until recently, coal has been more expensive than oil as a means of generating electricity in the United Kingdom, and the continuing increase in coal consumption for this purpose reflects political decisions on the part of the government.

(2) Coal gasification reached a peak of consumption in the mid-fifties, and thereafter was steadily replaced, first by oil and then by natural gas. The North Sea gas fields are expected to supply the United Kingdom market for the next 25 years or so, and it is unlikely, therefore, that coal will win much of this market back for some decades.

(3) The figures for coke ovens cover both metallurgical coke production

for steelworks, and other types of coke production including smokeless fuels but excluding gasworks. By far the most important consumer in this category is the steel industry. The decline in consumption in recent years is not due to a decline in steel production, but to increasingly efficient use of coke. This is partly due to the use of other fuels, especially oil, as a supplementary source of heat in blast furnaces, thus conserving coke for its main function of reducing the ore. It is also due to changes in blast furnace technique, such as oxygen enrichment, high top pressure, sintering of the charge before loading into the furnace, and the use of larger furnaces with higher thermal and chemical efficiency. This trend has slowed down lately, and further major improvements in blast furnace practice are unlikely in the near future. There is a continuing shortage of good coking coal, and the steel industry is likely to turn to alternative technologies, such as direct reduction, which do not require such good-quality coal. It will still require carbon as a reducing agent, however, and it is likely to continue to use coal as its source of carbon.

(4) Collieries appear to have ceased to use their own product as a major fuel. This is largely illusory, however, since many of the operations which were formerly coal-fired are now done electrically, and since much of the United Kingdom's electricity is coal-fired, the energy comes ultimately from coal.

(5) Railways have only very recently ceased to use coal; the steam locomotive is still part of most adults' memories. This market was almost eliminated by the introduction of diesel locomotives, but the increasing use of electricity on main lines has effectively won some of it back for coal.

(6) The domestic market for coal has declined since the mid-fifties, partly because of smokeless-zone legislation, and partly with the spread of central heating, which has usually been fired by oil, gas or electricity. Despite recent rises in the price of oil, this trend is likely to continue.

(7) The United Kingdom has ceased to be a major exporter of coal—tonnage has dropped by a factor of 45. It is conceivable that new discoveries such as the Selby seam (see Section 5.5) might alter this situation, but it might be that conservation of national reserves will be given higher priority than any financial advantage to be gained by export.

(8) The last column, marine fuel, has been eliminated by oil, which apart from greater convenience has the advantages of greater energy content per unit mass and per unit volume, both of which are important to ship designers and operators.

A set of comparable figures for the United States and Canada is shown in Table 5.3. Broadly similar trends are evident; differences are due mainly to the availability of native reserves of oil and natural gas almost since the beginning of industrialization, a luxury which has been denied to the United Kingdom until very recently.

Now that we have considered the quantitative aspects of coal consumption, we shall devote the rest of the chapter to methods of production and use of coal and peat. The reader should bear in mind that there are variations in

Table 5.2 Coal production and consumption for the United Kingdom

Year	Supply (Total)	Electricity	Gas	Coke ovens	Collieries	Railways	Other Industries	Domestic	Miscellaneous	Export	Overseas marine bunkers
1923	281·5	7·3	17·5	20·1	17·2	14·2		103·6		80·6	18·4
1924	268·9	7·8	18·4	19·2	16·9	14·4		109·5		62·5	18·0
1925	247·4	8·2	18·1	16·7	15·6	14·3		104·2		51·4	16·8
1926	151·8	8·4	17·6	7·2	7·7	12·3		69·0		20·8	7·7
1927	255·1	9·1	18·8	17·7	14·8	14·5		109·1		51·7	17·1
1928	241·7	9·4	18·6	17·7	13·7	14·0		98·6		50·7	17·0
1929	262·1	10·0	18·9	20·3	13·9	14·3		104·7		61·1	16·7
1930	246·1	9·9	18·7	17·5	13·7	13·8		99·0		55·6	15·8
1931	220·1	9·8	18·4	12·9	12·8	13·2		92·9		43·2	14·8
1932	212·9	10·0	18·0	12·9	12·2	12·6		91·4		39·3	14·4
1933	212·0	10·5	17·7	13·3	11·8	12·6		91·1		39·4	13·6
1934	224·2	11·4	18·2	17·2	11·9	13·1		96·6		40·0	13·7
1935	227·4	12·4	18·3	17·7	11·8	13·2		100·1		39·0	12·7
1936	232·2	13·8	19·4	20·4	12·0	13·7		103·4		34·8	12·2
1937	244·8	15·0	19·7	22·3	12·4	14·0		106·4		40·7	11·9
1938	227·7	15·1	19·4	19·4	12·1	13·4		99·2		36·2	10·7
1939	236·6	16·2	19·2	20·7	12·3	13·1		105·7		37·3	9·8
1940	228·6	18·4	18·1	22·7	12·5	13·7		111·0		19·8	7·1
1941	210·7	20·7	19·4	21·4	12·3	14·0		109·3		5·2	4·4
1942	208·4	22·7	21·0	21·9	12·2	14·7		105·6		4·4	3·6
1943	200·6	23·0	21·1	21·2	11·8	15·0	43·1	39·6	16·0	4·9	3·3
1944	194·9	24·5	21·0	20·4	11·3	15·2	40·2	37·6	17·5	3·8	2·4
1945	187·2	23·9	21·3	20·4	10·7	14·8	37·0	35·6	16·1	5·6	3·1

Year											
1946	194·2	26·6	23·1	20·4	10·8	15·0	35·9	38·6	16·4	4·2	4·8
1947	201·5	27·5	23·1	20·1	11·2	14·5	36·4	36·6	15·5	0·9	4·5
1948	211·2	29·3	25·0	22·7	11·4	14·5	37·0	38·6	16·2	10·9	5·5
1949	218·2	30·5	25·7	23·0	11·0	14·6	36·6	38·9	16·2	14·3	5·2
1950	220·4	33·4	26·6	23·0	10·9	14·4	37·9	40·7	16·3	13·1	4·1
1951	227·6	36·0	27·8	23·8	10·8	14·3	37·7	41·6	16·4	7·8	3·8
1952	225·6	36·1	28·1	25·5	10·5	14·1	37·4	39·5	16·3	11·8	3·4
1953	230·4	37·3	27·5	26·3	10·1	13·6	37·7	40·0	15·8	13·9	2·9
1954	231·2	40·2	27·7	27·0	9·7	13·2	38·8	41·5	16·4	13·7	2·5
1955	236·1	43·6	28·3	27·4	8·8	12·4	37·7	41·4	16·2	11·9	2·2
1956	230·6	46·3	28·2	29·8	8·0	12·3	38·1	40·1	15·3	8·2	1·6
1957	224·9	47·1	26·8	31·2	7·3	11·6	36·2	38·1	15·1	6·8	1·2
1958	209·4	46·8	25·2	28·2	6·6	10·5	36·8	34·2	14·5	4·1	0·9
1959	194·0	46·7	22·9	26·1	5·7	9·7	33·9	32·2	12·7	3·7	0·7
1960	204·4	51·9	22·7	29·0	5·1	9·0	35·0	31·8	12·7	5·3	0·3
1961	203·0	55·6	22·6	27·2	4·6	7·8	32·9	29·7	11·9	5·7	0·1
1962	198·6	61·4	22·5	23·9	4·3	6·2	33·2	27·8	12·4	4·8	0·1
1963	206·2	67·9	22·5	23·9	4·0	5·0	32·5	26·3	12·5	7·6	
1964	197·2	68·5	20·5	25·9	3·8	3·9	28·3	25·1	11·9	6·1	
1965	190·5	70·4	18·3	26·1	3·5	2·8	27·7	24·5	11·7	3·8	
1966	182·0	69·0	17·0	24·7	3·1	1·7	25·9	22·6	11·0	2·8	
1967	177·6	68·3	14·8	24·0	2·9	0·8	23·3	20·8	10·8	1·9	
1968	169·9	74·4	10·9	25·3	2·4	0·2	23·0	20·3	9·7	2·7	
1969	155·7	77·1	7·0	25·7	2·0	0·2	21·7	19·0	9·0	3·6	
1970	147·1	77·2	4·3	25·3	1·9	0·1	19·6	17·5	8·8	3·3	
1971	149·5	72·8	1·8	23·6	1·6	0·1	15·8	14·7	7·5	2·6	
1972	121·8	66·6	0·6	20·4	1·4	0·1	11·7	12·3	6·4	1·8	

(Derived from U.K. Energy Statistics 1973, Department of Energy, 1973 and from Statistical Digest 1966, Ministry of Power, 1966. Reproduced with the permission of the Controller, H. M. Stationery Office.)

102

Table 5.3 Coal production and consumption for North America (Canada and USA) in million metric tonnes

Year	Production	Patent fuel plants	Coke ovens	Gas works	Electricity	Coal mines	Railways	Industry	Domestic and other
1950	520·7	3·0	99·3	5·9	102·6	13·3	66·2	84·7	108·9
1951	534·7	2·6	108·0	5·5	116·2	15·0	59·8	89·8	97·2
1952	471·7	2·5	93·7	3·9	116·7	12·9	44·5	82·3	88·1
1953	452·9	2·1	107·9	3·4	126·9	12·2	33·4	78·2	78·8
1954	390·8	2·0	81·9	1·7	127·9	8·3	22·1	67·4	67·7
1955	453·8	2·0	102·6	1·5	153·9	9·5	19·5	73·8	68·1
1956	489·4	1·8	101·7	1·5	168·1	10·1	16·6	78·3	64·5
1957	477·5	1·3	103·5	1·3	170·4	10·3	11·0	71·4	49·2
1958	397·9	1·1	74·1	0·9	162·9	9·8	4·9	64·9	49·5
1959	398·0	0·9	77·3	0·8	174·6	12·0	3·1	56·3	42·3
1960	399·6	0·7	78·7	0·6	182·5	11·4	2·5	59·4	41·6
1961	386·1	0·6	72·1	0·5	186·3	11·2	2·2	58·6	37·6
1962	412·8	0·5	72·6	0·5	198·7	11·1	1·8	58·1	36·5
1963	438·4	0·5	76·0	0·5	217·6	12·4	1·6	59·7	32·0
1964	463·2	0·3	86·3	0·6	232·6	11·8	1·4	62·2	27·7
1965	483·9	0·3	91·8	0·6	252·9	11·1	1·3	64·4	25·9
1966	501·0	0·3	92·8	0·6	272·4	16·5	1·3	61·3	25·5
1967	516·9	0·2	89·5	0·5	278·9	17·7	1·2	53·9	21·8
1968	508·6	0·2	89·5	0·5	300·4	18·4	0·9	50·8	19·7
1969	521·3	0·1	91·1	0·5	313·5	23·3	0·8	40·3	16·6
1970	562·0	0·1	94·7	0·5	322·6	3·8	0·7	56·2	15·5
1971	516·8	0·1	82·0	0·5	329·2	2·9	0·5	44·0	14·5
1972	551·4	0·1	86·3	0·4	348·4	2·8	0·4	42·8	13·5

(Figures derived from Statistics of Energy 1958–1972, OECD, Paris 1974 and from Statistics of Energy 1950–1964, OECD, Paris 1966. Reproduced by permission of OECD.)

practice from one country to another, and that particularly in the case of coal mining it is difficult to generalize adequately.

5.2 Coal Mining

Prospecting for coal has been relatively easy in the past. In most of the World's important coal-bearing regions some of the seams outcrop at the surface, so that a simple geological map suffices as a guide to the prospector. Such coal-fields are described as *exposed*. Coalfields in which the seams are not visible at the surface are described as *concealed*, and in these cases drilling is necessary to locate the limits of the field. Most of the exposed coalfields of the industrial nations are now fully developed, so the discovery of new reserves depends on geological surveys conducted by trial borings, to assess not only the position but also the thickness and extent of the seams. Such surveys have been done offshore as well as on land, notably off the coast of Durham in North-East England, in water depths of up to 30 metres or thereabouts. In some cases

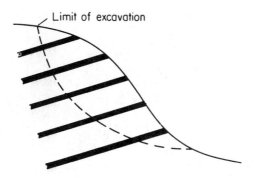

Figure 5.1. Open-cast mining of seems outcropping on a hillside

seismic surveys (see Chapter 6) have been used to locate likely areas prior to drilling.

When the coal seams are relatively near the surface they can be mined by *opencast* or *strip* working. The material above the coal seam, known as the overburden, is removed by excavation with a large dragline, and is dumped at a convenient place near the workings. The coal is dug out and carried away, after which the overburden is returned to its original location and the land is restored. In the past the opencast technique has been confined to seams which are quite close to the surface, but the development of very large earth-moving equipment has made opencast mining possible at greater depths, especially when several seams outcrop at the side of a hill (Figure 5.1). Some of the largest machines in the world are employed in this type of work. The scale of typical plant can be judged from Figure 5.2.

From the point-of-view of high productivity and low cost, opencast mining is very satisfactory. It also has a good safety record, and the working conditions are inherently more healthy than in underground mines. There are two main problems. Firstly, the whole operation is done where it can be seen, and as it is not visually attractive there are strong environmental objections to it, especially when done near centres of population or in areas of great natural beauty. Secondly, the reclamation of the land, although perfectly feasible, iš expensive, and the cost of doing so has to be set against the low cost of the mining operation itself. It has often been difficult to convince interested parties that reclamation will be done to the required standard, especially in those countries such as the United States where mining is carried out by privately-owned companies. It must be admitted that the mining companies have a great deal to answer for in some of the older mining areas, but it must be said in fairness that reputable mining organizations are nowadays most concerned about the importance of satisfactory reclamation.

Opencast mining is fairly common in the United States but in the United Kingdom, where the population density is higher and the coal seams are in general deeper, it is less attractive. Nevertheless, between 1952 and 1973 about

Figure 5.2. Large dragline on an open-cast site in North East England. The scale of the machine can be judged from the size of the bulldozer at the right of the machine, and by the 100-ton capacity of the bucket. (Reproduced by permission of the National Coal Board)

180 million tonnes of opencast coal were mined in Britain, at a profit of £130 million after allowing for all costs including restoration of the land, and in 1972/1973 the output was 10·5 million tonnes (National Coal Board, 1974).

When the seams are too deep for opencast mining, underground mining is necessary. We can classify such mining techniques roughly according to two criteria: the method by which the seam is approached, and the method of working the seam itself.

If a coal seam of substantial thickness outcrops at the surface, a simple method of approach known as *drift mining* is possible, in which a tunnel or *adit* is cut into the seam along the dip of the strata (Figure 5.3). Such an approach is also possible when the seam is concealed, provided that it is not too deep. The advantage of an adit is that both coal and miners can be transported on a railway, which is much more efficient in both time and fuel than a lift. In cases where the adit is very long, it may be quicker to raise and lower the men by lift, but a railway is still preferred for the coal. However, in many cases it is impracticable to gain access to the mine via an adit, and a vertical *shaft* is

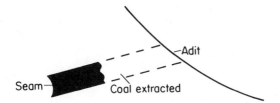

Figure 5.3. Drift mining of a seam outcropping at
the surface

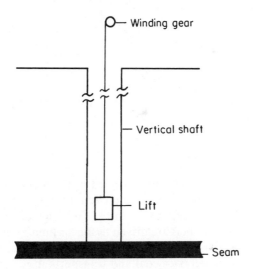

Figure 5.4. Shaft mining

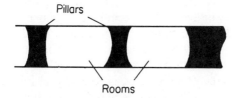

Figure 5.5. Room-and-pilar mining

used instead (Figure 5.4). Both men and coal must then travel in lifts, which
slows down the removal of coal from the mine.

There are two basic methods of working the seam, which differ in the method
used to support the roof. In *room-and-pillar* mining (Figure 5.5) the coal is
removed, to form a series of rooms, and pillars of coal are left between the
rooms to hold up the roof. This minimizes the amount of subsidence of the
overlying land, and it also saves labour because relatively few artificial
roof supports are needed, but it clearly results in a massive waste of coal.

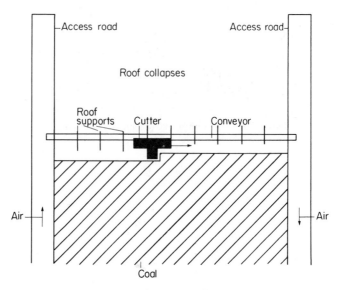

Figure 5.6. Longwall mining

For this reason it has been largely avoided in British coalfields since the 17th century, although it is useful in certain circumstances where subsidence must be minimized, as will be discussed later. The *longwall* method is generally used nowadays. Here tunnels are driven to the far limit of the mining operation (Figure 5.6), and a coal-cutting machine (Figure 5.7) moves to and fro between pairs of tunnels. This removes coal and transports it on a conveyor to one of the tunnels, whence it is removed by railway. As the coal-cutting machine works, the roof is held up by powered roof supports (Figure 5.8), which move forward after the cutter has passed. Behind the supports the roof is allowed to collapse. The advantage of this technique is that it removes all the coal. The principal disadvantage, namely the labour required to deal with the roof supports, has been partly solved with the introduction of self-advancing powered roof supports, but there remains the problem of subsidence.

Usually some degree of subsidence is tolerable in mining areas, but there are circumstances in which it must be avoided, notably when large industrial buildings or historic monuments are located directly above the mine workings. In these circumstances it is usual to leave the coal beneath the building unmined. A particularly interesting subsidence problem occurs with the newly-discovered Selby seam in Yorkshire. Here the overlying land is rather flat, and it is crossed by the River Derwent, so there is a danger that uncontrolled subsidence would cause serious flooding. The problem is accentuated by the unusually large thickness of the seam (by British standards) of about 3·4 metres. To alleviate the problem it is proposed to control the subsidence to within acceptable amounts by leaving coal unmined beneath the river, and by gradually increasing the fraction of coal removed as the workings get further away. This is shown

Figure 5.7. Power shearer. (Reproduced by permission of the National Coal Board)

Figure 5.8. Self-advancing roof supports. (Reproduced by permission of the National Coal Board)

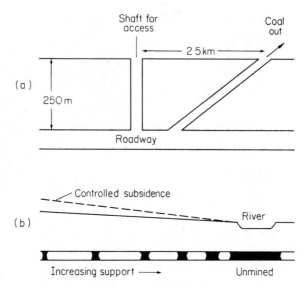

Figure 5.9. Sketch of proposals for mining the new Selby seam in Yorkshire, England

in Figure 5.9, which also gives a general impression of how the seam will be approached.

Among the principal problems of underground mining are the thickness of the seam, ventilation, drainage, dust and geological faults.

Thickness

Except in the case of opencast mining, there is a lower limit of about one metre to the thickness of seams which can be mined by present-day techniques. This is set by the problems of access for men and machinery. At this thickness the headroom for the miner is probably little more than 80 cm because of the thickness of the roof supports; thus the working conditions cannot be described as pleasant. It is conceivable that an entirely remote-controlled machine could work thinner seams, and also give the operator a more agreeable working environment, but up to the present the variability of coal seams has rendered full automation impossible. Nevertheless, experimental remote-controlled coalface equipment has been used by the British National Coal Board since 1956, an interesting example of which is the nucleonic-controlled power loader (Figure 5.10), which uses gamma radiation to measure the thickness of the seam and thus prevent the cutter from cutting rock from the roof or the floor of the workings. In 1972 there were six such machines working in the United Kingdom, and, although this particular machine is limited in its application to cases where a certain amount of coal must be left to support the roof, it is doubtless the fore-runner of other automatic machines, perhaps enabling narrower seams to be worked routinely.

Figure 5.10. Nucleonic-controlled shearer. (Reproduced by permission of the National Coal Board)

Ventilation

Adequate ventilation is essential, not only to provide enough air and reasonable working conditions for the miners, but also to keep down the concentration of inflammable gases such as methane. The usual arrangement is to sink two shafts from the surface, and to draw air up one of them with fans, thus drawing air down the other shaft and through the workings. An elaborate arrangement of underground doors ensures that the air passes through all the passages and coal faces. Individual faces are ventilated by passing air along one of the access tunnels, across the face and back along the other access tunnel.

Drainage

The coal seam is usually underlain by a layer of clay which is largely impervious to water. Inevitably this necessitates drainage, and before the invention of the steam engine it was very difficult to ensure adequate drainage in deep workings. Nowadays, of course, the actual pumping is not difficult, but if the pumping should cease for any reason, such as roof falls or industrial disputes, serious flooding can occur which may permanently damage mining equipment.

Dust

Coal dust is a hazard both to health and to safety. Careful monitoring of the particle concentration in the mine air, together with dust control measures such as spraying the coal with water jets as it passes along the conveyors, have helped to reduce the incidence of lung complaints, but the rate of disease attributable to dust is still somewhat high. This topic, and other matters of health and safety, is discussed in Chapter 18.

Faults

Geological faults, in which there is a sharp vertical discontinuity in the coal seams, are a serious problem to mining engineers. Severe faulting can halt production completely, and since the capital cost of the machinery at the coal face is high, a stoppage of production represents a serious financial loss.

Despite all the problems enumerated above, there is no doubt that coal mining is a thriving industry, particularly following the recent rise in the price of oil. The competitiveness of coal is very sensitive to economic circumstances, however, and this will be discussed further in Chapter 18.

5.3 Coal-fired Electricity Generation

The electricity industry is by far the largest single consumer of coal. In the United Kingdom in 1972/1973 the electricity industry consumed 69 million tonnes, representing 53 per cent of the National Coal Board's output, and in the United States the situation is similar. Moreover, the proportion of coal used in this way seems likely to continue to rise, since coal is in general a difficult fuel to work with on all but the largest scale, while electricity offers the ultimate in convenience for most purposes. We shall, therefore, examine the technology of coal-fired electricity generation in some detail.

From economic necessity, it is essential that the coal used in electricity generation is the cheapest obtainable, because in a typical plant the cost of fuel represents about 80 per cent of the cost of the electricity sent out, and any reduction in fuel cost represents a major saving. Consequently power stations are designed to use poor quality coal, and particularly small coal which is produced in large quantities by modern mechanical mining methods and is almost unusable except in large industrial plant, so that it commands a comparatively low price. It is also important to minimize transport cost by siting the power station fairly close to the coal mines, because the cost of carrying the coal is greater than the cost of transmitting the electricity produced. For this reason, there is a massive concentration of power stations along the River Trent in the English Midlands, sometimes called the Trent Bus Bar, where both coal and cooling water are readily available, and there is a similar concentration at the Four Corners power station in New Mexico, U.S.A., in proximity to the Navajo open-cast mine.

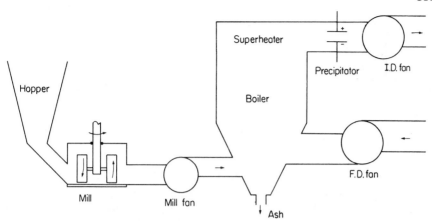

Figure 5.11. Pulverized fuel plant for large scale combustion of coal. The forced draught (F.D.) fan supplies combustion air, the mill fan supplies fuel and primary air, and the induced draught (I.D.) fan keeps the pressure in the combustion chamber just below atmospheric

Although mechanical conveyors are sometimes used to bring coal from nearby mines, most of the coal arriving at a generating station is transported by rail, usually in trains of about 50 wagons, each holding about 20 tonnes. Special equipment is used to ensure easy and rapid handling. From the wagons the coal is taken by conveyor either to storage or to the plant's bunkers. Usually the choice of storage or immediate use is determined by the urgency of the station's need, but in some cases the nature of the coal is of importance. Some coals are rather high in sulphur, and may be spontaneously inflammable if stored, and in these circumstances immediate use is essential.

From now on the fate of the coal is as shown in Figure 5.11. The bunkers feed the coal into a mill where it is crushed into a fine powder known as *pulverized fuel*. This is done partly to make it easy to handle—it can be borne along in an air stream and treated as a fluid—and partly to maximize the surface area so that combustion is rapid and complete. Mills are notoriously unreliable, and it is usual to equip a boiler with several small mills rather than with one larger one, so that a failure of one of them does not restrict the boiler's output.

The powdered coal is blown from the mill into a boiler, where it meets a supply of air for combustion. Ignition is achieved initially with an oil flame, but once the air-borne coal is ignited, it burns continuously without assistance. Heat from the combustion process is extracted in the boiler and used to raise steam to drive a turbine. The details of this part of the process are considered in Chapter 7, and for the present it is sufficient to note that for maximum efficiency it is necessary to achieve the highest possible steam temperature and also to transfer as much heat as possible from the combustion products to the steam.

Coal always contains some ash, and the low-grade coal used in power stations often has quite a high additional incombustible content because

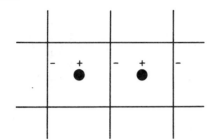

Figure 5.12. Principle of electrostatic precipitation

it is not cleaned before use. All the incombustible content ends up as ash, some of which falls to the bottom of the boiler, but most of which is carried along in the gas stream and must be removed before discharge into the atmosphere.

There are several methods by which solid particles can be removed from a gas stream, of which the most obvious is filtration, but on the very large scale of a power station the most effective system is electrostatic precipitation. Because they are still quite hot, many of the particles will be ionized. If they are passed through a mesh (Figure 5.12) between the bars of which there is a strong electric field, ions will be attracted to the electrode of the opposite polarity, where they will be discharged, and will remain adhering to the electrode because of polarization induced electrostatic forces. Neutral ash particles will tend to become ionized in the strong electric field, and they too will be removed. When the mesh is full, the gas stream is diverted to a second precipitator, while the first one is switched off and shaken vigorously to remove the ash.

Ash from the precipitators and also from the bottom of the boiler is collected in a stream of water, in which it forms a thin slurry. The slurry is pumped out to shallow lakes known as *ash lagoons*, in which the ash precipitates as a grey sand-like deposit. Until fairly recently the disposal of this ash was a serious and expensive problem, and it was usually left in the lagoons, which steadily accumulated around the larger power stations. However, the British Central Electricity Generating Board has conducted extensive research and marketing campaigns to find useful purposes for the ash, and most of its output is now used, either by the Board itself or by other concerns which buy it from the Board.

The remaining gases leaving the precipitators are sent up a tall flue and released into the atmosphere. For the most part they are clean, consisting of water vapour and carbon dioxide, which are tolerable under normal circumstances (see Chapter 2), but there is a small amount of sulphur dioxide (see Chapter 15), which arises from the sulphur in the coal and cannot be eliminated from the flue gases without great expense. Limitation of SO_2 emissions is therefore possible only by choosing a low-sulphur coal in the first place. Some concern has been expressed about the sulphur content of rain falling over Scandinavia, and it has been suggested that the use of high-sulphur coal for power generation in Western Europe is the major source of the problem.

Methods of burning high-sulphur coal without causing SO_2 emission will be discussed later in this chapter.

5.4 Small-scale Coal Burning Equipment

At the other end of the size range, but of considerable economic importance, are domestic and small industrial heating plants. Although most of this market has been invaded by oil, especially in industry, there is still an extensive market for coal.

The traditional open fire is inefficient both thermally and chemically. The thermal efficiency is often as low as 10 per cent, most of the heat being lost up an excessively large chimney, and the inefficient conditions of combustion, especially when newly stoked, lead to a loss of much of the volatile content of the fuel as smoke. Modern open fires have improved both thermal and combustion efficiencies, mainly by control of the primary air supply (the draught reaching the fire from beneath the grate) and by a reduction in the aperture at the throat of the chimney, and the British National Coal Board now quote a thermal efficiency of 37 per cent (Bradbury, 1973) for a good open fireplace, but there is still the problem of smoke. Since the mid-1950's, anti-pollution legislation has empowered local authorities in many parts of the World to restrict the emission of smoke from chimneys, and in areas affected by smoke control orders it is now necessary to use smokeless fuels (considered in detail in the next section). Most of these fuels require rather vigorous draughts of primary air, and fireplaces intended for such fuels often have a controllable under-floor flue or even an electrically-driven fan.

Most of the output of an open fire is by radiation, and this creates a problem well known in households which rely on open fires as their main source of heating, namely that only the area around the fireplace is heated adequately. Moreover, there is a considerable loss of potentially useful heat in the flue gases. An improvement in thermal efficiency, and in the comfort in the further corners of the room, can be achieved by passing the flue gases through a heat exchanger to generate convected heat and also through a boiler to supply hot water, and this leads to a design typefied by Figure 5.13. The glass front allows radiated heat to reach the room, slightly improving thermal efficiency and contributing greatly to the acceptability of the appliance as a feature of the room. Smokeless fuels are usually necessary in such appliances, for, although the combustion efficiency is better than in an open fire, there would be some loss of volatiles from bituminous coal and the glass doors would quickly become opaque.

For several reasons smokeless fuels are not entirely acceptable to many consumers, and considerable effort has gone into designing a domestic appliance which will burn ordinary bituminous coal smokelessly. A diagram of such an appliance, which rejoices in the name of 'the smoke eater', is shown in Figure 5.14. The operation is as follows. A small electric fan forces primary air through the firebed of burning coal so that smoke and volatiles are drawn downwards. Unburnt volatiles pass to the back of the appliance where they

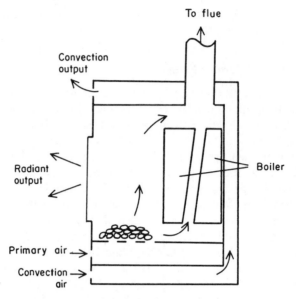

Figure 5.13. Solid fuel room heater (Simplified for clarity)

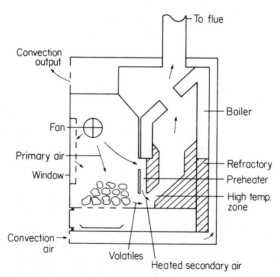

Figure 5.14. Bituminous coal fired smokeless room heater. (Modified from Bradbury (1973). Reproduced by permission of the Women's Solid Fuel Council)

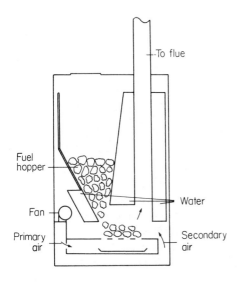

Figure 5.15. Gravity-feed boiler. Modified from Bradbury (1973). (Reproduced by permission of the Women's Solid Fuel Council)

burn in a special high-temperature combustion zone lined with hot refractory material. The fire is visible through a double-glazed window, and part of the secondary air for combusion passes through louvres in the inner pane of the window, keeping the glass clean. The flue gases pass over a boiler to supply hot water, and there is also a heat exchanger to supply convected heat to the room. A typical appliance has a rating of 3·5 kW for direct space heating and 6·4 kW for water heating, so it has some spare water-heating capacity for central-heating of other rooms in the house.

The appliances described above suffer from the major disadvantage that they require stoking by hand at fairly frequent intervals. A genuine central heating boiler requires some form of automatic stoking, the simplest of which is gravity feeding. A typical design is shown in Figure 5.15. The fuel falls down automatically from a hopper, and air is supplied to the fire by a fan, producing a sufficiently high temperature to fuse the ash into clinker, which can be removed manually at infrequent intervals. A thermostat measures the water temperature and switches off the fan when the required temperature is reached. The fuel for such an appliance is usually small pieces of anthracite, which is a naturally smokeless fuel having a higher density than most other smokeless fuels, allowing a smaller size of hopper to be used without the inconvenience of frequent stoking. Normally stoking and clinker removal are done once daily.

Larger coal-fired boilers often use an underfeed stoking system in which the fuel is fed to the under-side of the firebed by a screw (Figure 5.16). The system works surprisingly well, especially with friable fuel. Some of the last coal-fired railway engines were equipped with similar stoking devices.

5.5 Smokeless Fuels

Smoke arises mainly from incomplete combustion of the volatile components

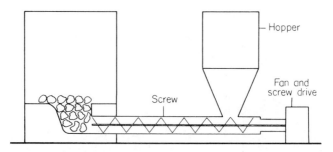

Figure 5.16. Underfeed screw stoker

of coal. A smokeless fuel must, therefore, be low in volatiles, but this creates the problem of difficult ignition and necessitates a strong draught of primary air for satisfactory burning. Developments in smokeless fuels have concentrated on minimizing these problems. In this section we shall study the characteristics and the methods of manufacture of several different types of fuel.

Natural Smokeless Fuels

There are two general classes of natural smokeless fuel. *Anthracite* has a low volatile content (less than 10 per cent), a high calorific value of about 33 MJ kg^{-1} and burns with a short blue flame. Its relatively high bulk density of about 720 kg m^{-3} is about the same as that of bituminous coal, so it is convenient to store and is very satisfactory for hopper-fed boilers. *Dry steam coal*, known as *semi-anthracite* in the United States, has a rather higher volatile content than anthracite, so it is easier to ignite and burns with a rather longer flame. Its calorific value and bulk density are similar to those of anthracite. The supply of these two natural smokeless fuels is limited, and neither is really suited to open fires, so it is necessary to manufacture artificial fuels to supplement them.

Manufactured Ovoids

These are made from selected dry steam coals blended with some anthracite. The mixture is ground to powder, bound with pitch and moulded into ovoids. They are then heated in retorts, driving off most of the volatile matter which was contained in the pitch. The result is a consistent, dense fuel used for cookers, room-heaters and boilers. Similar ovoids, heated at a rather lower temperature during the retorting stage, retain some of their volatile content and are suited to open fires.

Gas Coke

This is the solid by-product of gasworks which manufacture coal gas by distillation of coal. Good-quality coking coals are carbonized by heating at

1000°C for ten to twelve hours in closed retorts. The gas evolved, which is mostly hydrogen and methane, is sent out to consumers, and it will be discussed more fully later in this chapter. The coke which remains in the retorts is low in volatiles (typically 2 per cent) and is, therefore, of rather low reactivity. It also has a low bulk density of about 480 kg m^{-3}. Now that coal gas has been replaced by natural and petroleum-based gases, this form of coke has decreased in importance.

Hard Coke

This is produced by carbonization of coal at a high temperature. Its main use is as a reducing agent in blast furnaces, but it is also used as a domestic smokeless fuel. It is made from coking coals by blending, crushing and carbonizing at over 1000°C for about 18 hours. The volatile content is even lower than that of gas coke, about 1 per cent, and the pore size is smaller, but the bulk density is much the same. Hard coke has insufficient reactivity for open fires, but is useful in closed appliances.

Low-temperature Carbonization

There are several proprietary brands of semi-cokes produced by carbonization at low temperatures. In one particular process, selected small coal with medium and weak coking properties is blended and heated in closed retorts at 630°C for four hours, during which some of the volatiles are driven off and condensed, while the gas evolved is burnt to keep the retorts hot. The carbonized solid residue is cooled slowly in sealed chambers in the absence of air. The product has a much higher reactivity than other smokeless fuels, and is very satisfactory on open fires, where, because it is a better radiator than coal, it is a more efficient fuel, largely offsetting its higher cost. Like other cokes its density is low, about 480 kg m^{-3}

Fluidized-bed Carbonization

Some new smokeless fuels have appeared in recent years made by an interesting process using a fluidized bed. Coal is crushed to particles of about 1·5 mm diameter, and is fluidized in a stream of air and steam, the relative proportions of which are regulated to control the temperature of the bed. Heat is supplied to the bed by burning off some of the coal in the gas stream. Carbonization occurs at 430°C for 20 minutes, after which the residue, known as a *char*, is pressed or extruded into the required shape. The briquettes are cooled slowly to 150°C in a cooling chamber and are then sprayed with water. The products are suitable for open fires, and the major advantage claimed for this process over other low-temperature carbonization methods is its ability to use relatively low-quality non-coking coal as feedstock.

A summary of the various carbonization processes is shown in Table 5.4,

Table 5.4 Carbonization processes and products

Process	Conditions	Coke quantity	Coke reactivity	Products from 1 tonne of coal				
				Benzole	Tar	Ammonium sulphate	Gas
Hard coke	> 1000°C 18 hours	725 Kg	Low reactivity; 1 per cent volatiles	13·6 litres	40·9 litres	9·5 Kg	6330 MJ
Gas coke	1000°C 10–12 hours	700 Kg	Low reactivity; 2 per cent volatiles	13·6 litres	45·5 litres	9·5 Kg	7600 MJ
Low-temperature carbonization	630–700°C 4 hours	700–750 Kg	High reactivity; medium volatile content	Mixed oil products 95 litres approx.			3000 MJ approx.

(Reproduced by permission of the Women's Solid Fuel Council.)

where some indication is also given of the by-products which are obtained by each technique. These by-products can be extremely valuable as chemical feedstock.

5.6 Coal-based Chemicals

Smokeless fuels are examples of chemical production based on coal. Many other chemical products can be made from bituminous coal, as is shown in Figure 5.17. Brown coal is similarly versatile, but anthracite, because of its low volatile content, is not useful for chemical technology. Although not strictly part of a study of energy and power, it is important to examine the

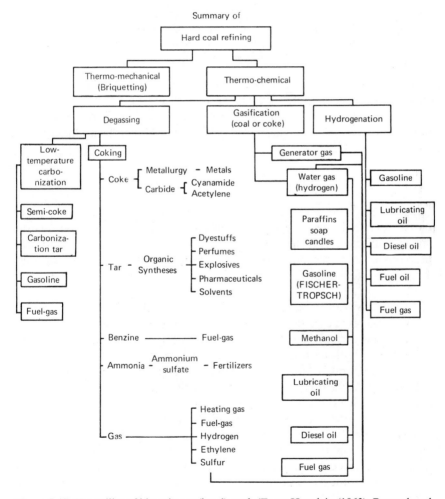

Figure 5.17. Versatility of bituminous (hard) coal. (From Henglein (1963). Reproduced by permission of Verlag Chemie GmbH)

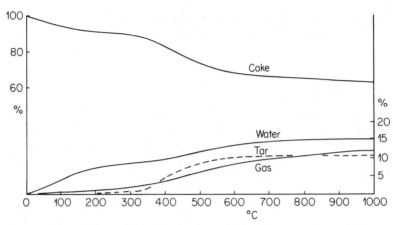

Figure 5.18. Products of carbonization of bituminous coal at different temperatures. (From Henglein (1969). Reproduced by permission of Verlag Chemie GmbH)

chemical possibilities of coal, because it is conceivable that its value to mankind as a source of chemicals might exceed its value as a fuel.

There are four classes of chemical treatment of coal: carbonization, partial combustion, hydrogenation and solvent extraction. All but the last have proved useful commercially.

Carbonization

This is the process used in coke manufacture, which we have already studied. As well as coke, it yields other useful products, and the nature and quantity of these is affected, as is the nature of the coke, by the temperature and other conditions of carbonization. Table 5.4 summarizes the conditions and main products of different processes, and Figure 5.18 summarizes the quantitative aspect.

The coal gas from carbonization retorts is a satisfactory fuel for domestic and industrial purposes. Its composition is approximately as shown in Table 5.5, and it has a calorific value of about $1 \cdot 5 - 1 \cdot 8 \times 10^7$ J m^{-3}, which is fairly low compared with natural gas but is acceptable. It burns with a high flame speed, and requires a fairly small admixture of primary air, so the burner design in simple. Coal gas was used very widely in the United Kingdom from the early 19th century, first for lighting and later for heating and cooking, but it has been superseded as a fuel, first by a rather similar gas made from oil and later by natural gas (see Chapter 6) because of its high production cost. Other disadvantages in comparison with natural gas are high transmission costs due to its low calorific value, and toxic effects due to the carbon monoxide content. As a chemical feedstock, coal gas is useful for its hydrogen and ethylene content.

The gas from the retorts is contaminated with tar vapours, and these are

Table 5.5 Composition of coal gas

H_2	53 per cent
CH_4	23 per cent
N_2	12 per cent
CO	7 per cent
CO_2	2·5 per cent
C_2H_4	2·0 per cent
O_2	0·5 per cent

removed by spraying dilute ammonia solution into the gas stream, followed by cooling. The tar is separated from the ammonia by settling, and provides a raw material for production of dyes, pharmaceuticals and other organic chemicals. Another product obtained in the same way is benzole, which is refined to benzene (C_6H_6). Some of the benzene is sold as such, and some is hydrogenated to cyclohexane (C_6H_{12}) which is a feedstock in the manufacture of certain types of nylon.

The coal gas is also contaminated with sulphur, which is present as hydrogen sulphide and which is undesirable in fuel gas for several reasons. It is removed by passing over iron hydroxide, or by more modern processes, and has been used as a raw material for sulphuric acid, sulphate fertilizers, etc.

Partial Combustion

This includes the well-known producer-gas and water-gas processes. Producer gas is a mixture of carbon monoxide and nitrogen made by blowing air over hot coal or coke. If the carbon and air are hot enough, the reaction

$$2\,C + O_2 \longrightarrow 2\,CO$$

is favoured for thermodynamic reasons rather than the alternative reaction

$$C + O_2 \longrightarrow CO_2$$

and as the reaction is exothermic, there is little difficulty in keeping the temperature sufficiently high. The nitrogen content of the air is unaffected and so forms a high proportion (about 65 per cent) of the final product, making it rather low in calorific value. Its main use is as an industrial fuel, and it is made where it is needed, so eliminating the high cost of transmitting a low-quality fuel. Its advantage over the coal from which it is made lies in its cleanliness and in its great controllability, an advantage which it shares with all gaseous fuels.

Water gas is a mixture of hydrogen and carbon monoxide made by blowing steam over very hot coal or coke; the reaction is

$$C + H_2O \longrightarrow CO + H_2$$

As all the gases in the product are combustible, it is of higher calorific value than producer gas, around $1\cdot1 \times 10^7$ J m^{-3}, which is comparable with coal

122

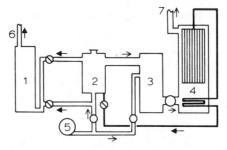

Figure 5.19. Alternating water-gas plant: 1. scrubber; 2. generator; 3. combustion chamber; 4. waste heat boiler; 5. fan; 6. gas outlet; 7. outlet to flue
The operation is as follows: In the hot-blowing phase, valves shown by open circles (O) are open, and those shown by crossed circles (Ø) are closed. Gas flows are indicated by skeleton arrows. The fan blows air through the coke in the generator, and the coke gets hotter. Producer gas is evolved and enters the combustion chamber, where it burns in a supply of air. Heat is released which raises steam in the waste heat boiler. In the water-gas phase, the valves are changed over, and gas flows are indicated by bold arrows. Steam from the waste heat boiler is blown into the generator, first from the bottom and then from the top. Water gas evolved is cleaned in the scrubber. The coke cools owing to the endothermic reaction, and when necessary, the hot-blowing phase is repeated. (Modified from Henglein (1969). Reproduced by permission of Verlag Chemie GmbH)

gas. By spraying a petroleum oil known as *gas oil* (see Chapter 6) into the reaction vessel, the water gas is enriched to a calorific value of about 1.9×10^7 J m^{-3}, which is as good as the best coal gas, and this enriched water gas was the usual city gas supply in the United States until natural gas became important.

The problem with water gas is that the reaction is endothermic, and the carbon cools unless it is externally heated. A neat way of overcoming this is to run the exothermic producer-gas and the endothermic water-gas reactions in the same plant, either in turn or simultaneously, so as to keep the coal or coke at the desired temperature. A typical plant for alternating operation is shown schematically in Figure 5.19. Producer gas is made by blowing air through the coal in the generator; the exothermic reaction occurs, and the combustible products are burnt in the combustion chamber. The heat produced raises steam in the boiler, and the steam is used to make water-gas by passing it through the hot coal in the generator and out via the scrubber.

Intermittent processes are unpopular in industry, and an interesting continuous process has been devised, known as the Winkler process, in which the two reactions run simultaneously. A schematic diagram of the Winkler process is shown in Figure 5.20. Granulated coal is fluidized in a blast of oxygen and

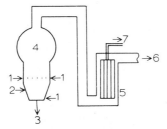

Figure 5.20. Winkler fluidized-bed water-gas process. A blast of steam and oxygen (1) and a supply of powdered coal (2) form a fluidized bed. Ash leaves at the bottom of the vessel (3) and the water gas evolved (4) proceeds to a waste heat boiler (5) and thence out at 6. Steam (7) evolved in the boiler is returned to the reaction. (Modified from Henglein (1969). Reproduced by permission of Verlag Chemie GmbH)

steam at a temperature of 1000°C, and the products pass through a boiler (in which the steam is raised) and a dust-extractor. The process was developed for brown coal, but can be used with bituminous coal, and has even been tried with peat.

As well as its use as city gas, water gas is a useful fuel for reciprocating engines and gas turbines, in which role it may provide a means of using high-sulphur coal without causing atmospheric pollution (see later in this chapter). Its most important use, however, is as a feedstock for chemical syntheses, especially of petroleum-type products. These have been developed from the Fisher–Tropsch process in which water gas is passed at a temperature of 200–300°C and a pressure of 20–30 atmospheres over an iron or cobalt catalyst, to yield a range of hydrocarbon products which can be separated by the methods of the petroleum industry (Chapter 6). Figure 5.21 shows the great range of products which can be made, and in view of the threatened shortage of oil, the Fischer–Tropsch process and its more modern derivatives may well become of great importance in the future.

Hydrogenation

Coal can be regarded as a hydrogen-deficient hydrocarbon, and if hydrogen is added to coal under suitable conditions, petroleum-type oils are produced. The Bergius process, invented in 1913 and used commercially in Germany from 1927, makes use of high-pressure catalytic hydrogenation to transform coal into oil. The process is considered obsolete, but it is of historic interest because it supplied much of the demand for petroleum in Germany during the Second World War.

A schematic representation is shown in Figure 5.22. Coal is ground to a paste with recycled oil containing the catalyst, and it is injected as a liquid

124

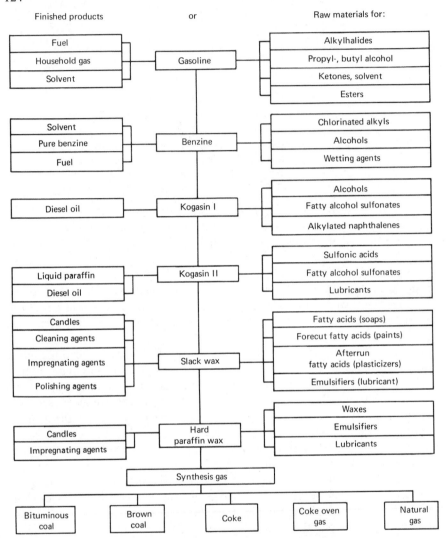

Figure 5.21. Possibilities of the Fischer-Tropsch process. (From Henglein (1969). Reproduced by permission of Verlag Chemie GmbH)

at 200 atmospheres into a preheater and thence into reactors where the hydrogenation occurs, the temperature rising to 460°C. The products are passed to a separator where the heavy and volatile fractions are separated. The volatiles pass through a heat exchanger, giving up their heat to the feed materials. Gases and liquids are separated; the gases are recycled, and the liquids are distilled, yielding gasoline (petrol), middle oil and heavy oil. The heavy oil is recycled, and the middle oil is converted into gasoline by gas-phase hydrogenation over a solid catalyst. One tonne of coal yields about 600 kg of gasoline,

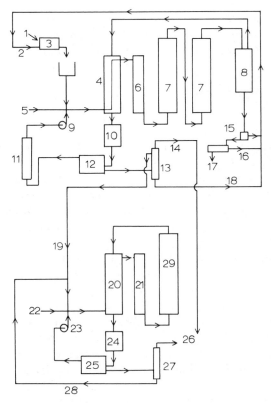

Figure 5.22. Bergius process for high-pressure
hydrogenation of coal. 1. coal; 2. grinding oil;
3. coal grinder; 4. heat exchanger; 5. fresh
hydrogen; 6. pre-heater; 7. furnaces; 8. separator;
9. gas circulating pump; 10. cooler; 11. gas purifica-
tion; 12. scraper; 13. distillation; 14. gasoline;
15. centrifuge; 16. carbonization furnace; 17. ashes;
18. grinding oil; 19. middle oil; 20. heat exchanger;
21. pre-heater; 22. fresh hydrogen; 23. gas circulating pump; 24. cooler; 25. scraper; 26. gasoline;
27. distillation; 28. middle oil; 29. furnace.
(Modified from Henglein (1969). Reproduced by
permission of Verlag Chemie GmbH)

but since the hydrogen required by the process is also obtained from coal,
the overall conversion rate falls to nearer 300 kg per tonne of coal. By-products
of the process include phenols, cresols and gaseous hydrocarbons which are
used as feedstocks for the plastics industry.

Solvent Extraction

As a technical process solvent extraction has not yet proved viable, but as a means of fundamental research on the composition of coal it has proved invaluable. The techniques and results have been reviewed by van Krevelen (1961), to which the reader is referred for further details.

5.7 Future Technologies based on Coal

Fluidized Beds

We have already alluded to possible means of burning high-sulphur coal without causing atmospheric pollution. Two methods of doing so have been proposed, one of which is a combustion process, and the other a variant of the water-gas reaction. In *fluidized-bed combustion,* powdered coal is fluidized by hot air jets, and burns; the heat of combustion is transferred to a nest of pipes in the bed, which forms a boiler, in which steam is raised to power a conventional steam turbine. The advantages are twofold. Firstly, the temperature of combustion is lower, and the sulphur remains in the ash rather than being released as SO_2. Secondly, the heat transfer from the burning coal to the water tubes is improved greatly by the agitation of the fluidized bed, so the size of boiler can be reduced considerably without losing efficiency. The British National Coal Board is working on this system, and has negotiated a joint programme with the United States Environmental Control Agency to develop it. In *fluidized-bed water gas production,* reviewed recently by Squires (1972), coal is gasified in a fluidized bed maintained by a blast of mixed air and steam, as in the Winkler system described earlier. The gas contains sulphur as hydrogen sulphide, and it is desulphurized by passing through a bed of partially calcined dolomite, where the following reaction occurs:

$$CaCO_3 + H_2S \longrightarrow H_2O + CO_2 + CaS$$

The cleaned water gas is used as fuel for a gas turbine, the exhaust gases of which pass to a conventional boiler where steam is raised to drive a steam turbine. Such a cycle has a higher 'top' temperature than one involving only a boiler and steam turbine, so it can offer a higher thermal efficiency (see Appendix 2), in principle about 50 per cent compared with the highest at present of about 40 per cent. The reason for this is basically that in a boiler the maximum temperature is restricted by the materials of which the boiler is built to a value considerably less than the theoretical maximum flame temperature for the reactions involved. In a gas turbine there are still restrictions caused by the materials of construction but they are much nearer to the theoretical maximum.

Underground Gasification

In order to reduce the amount of labour needed in coal mining, and to make good use of coal seams which are inaccessible to present-day mining technology, it has been proposed to gasify the coal underground. Such a process would extract a fair proportion of the thermal energy and might extract most of the useful chemicals from the coal. The Soviet Union began trials in 1931 (Henglein, 1969), and the United Kingdom conducted a series of experiments from 1949 (Ministry of Fuel and Power, 1956).

In the British experiments two bore-holes were drilled into the coal seam, and they were linked through the coal to create a pathway in which the gasification reaction could occur. Three methods of linkage were tried: mechanical fracture by high-pressure air, electrical carbonization by passing a heavy current through the coal and directional drilling. The first two methods had the advantage that all the work could be done from the surface, but there were many practical difficulties. Directional drilling involved sinking a shaft and sending a drilling crew down, for methods of turning the drill through a right angle were considered too difficult, but this method of linkage gave the best results. Air was passed down one of the bore-holes, and the coal was ignited. A reaction proceeded, which was probably a mixture of the coal-gas, producer-gas and water-gas reactions described earlier in this chapter, and a gas was evolved which emerged from the second bore-hole. The gas produced was of low calorific value (usually less than $3 \cdot 5 \times 10^6$ J m^{-3}) and was, therefore, unsuitable for long-distance transmission, but it was thought to be usable for electricity generation, and was in fact used to power a gas-engine on an experimental basis. However, the economics of the process were not considered sufficiently attractive to warrant commercial exploitation, and the experiments were discontinued.

Since then there has been renewed interest in this process, and the latest proposal, from the United States Atomic Energy Commission, involves shattering the coal with chemical explosives, retorting it *in situ* by starting a methane flame, sustaining combustion with a supply of oxygen and water and piping the resulting gas (mostly methane) to the surface. Sulphur and other pollutants would remain underground, which is claimed as an advantage, but clearly represents a loss of valuable chemicals. It is not known how feasible the method might prove, but there is little reason to expect it to be much more successful than the methods tried several decades earlier by other countries. However, the use of tonnage oxygen may be an important factor.

Nuclear Gasification

Another United States Atomic Energy Commission proposal is to use heat from a nuclear reactor to heat the coke ovens or retorts of a conventional gasification plant. Temperatures of the order of 1000°C would be needed for high-temperature gasification, and this is somewhat beyond present-day reactor technology, but the lower-temperature gasification systems may be feasible. The economics of the proposal may be rather doubtful.

District Heating

The disadvantage of coal as a source of domestic heating, namely its inconvenience, can be circumvented by burning it at a central boiler-house, which also allows the use of cheap small coal instead of high-grade domestic fuel. According to the National Coal Board there were about 75 solid-fuelled heating schemes operating or under construction in Britain in 1972. District heating is considered more fully in Chapter 15.

5.8 Peat

Peat is widely used as a domestic fuel on a small scale, but its use in large quantities for electricity generation is confined at present to the U.S.S.R. and the Irish Republic, although other countries including Germany, Canada and Finland have shown some interest in recent years. The Russian enthusiasm for peat-fired electricity generation undoubtedly arises from the enormous amount of peat available in the Soviet Union (see Chapter 4). Because of the nature of the Soviet economic system it is difficult to assess the economic viability of their use of peat for this purpose. In Ireland the use of peat is related to the absence of any appreciable quantities of coal and (at least at present) oil. There is no doubt that peat has been more expensive up to now than imported fuels, but the use of native fuel saves foreign exchange and reduces dependence on the goodwill of foreign governments. Recently, of course, the rise in oil prices has made peat more competitive, and O'Donnell (1974) has gone so far as to suggest that already peat is cheaper than imported oil, but the situation may change to the detriment of peat if the promise shown by offshore oil prospecting along the southern coast of Ireland is fulfilled. Whatever the economic position, however, there is a much more important reason for the Irish interest in peat, which is partly an economic and partly a social consideration. The peat bogs occur in remote rural regions, which for centuries have suffered from depopulation and from the lack of satisfactory employment other than small-scale farming. By developing the peat industry, a new source of employment is introduced which has an effect on the local economy out of proportion to the relatively small number of men employed. This important aspect of energy policy is discussed in more detail in Chapter 18.

Traditionally, the winning of peat is done by hand, using a specially-adapted spade. The process looks (and is) laborious, but it is surprisingly productive in terms of the fuel value produced per man-day. Most of the peat won in this way is produced on an 'amateur' basis by individual families for their own use, but there is still some hand-won peat used in electricity generation.

The mechanized winning of peat suffers from some difficult problems. Peat bogs have a very high water content, which, even in summer, can be as high as 92–94 per cent (Green, 1963). The first step in mechanical winning is, therefore, to drain the bog. To accomplish this difficult task, specialized machines have been developed with very wide tracks which reduce the loading on the surface of the bog to about 880 kg m^{-2} (Green, 1963). Even when the bog is drained, the water content is about 90 per cent, and the maximum tolerable surface loading is still very low at about 1550 kg m^{-2}. The figures quoted are applicable only in summer; in winter the bog is so wet as to be unusable. Even in summer an unusually wet spell can seriously disrupt work.

Another major problem in peat winning is the unpredictable occurrence of timber and similar obstructions. While occasional lumps of timber foul machinery they are otherwise not serious, but in some bogs the occurrence of timber is so frequent that is makes the use of machinery very difficult.

A final problem is that not all peat bogs are level enough for mechanical winning. Basin peat deposits, such as those in the Irish Midlands, lie in poorly-drained hollows, are fairly flat at the surface and have a reasonable thickness (typically 10 metres or more). This type of bog is amenable to mechanization using existing techniques. However, many of the peat deposits which are at present worked by hand on an amateur basis are upland blanket bogs which are thin and anything but flat. Mechanical winning is impossible with existing machinery.

Despite the problems, mechanized winning of peat is common in several parts of the World, notably the U.S.S.R., Germany and Ireland. The following account is of Irish practice, but the operations to be described are typical of the rest of the World, insofar as any description can be typical when the variability of the conditions of operation is so great.

The first operation is drainage of the bog, which is normally done with a dragline excavator. An alternative machine known as a disc ditcher (Figure 5.23) has been developed, and it is rather quicker than a dragline and also

Figure 5.23. Disc ditcher. Note the extremely wide tracks. (Reproduced by permission of Bord na Móna)

130

more satisfactory in operation since the rotating disc spreads the excavated peat out over a wide area. When timber is encountered the disc can be replaced by a saw-blade. It may take several years from the beginning of drainage before the bog is ready for production. When this stage is reached, the bog is stripped down to the peat and levelled.

The bog is now ready for harvesting. Two different techniques are used, one producing sods which superficially resemble the traditional product of hand digging, and the other producing milled peat which is of a loose powdery texture.

Sod Peat

The winning of sod peat (otherwise known as machine turf) begins with the excavation of the peat bog by means of a bucket chain (Figure 5.24). The excavated peat is macerated in what is essentially a large mincing machine, and it is then moulded and spread out in the form of sods of uniform size and shape (Figure 5.24). The operations of excavation, maceration, moulding and spreading are all done by one machine, misleadingly known as a *bagger*.

Figure 5.24. Sod peat bagger. Spreader arm and sods in foreground, bucket chain and control cab in middle ground and peat-fired power station in background. (Reproduced by permission of Bord na Mona)

The sods are very wet at this stage, and in order to hasten the drying, they are exposed to the air by gathering them into *windrows* by means of a tractor fitted with a snowplough blade. During drying they are turned at intervals, and eventually collected into ricks (Figure 5.25). The normal drying period is six to ten weeks, depending on the weather.

The sods produced by this process are typically $300 \times 75 \times 50$ mm and have a moisture content of around 25–40 per cent (Lunny, 1963). The calorific value is about 10^7 J kg^{-1} (Miller, 1963) although this depends greatly on the moisture content. They are suitable not only for domestic heating but also for larger-scale industrial plant such as boilers and power stations, although for the latter purpose milled peat is rather more satisfactory. Machine turf has three important advantages over traditional hand-cut turf. Firstly, the surface is covered with a skin which is resistant to wetting during storage. Secondly, it is of more uniform quality than the hand-cut product, for during the maceration a mixture is formed containing peat from all levels of excavation. Thirdly, it is cheaper, since the labour content is reduced, and indeed the productivity of sod-peat operations in Ireland is increasing. In 1960 it was 300 tonnes per man per annum, and by 1971 increased mechanization and improvement in techniques had raised the figure to 550 tonnes (Callanan, 1972). Incidentally, this is comparable with the 1971 British coal production of 475 tonnes per man-year.

Figure 5.25. Mechanical loading of sod peat into railway wagons. (Reproduced by permission of Bord na Móna)

Figure 5.26. Triple drum miller used for milled peat production. (Reproduced by permission of Bord na Móna)

Milled Peat

A peat miller (Figure 5.26) scrapes about 10 mm of peat from the surface of the bog at each pass, and causes a certain amount of air drying at the same time. The milled peat is harrowed at intervals until the moisture content is down to about 50 per cent, which in good conditions takes two to five days (Rapple, 1968). It is then moved into ridges (Figure 5.27) and harvested into long stacks placed beside a temporary railway track. A mechanical loader transfers the light, powdery material into wagons

Milled peat offers to the user two main advantages over sod peat, both of which are of greater importance for large-scale users than for the domestic market. The production cost is lower, and this is reflected in figures quoted by the electricity industry (Electricity Supply Board, 1973) which show appreciably lower fuel cost per unit sent out for milled peat stations than for sod peat stations. Furthermore, the technology of large-scale combustion is simpler for milled peat than for sods; this is especially true of the stoking systems. There is also an advantage to the producer in that the drying time is ideally only a few days rather than weeks, so that the production rate is higher. However, these advantages are not obtained without some problems. A major difficulty with milled peat is that the material is easily blown away by strong winds. On several occasions this has caused large losses, but the problem has been

Figure 5.27. Ridging of milled peat. (Reproduced by permission of Bord na Móna)

solved by covering the ridged peat with exceptionally large areas of polythene sheet. Another major difficulty, particularly in Ireland with its variable climate, is the susceptibility of the process to rainy conditions. The moisture content of the top layer of a working peat bog is very sensitive to the weather, and as a peat miller removes only the top 10 mm or so, the dampness of the newly-milled product can vary enormously. Moreover, the rate of drying of the peat is dependent on the atmospheric humidity. If the 'normal' drying period of two to five days is to be achieved, it is necessary for the rainfall to be less than 1 mm during the drying (Rapple, 1968), so production at the maximum rate is only possible during periods of prolonged anticyclones. This has proved to be such a severe restraint on milled peat production that a modified sod-peat process, known as the *foidin* system, has been introduced as a standby. Normal milling is used when possible, and foidin production is used when the weather is unsuitable for milled peat production. A foidin is a small sod, and usually takes about 30 days to dry to 50 per cent moisture content. They can be burnt by normal milled-peat plant, and are, therefore, interchangeable with milled peat as a fuel (Rapple, 1968).

5.9 Peat as a Fuel

Peat has a very high volatile content compared with coal. A typical analysis (Lunny, 1963) is: Fixed (non-volatile) carbon 31 per cent, volatiles 67 per cent, ash 2 per cent. To this must be added the moisture content, which is also volatile. During combustion the volatile fraction is driven off, producing a very long flame. Further, the volatiles have a very high ignition temperature (Lunny, 1963), and unless they are ignited, they will escape unburnt, which is a waste of fuel as well as a pollution problem. However, the remaining non-volatile component ignites at a very low temperature. The design of peat-burning equipment must take these characteristics into account. It is possible to modify the characteristics of natural peat to advantage by the production of peat briquettes, which are made by artificially drying and compressing milled peat. The production of briquettes is of considerable commercial importance, and it offers the twofold advantages of yielding a fuel which is clean to handle and easy to use, and of reducing the moisture content to about 10–12 per cent so as to reduce transport costs. Briquetting is of importance in the production and sale of brown coal as well as of peat.

Small Scale Peat-fired Plant

The most important factor in the design of small-scale peat burning equipment is the long flame length. It is perfectly possible to burn peat in a conventional solid-fuel heater designed for coal, but much of the flame may well disappear up the flue, leading to waste of heat and to dangerously high flue temperatures, often in the region of 630–700°C (Lunny, 1968). An efficient peat-burning room heater must incorporate a heat exchanger by means of which heat can be transferred from the flue gases to the air in the room, and this tends to make the appliance rather large unless the flues are convoluted. This in turn necessitates a strong chimney draught, which is not available when the chimney is cold, so the appliance tends to smoke during lighting up. The problem can be solved with a manually operated flue damper, but domestic users tend to forget to adjust it after lighting, leading to excessive and inefficient use of fuel. A better arrangement uses a thermostatically-controlled damper.

The methods of stoking peat-fired appliances are broadly similar to those used for coal. Apart from the obvious method of hand-stoking, domestic-scale equipment can be fuelled from a gravity-fed hopper, and the free-burning, non-caking properties of peat, together with its low ash content, make this simple method of fuel feeding very satisfactory. Screw-operated underfeed stokers have also been used with limited success (Lunny, 1968), but the burning characteristic of peat are not ideally suited to this type of feeding.

Large-scale Peat-fired Electricity Generation

Peat-fired power stations are of considerable importance in the U.S.S.R., where there are said to be some very large stations is excess of 300 MW. In Ireland the absolute size of the station is less impressive, the largest being only 90 MW. However, in terms of relative importance of peat and other fuels the Irish effort in peat is indeed impressive, for in the year 1972/1973 peat-fired plant provided 24 per cent of the Irish Republic's electricity (Electricity Supply Board, 1973).

Both sod and milled peat are used for electricity generation; in general the older and smaller stations use sod peat, because the milled-peat technique is of relatively recent importance in Ireland. Sod peat is burnt in chain-grate furnaces, in which sods are loaded on to an openwork conveyor which moves slowly under the boiler and conveys the fresh sods to the combustion zone. Air is supplied through the bars of the conveyor, and the position at which air is supplied, and the speed of the conveyor, are controlled so as to keep the combustion zone in the right place and to ensure complete combustion. Milled peat is burnt in a manner largely similar to pulverized coal, which was described earlier in this chapter. The milled peat arrives in a rather wet condition (about 50 per cent moisture), so the first operation is power drying, which is done by supplying a portion of the flue gases from the boiler to the drying plant. Flue gases are used in preference to hot air to minimize the risk of ignition, for peat is highly reactive. This drying operation is known as milling by analogy with the corresponding step in a coal-fired boiler, but in fact the drying is rather more important than any reduction of size which may occur.

In both types of plant the boiler design resembles that for coal-firing. Typical steam conditions are 440°C and 29 bars. The nature of the ash involves some slight modifications in the arrangements for cleaning the boiler tubes (Wasserohrkessel-verband, 1963). Ash removal is done very simply by means of tipping railway trucks, and because of the remoteness of the power stations it is not considered necessary to use electrostatic precipitators to remove ash from the flue gases. The steam turbines and alternators are similar to those used in coal-fired plant, although they are, at 40 MW, considerably smaller than the most modern coal-fired sets.

Unusual Uses of Peat as a Fuel

This chapter ends with two interesting examples from Scotland. An experimental peat-fired gas turbine was set up in the late 1950's at Altnabraec, Caithness (Scottish Peat Committee, 1962), with a view to testing its potential as a means of generating electricity from peat. Two approaches were tried: a closed cycle, in which air was heated via a heat exchanger and used as the working fluid, and an open cycle, in which peat was actually burnt in the combustion chamber of a modified industrial turbine. Ash accumulation was a problem in the open-cycle system, and the heat exchanger was not entirely satisfactory in the closed-cycle system. It was concluded that there was no advantage in practice over a conventional steam turbine with a peat-fired boiler, and the experiment was eventually terminated. In contrast, many of the better-known distilleries in Scotland continue to use peat-fired stills in the manufacture of their traditional malt whiskies, and as far as the authors are aware they have no plans to discontinue this practice.

136

References

Bradbury, K. A., *Solid Fuel in the Home*, Women's Solid Fuel Council, London, 1973.

Callanan, P. F., *Transactions of 4th International Peat Congress*, Otaniemi, Finland, Vol. 2, p. 29, 1972.

Electricity Supply Board, Ireland, *45th Annual Report*, 1973.

Green, W. A. R., *1963; Transactions of 2nd International Peat Congress*, Leningrad, Vol. 1, p. 291, H. M. S. O., Edinburgh, 1968.

Henglein, F. A., *Chemical Technology*, Pergamon, Oxford, 1969.

Lunny, F., *1963; Transactions of 2nd International Peat Congress*, Leningrad, Vol. 1, p. 409, H. M. S. O., Edinburgh, 1968.

Lunny, F., *Transactions of 3rd International Peat Congress*, Quebec, p. 307, National Research Council of Canada, Ottawa, 1968.

Miller, H. M. S., *1963; Transactions of 2nd International Peat Congress*, Leningrad, Vol. 1, p. 471, H. M. S. O., Edinburgh, 1968.

Ministry of Fuel and Power, *British Trials in Underground Gasification, 1949–1955*, H. M. S. O., London, 1956.

National Coal Board, *Facts and Figures*, 1974.

O'Donnell, S., *New Scientist*, **63**, (904), 18 (1974).

Rapple, E. A. J., *Transactions of 3rd International Peat Congress*, Quebec, p. 236, National Research Council of Canada, Ottawa, 1968.

Scottish Peat Committee, 2nd Report, H. M. S. O., Edinburgh, 1962.

Squires, A. M., *Scientific American*, **227**(4), 26 (1972).

van Krevelen, D. W., *Coal—Typology, Chemistry, Physics, Constitution*, Elsevier, Amsterdam, 1961.

Wasserohrkessel-verband, 1963; *Transaction of 2nd International Peat Congress*, Leningrad, Vol. 1, p. 387, H. M. S. O., London, 1968.

Further Reading

Bateman, A. M., *Economic Mineral Deposits*, Wiley, New York, 1950.

Bord na Mona, *The Moving Bog*, Bord na Mona, Dublin, 1972.

Bradbury, K. A., *Solid Fuel in the Home*, Women's Solid Fuel Council, London, 1973.

6 The Use of Oil and Natural Gas

Unlike coal, oil is a relative newcomer to the World's fuel markets. Coming after the industrial revolution, the oil industry has always used advanced methods of exploration, production and marketing. It has come to be identified with much that is bad as well as good in modern industrial civilization. In this chapter we are concerned, not with business ethics, but rather with the techniques of the oil industry, which, both because of their scientific nature and because of their enormous scale, are of considerable interest.

An analysis of world oil production is shown in Table 6.1. At first sight it would appear that oil production is scattered over the World in much the same manner as coal; however, closer inspection reveals that the bulk of the World's oil comes from relatively few areas, principally the Middle East, North and South America and the Soviet Union. Furthermore, the Middle Eastern oil producers are able to use only a very small proportion of their output, and their large surplus production, together with the complete absence from Table 6.1 of many of the World's industrial nations, generates a very large international trade in oil and petroleum products. This is quite unlike coal, where, as we have noted in Chapter 5, the principal consumers are in general the principal producers.

Consumption of oil products is analysed for the United Kingdom in Table 6.2 and for the United States and Canada in Table 6.3. It should be noted that the figures are tabulated in terms of different types of product, each having a fairly narrow range of uses. The oil industry supplies a great range of chemically processed products, each tailored to suit the requirements of the consumer, unlike the coal industry where, although the product is graded and washed, there is relatively little chemical processing.

The following trends are of interest:

(1) In the United Kingdom there has been a relative increase over the 20 years from 1950 to 1970 in the importance of the heavier fractions (for example fuel oil) compared with the lighter fractions (for example motor spirit). This is probably accounted for by the major shift of industrial fuel consumption from coal to oil, which was noted in Chapter 1. In the figures for the United States there is no comparable effect, because oil has always been important to American industry.

(2) The history of the British gas industry is written in the figures for naphtha. In 1950 all the United Kingdom's gas supplies came from coal. Within a decade

Table 6.1 Crude petroleum production (in millions of tonnes)[a]

Year	Algeria	Argentina	Canada	China	Colombia	Egypt (U. A. R.)	Indonesia	Iran	Iraq	Kuwait	Libya	Mexico	Neutral zone	Nigeria	Qatar	Romania	Saudi Arabia	U.S.S.R.	U. Arab Emirates (Trucial Oman)	United States of America	Venezuela	World Total
1950		3	4	0·2	5	3	7	32	7	17		10			2	5	27	38		267	80	523
1951	0·5	4	6	0·3	5	3	8	17	9	28		11			2	6	37	42		304	91	593
1952	0·9	4	8	0·4	5	3	9	1	19	38		11	0·8		3	8	41	47		309	97	624
1953	0·8	4	11	0·6	5	3	10	1	28	43		10	1		4	9	42	53		319	94	660
1954	0·6	4	13	0·8	6	2	11	4	31	48		12	2		5	10	47	59		313	101	691
1955	0·3	4	17	1	5	2	12	16	34	55		13	3		5	11	48	71		336	115	774
1956	0·2	4	23	1	6	2	13	26	31	55		13	4		6	11	49	84		354	132	842
1957	0·5	5	25	1	6	2	16	35	22	57		13	4		7	11	49	98		354	148	886
1958	1	5	22	2	6	3	16	40	36	70		14	4	0·2	8	11	50	113		331	139	910
1959		6	25	4	7	3	18	46	42	70		14	6	0·5	8	11	54	130		348	145	978
1960	9	9	26	5	8	3	21	52	47	82		14	7	0·9	8	12	62	148		348	149	1054
1961	16	12	30		7	4	21	59	49	83	0·9	15	9	2	8	12	69	166		354	153	1122
1962	20	14	33		7	5	23	66	49	92	9	16	13	3	9	12	76	186	0·8	362	167	1217
1963	24	14	35		8	6	22	74	57	97	22	16	17	4	9	12	81	206	2	372	170	1305
1964	26	14	37		9	6	23	85	62	107	41	17	19	6	10	12	86	224	9	377	178	1410
1965	26	15	39		10	6	24	94	64	109	58	17	19	14	11	13	101	243	14	385	182	1511
1966	33	16	43		10	6	23	105	68	114	72	17	22	21	14	13	119	265	17	409	176	1641
1967	38	16	47	15	10	6	25	129	60	115	83	19	22	16	15	13	129	288	19	435	185	1759
1968	42	18	50	20	9	9	30	142	74	122	126	20	22	7	16	13	141	309	24	450	189	1923
1969	45	18	54		11	12	37	168	74	130	150	21	22	27	17	13	149	328	29	456	188	2070
1970	48	20	61	24	11	16	42	192	76	137	162	22	25	54	17	13	177	353	38	475	194	2269
1971	38	22	64	26	11	15	44	224	84	147	132	21	27	76	20	14	223	377	52	467	186	2396
1972	50	22	73	30	10	15	54	248	71	151	106	22	29	91	23	14	286	400	58	467	168	2527

(Source:: United Nations Statistical Yearbooks 1972 and 1973. Copyright United Nations, 1972 and 1973. Reproduced by permission.)

[a] Countries producing less than 10 M tonnes per annum throughout the period 1950–1972 are excluded.

Table 6.2 United Kingdom petroleum deliveries (in tonnes × 10³)

Year	LPG (including natural gas imports)	Petrochemical feedstock	Gasworks naphtha	Aviation spirit	Motor spirit	Wide-cut gasoline ('Avgas')	Aviation kerosene ('Avtur')	Burning kerosene	Vaporizing oil (TVO)	Gas oil (engines)	Fuel oil (including light oil)	Lubricating oil	Bitumen	Refinery fuel
1950	31	218	0	287	5278		179	571	814	2671	3142	761	631	660
1955	65	575	0	361	6340	798	567	703	700	3821	5470	903	879	2093
1956	71	677	0	366	6425	812	605	844	583	4252	6575	908	956	2192
1957	75	806	0	393	5837	731	532	809	528	4202	7042	839	862	2151
1958	164	844	42	351	6730	731	509	1142	452	5117	10751	886	882	2573
1959	279	1084	229	293	7238	695	666	1222	375	5486	14033	943	1007	3127
1960	401	1587	406	267	7747	733	792	1316	294	6220	17717	980	1079	3396
1961	482	1603	505	272	8273	1026	892	1251	265	6895	19118	996	1185	3571
1962	648	1980	726	226	8702	820	1096	1454	215	7783	21670	983	1258	3758
1963	911	2749	914	244	9189	523	1508	1700	187	8898	23067	1019	1336	3938
1964	1272	3177	1369	196	10172	458	1686	1491	150	9791	25202	1098	1515	4083
1965	1553	3382	2098	170	10911	326	1929	1610	122	10871	27733	1124	1483	4234
1966	1904	3685	3041	149	11503	283	2190	1676	101	11813	29561	1156	1564	4640
1967	1867	4685	4035	130	12277	311	2517	1776	89	12557	31098	1118	1789	4766
1968	1731	5427	5176	105	13013	347	2765	1998	75	13820	31101	1152	1857	5164
1969	1673	6263	5400	101	13443	282	2968	2243	65	15382	33927	1227	1841	5653
1970	1497	6336	3542	75	14234	152	3253	2481	54	17143	38583	1176	2069	6028
1971	1424	6053	1898	64	14963	83	3667	2565	48	17754	39392	1147	2208	6183
1972	1706	6314	1469	64	15898	77	3929	2928	41	20367	41304	1113	2204	6420

Source: United Kingdom Energy Statistics 1973 (Department of Energy) and Statistical Digest 1966 (Ministry of Power). Reproduced with the permission of the Controller, Her Majesty's Stationery Office.

Table 6.3 Petroleum consumption for North America (Canada and U.S.A.) (in million metric tonnes)

Year	Liquefied gases	Aviation gasoline	Motor gasoline	Jet fuel	Kerosene	Gas oil for diesel engines	Fuel oil including gas oil for burning
1950	7·3	5·2	117·4	1·5	16·4	11·8	106·0
1951	8·7	7·1	127·3	2·0	17·2	14·2	110·6
1952	9·3	7·5	135·5	2·6	16·9	16·4	111·1
1953	10·2	8·5	140·9	4·4	16·0	18·0	112·4
1954	11·1	8·8	143·8	5·9	16·8	18·9	113·5
1955	12·6	9·9	155·4	7·6	16·6	21·1	124·5
1956	13·7	10·3	160·6	9·4	16·9	22·6	129·8
1957	14·0	10·6	163·3	9·4	15·6	22·8	126·9
1958	14·9	12·5	166·5	12·1	16·5	25·0	131·2
1959	17·5	12·9	172·6	15·0	14·4	26·4	137·2
1960	19·3	11·9	177·1	17·4	15·0	26·6	142·2
1961	19·8	12·8	178·8	19·5	14·6	26·9	141·5
1962	21·6	13·6	184·9	23·0	15·0	29·2	147·2
1963	20·1	13·8	190·7	24·6	14·9	31·9	147·9
1964	21·0	5·4	203·2	25·7	14·3	33·7	150·4
1965	22·3	5·1	211·7	27·7	15·1	35·0	162·4
1966	23·0	4·5	221·9	30·9	15·5	37·6	169·1
1967	21·3	3·9	229·1	37·9	15·4	38·4	175·8
1968	21·8	3·6	243·7	43·8	15·8	41·2	185·6
1969	30·1	3·1	255·2	46·9	15·5	42·1	205·0
1970	30·7	2·4	267·3	46·1	15·1	44·2	226·2
1971	29·7	2·2	278·2	47·9	14·4	48·1	236·3
1972	33·2	2·0	295·6	50·1	13·8	52·4	264·8

(Figures derived from Statistics of Energy 1958–1972, OECD, Paris 1974 and form Statistics of Energy 1950–1964, OECD, Paris 1966. Reproduced by permission of OECD.)

the gas industry had changed over to naphtha cracking, and also imported methane from Algeria. From 1970, natural gas from the North Sea became important, and naphtha consumption declined.

(3) Aviation spirit has declined recently in both nations because of the phasing out of piston-engined aircraft in favour of gas-turbine and jet aircraft. The only remaining market is the private and small commercial sector, including crop spraying and similar industrial uses. At the time of the 'Energy Crisis' in 1973 and 1974, restrictions were placed on the use of private aircraft, and there were complaints from persons affected that they were being unfairly singled out despite the fact that their consumption of fuel was relatively tiny. The figures show that they were amply justified in complaining, for even if

all private flying ceased completely, there would be an utterly negligible saving in fuel.

(4) Wide-cut gasoline (a fuel for aviation gas turbines otherwise known as JP. 4) has been phased out in the United Kingdom in favour of the safer alternative, kerosene (Avtur). Figures for the United States are composite, and an exact comparison is not possible, but it is believed that JP. 4 is rather more widely used in America, largely for climatic reasons.

(5) Tractor vaporizing oil (a variety of kerosene) has ceased to be important now that the farming and inshore fishing industries have changed over to diesel engines as their motive power.

Natural gas is an even newer arrival than oil to the World's fuel supply. As can be seen in Table 6.4, only the United States has any long history of

Table 6.4 World natural gas production (in units of cubic metres $\times 10^9$)[a]

Year	Canada	West Germany	Iran	Italy	Mexico	Netherlands	Romania	U.S.S.R.	United Kingdom	United States of America	World total
1950	2			0·5	2		3	6		178	
1951	2			1	3		4	6		211	
1952	3			1	3		5	6		227	
1953	3			2	3		6	7		238	
1954	3			3	3		6	8		248	
1955	4			4	3		6	9		266	
1956	5			4	4		7	12		285	
1957	6			5	5		7	19		302	
1958	10			5	6	0·2	8	28		312	
1959	12	0·8	0·9	6	9	0·2	9	35		339	429
1960	15	0·9	1	6	10	0·3	10	45	0·1	360	468
1961	19	0·9	1	7	10	0·4	11	59	0·1	373	505
1962	27	1	1	7	11	0·5	13	74	0·1	391	552
1963	32	2	1	7	11	0·6	14	90	0·2	415	604
1964	37	3	1	8	14	0·8	15	109	0·2	438	658
1965	41	3	1	8	14	2	17	128	0·2	454	705
1966	44	4	1	9	15	3	19	143	0·2	487	765
1967	48	5	1	9	16	7	21	157	0·6	515	822
1968	48	7	2	10	16	14	22	169	2	547	887
1969	56	9	3	12	17	22	24	181	5	586	972
1970	64	12	11	13	19	32	24	198	11	621	1069
1971	71	15	14	13	18	44	25	212	18	637	1140
1972	83	17	17	14	19	58	26	221	27	638	1204

[a] Excludes amounts less than 10×10^9 m^3 throughout the period, and gas of C. V. less than 8000 Kcal m^{-3}. (From United Nations Statistical Yearbooks 1972 and 1973. Copyright United Nations 1972 and 1973. Reproduced by permission.)

natural-gas production. To some extent this is because gases are much more difficult to transport than liquids, and it has only recently become possible to use natural gas at places remote from the producing wells.

A breakdown of natural-gas consumption is less revealing than for liquid petroleum products because of the smaller range of uses. In the United Kingdom almost all the natural gas is burnt directly as fuel by industry and domestic consumers. A tiny proportion (0·35 per cent in 1972) is used by the chemical industry as feedstock. In the United States the situation is slightly different in that natural gas is consumed, not only directly as a fuel, but also for electricity generation. There has been pressure from the United Kingdom electricity authorities to allow the use of natural gas for electricity generation in Britain, but so far the government has resisted this pressure, arguing that a high-quality fuel should not be wasted in applications for which a lower-quality fuel is adequate.

6.1 Oil Prospecting

To locate oil is inherently more difficult than to locate coal, because it is unusual nowadays for oil to appear at the surface, and it is necessary to drill to great depths to find worthwhile quantities. Until quite recently it was not possible to tell directly from measurements at the Earth's surface whether or not oil was actually present underground, and the object of prospecting has been limited to the location of suitable sedimentary rocks and traps (Chapter 4) in which oil might be expected to accumulate. Recently techniques have been developed which give some indication at least of whether a trap actually contains oil or gas, or whether it is dry, but despite such advances, drilling is still the ultimate test. Deep drilling is expensive, so it is very important that the methods used to locate possible oil reserves are as effective as possible.

Prospecting begins with the production of a detailed geological map of the area. This is usually done by aerial photogrammetry, in which a series of overlapping aerial photographs is taken and then analysed in stereo pairs to yield a detailed map. Even though only the surface topography is revealed in this way, an experienced observer can learn much about the underlying strata, including the location of anticlines, faults and other geological phenomena which are likely to produce traps. Subsequently, a geological examination is made of the rocks at the surface in the vicinity of interesting areas so as to identify the rock types. Clearly the procedures just described are not applicable if the area of interest is under water.

The next stage, which is necessarily the first stage for underwater areas, is geophysical prospecting, in which the magnetic, gravitational and seismic properties of the area are measured. There has been a great deal of development in these techniques in recent years, so they will be considered in some detail.

Magnetic Survey

Matter can be classified according to its magnetic properties as diamagnetic, paramagnetic, ferromagnetic, antiferromagnetic or ferrimagnetic. Materials such as quartz, marble, rock salt and gypsum exhibit diamagnetism, while antiferromagnetism or ferrimagnetism occurs in several minerals containing iron. It appears that there are no ferromagnetic rocks, and that although paramagnetism is sometimes observed for example in pegmatite and dolomite, the effect is due to the presence of ferri- or antiferromagnetic impurities. Table 6.5 shows the magnetic susceptibilities of some important rock types, but it should be borne in mind that the actual values are highly variable.

Table 6.5 Magnetic volume susceptibilities of rocks (in units of 10^{-6} emu)

Quartz	-1.2	Gabbro	300–7200
Marble	-0.75	Basalt	120–2000
Graphite	-8.0	Magnetite ore	$10^5 - 1.2 \times 10^6$
Rock salt	-0.82	Ilmenite	25000–300000
Anhydrite	-1.12	Pyrrhotite	50–50000
Granite	80–1200	Haematite ore	200–3000
Pegmatite	250–6000	Pyrite	120
Dolomite	1600		

(From D. S. Parasnis, *Principles of Applied Geophysics*, Methuen, London, 1962.)

The presence of a substantial body of rock with a high susceptibility produces a distortion of the Earth's magnetic field. This can be detected by sensitive instruments and is used as a means of prospecting. The technique is best suited to locating bodies of magnetic ores, such as iron or chromium, but it is applicable also to oil prospecting in the measurement of the depth of interfaces between strata. This application relies on the fact that most of the non-sedimentary rocks are magnetic and can, therefore, create anomalies in the Earth's magnetic field, while sedimentary rocks are generally non-magnetic. The measured effect of a magnetic anomaly decreases as the distance is increased between the detector and the source of the anomaly, so it is possible to measure the thickness of sedimentary rocks overlying a non-sedimentary *basement*. Surveying is done from aircraft by means of a geomagnetic gradiometer, the output of which is processed numerically to compensate for extraneous effects. By flying a rectangular grid over the region of interest, it is possible to produce a magnetic map in which anomalies are measured and located. In this way a geological map can be drawn showing basement depth and structure.

Gravity Survey

Non-sedimentary basement rocks are more dense than the overlying sediments, so that when the surface sediments are thin, the basement rocks are close to the surface and the local gravitational field is high. If the surface sediments are thick, the gravitational field is low. The gravimeter used for the measurements is sensitive to movement, and it is unsuitable for airborne use. On land the use of a gravimeter is straightforward, though slow because of the necessity for careful adjustment before each reading. For marine surveys it can be lowered to the sea bed, but this is inconvenient, and ship-borne gravimeters in gyrostabilized mountings have been developed to improve the speed of marine survey.

An area to be gravity-surveyed is scanned in a rectangular grid, with measurements every kilometre or thereabouts. The observations are corrected for topography, elevation, latitude and tidal effects. They are then mapped. Interpretation involves assumptions about the densities of the subsurface rocks, and as these are not known exactly until drilling has started, there is inevitably some uncertainty in the interpretation of gravity measurements.

Seismic Survey

In seismic surveying a small explosion generates sound waves which travel through the rocks and are reflected or refracted at discontinuities in the geological structure. The reflected and refracted waves are received at a series of geophones and are recorded continuously over a period of time (usually a few seconds). Interpretation of the results is complicated, but provides more detailed information than either of the techniques previously described. Seismic surveying was originally developed for land but it has been modified very successfully for marine use.

There are four types of elastic wave in solids:

(1) Compression or P waves, in which the wave motion is in the same direction as the motion of the wavefront (as in sound waves in air).

(2) Shear or S waves, in which the wave motion is at right angles to the direction of motion of the wavefront (as in waves on a string).

(3) Surface waves at the interface between two media, in which the wave motion is elliptical in the vertical plane containing the direction of propagation (as in water waves).

(4) Love waves, which are surface shear waves.

Of these, the first two are of greatest importance in geophysical prospecting.

Under the conditions found in practice, the velocity of a P wave in a medium is given by

$$V_p = \sqrt{(k + 4n/3)/\rho}$$

where k is the bulk elastic modulus, n the shear modulus and ρ the density.

The velocity of a shear wave is given by

$$V_s = \sqrt{n/\rho}$$

and since k, ρ and n are always positive, $V_s < V_p$.

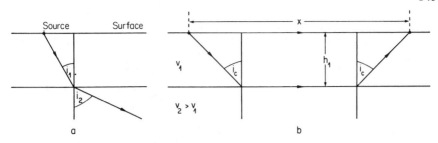

Figure 6.1. Refraction survey

In the *refraction method* of surveying, Snell's law is applicable. Consider a medium in which a P wave has a velocity V_1 overlying a medium in which a wave of the same type has velocity V_2. Then, using the notation shown in Figure 6.1a,

$$V_1/V_2 = \sin i_1/\sin i_2$$

If $V_2 > V_1$ there exists a critical angle i_c at which a wave leaving medium 1 does not enter medium 2, but rather travels along the interface (Figure 6.1b). Oscillatory stresses are set up which give rise to secondary waves emanating from all points along the interface, and a suitably placed geophone (Figure 6.1b) can pick up these waves, as well as the direct wave travelling entirely through medium 1. Information is obtained by timing the arrival of waves at the geophones.

The time of travel for the direct wave is $t_d = x/V_1$ (Figure 6.1b) while that for the refracted wave is easily shown by geometry to be

$$t_r = \frac{x}{V_2} + \frac{2\,h_1(V_2^2 - V_1^2)^{1/2}}{V_1 V_2}$$

Although it might appear that t_d must be less than t_r, it must be remembered that in order to get a wave travelling along the interface, V_2 must exceed V_1. Thus, if the distance x is great enough, it is possible for the refracted wave to arrive before the direct wave. It is easily shown that this occurs provided $x > x_c$, where

$$x_c = 2\,h_1\sqrt{\frac{V_2 + V_1}{V_2 - V_1}}$$

If a series of geophones is placed at various distances x from the point of the explosion, a graph of time versus distance for the first-arriving impulses will show two straight-line sections whose slopes are $1/V_1$ and $1/V_2$. This can be extended to any number of refracting interfaces provided there is an increase in wave velocity at each successive interface. In cases where this condition does not hold it is necessary to use S waves in addition to P waves as the source of information.

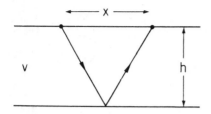

Figure 6.2. Reflection survey

Refraction surveying gives information about the depth of interfaces and also about the wave velocity in the strata.

In the *reflection method* of surveying, the echo from an interface between two strata is monitored. A P wave incident on an interface (Figure 6.2) is reflected partly as a P wave and partly as an S wave, and, for the P-wave component, the angle of incidence is equal to the angle of reflection. Once again, information is obtained by timing the arrival of the waves at the geophones, and, using the notation of Figure 6.2, it is easily shown that

$$t = \frac{2}{V}\sqrt{(h^2 + x^2/4)}$$

Hence h, the depth of the interface, can be measured. When refraction is involved there is a slight error, but the effect is negligible provided that the separation between the explosion and the geophone is small.

In order to use this equation for t it is necessary to know the velocity V. A direct method of obtaining this is to drill a hole and lower a geophone into it, and to measure the travel time to various depths for waves from a surface explosion. However, it is frequently necessary to know a value for V prior to drilling, and statistical methods have been developed for obtaining the information from reflection data, using a series of geophones.

In the practical application of seismic methods to surveying on land, the charge is exploded in a shallow hole and the geophones are placed on the ground surface. At sea, the charge is fired just below the water surface, and the geophones are trailed behind the ship, making rapid surveying possible.

The 'Bright Spot' Technique

The geophysical surveying methods just described suffer from the disadvantage that they do not locate oil itself, but only geological formations which may bear oil. Consequently, there is still a high probability that when a test drilling is made the hole will be 'dry'. A geophysical technique known as the 'bright spot' technique has been developed recently, however, which makes possible a distinction between oil-bearing and dry formations.

A normal reflection-type seismic survey measures the time delay between the signals received at the geophones, but does not make use of the relative amplitudes of the waves. The coefficient of reflection at the interface between two media depends on the densities and elastic properties of the two media, and, if the relative amplitudes of the direct and reflected waves are measured, a reflection at an interface between rock and gas, or between rock and oil, can in principle be distinguished from a reflection at other interfaces. Furthermore, the interface between gas and oil or oil and water is exactly horizontal, while one between rock strata is rarely so. By accumulating items of information such as these, it is possible to predict with some degree of certainty whether or not drilling is worthwhile. There is still some margin for error, but Harvey (1974) has suggested that the success rate of exploratory holes has increased fivefold by the use of this technique.

6.2 Drilling

When all the surveying has been done an exploratory well or *wildcat* is drilled. The main purpose of this well is not so much to find oil, although of course one is not displeased if one does so, but rather to investigate the nature of the rocks and to confirm or modify the results of the surveys. It is important, therefore, to analyse samples of the rocks taken at frequent intervals, either brought up by the lubricant (see later) or as *cores* (cylindrical samples of the rock through which the drill is passing). In addition a continuous record is kept of the electrical properties of the rock.

The drilling process is illustrated schematically in Figure 6.3. The bit (Figure 6.4) is rotated continuously, and the shaft is lowered so as to maintain a steady pressure. During drilling, the bit is cooled and lubricated by a stream of specially-formulated 'mud' which also serves to bring the cut pieces of rock to the surface for analysis. The shaft is lengthened at intervals until the hole has reached a predetermined depth, when the drill is withdrawn and a tubular casing is lowered into the hole and fixed with cement. A valve known as a blowout preventer is attached at the top of the well in case the drill should hit oil or gas unexpectedly. The drilling then continues with a smaller-diameter bit, and more casing is installed when necessary.

When the drilling bit enters an oil or gas reservoir, there may be an increase in pressure which is corrected temporarily with the blowout preventer while the density of the mud is increased to compensate for the pressure of the oil in the reservoir. Drilling can then continue while the geology is tested. If the well proves worthwhile, it is completed by cementing in the last section of casing.

A series of exploratory wells is drilled to establish the extent and the characteristics of the new oilfield, and when the exploration is completed, production wells can be drilled in much the same way.

The continental shelf has proved a very important new source of oil in recent years. The techniques of underwater drilling which were developed initially in shallow water have steadily moved deeper, until at present it is considered unexceptional to drill in 200 metres of water. The techniques of drilling under water are essentially similar to those just described for drilling on land, but the platforms from which the work is done call for some attention.

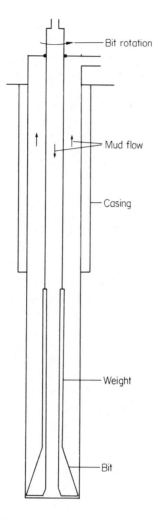

Figure 6.3. Sketch of drilling process. Vertical scale contracted for clarity

Self-contained Platforms

These are set up permanently on piles driven into the sea bed. They are relatively inexpensive, and although they are fixed, they can be used to develop a cluster of wells by directional drilling, in which the drill bit is forced to deviate from the vertical by the occasional insertion of steel guides into the hole. The depth of water in which this type of structure can operate is limited to about 30 metres, and since the platform itself must be about 15 metres above the mean sea level to allow for waves and tides, the overall length of the legs is typically 45 metres.

Submersibles

These are platforms which are floated into position and sunk on to the sea

Figure 6.4. Rock bit. (Reproduced by permission
of Hughes Tool Company, Belfast)

bed by the flooding of ballast tanks, leaving the deck above sea level. The limit of water depth for this type of structure is about 50 metres. A schematic diagram of a submersible, built in reinforced concrete, is shown in Figure 6.5.

Self-elevating Platforms

These consist of a barge-like hull with legs which can be lowered to the sea bed and then used to jack the hull above sea level. The limiting water depth is about 90 metres, beyond which there are excessive bending stresses in the legs.

Semi-submersibles

This type of craft, typified by *Sea Quest* (Figure 6.6), is possibly the most versatile of all the offshore drilling rigs. When the water depth is not too great, they can operate as submersibles, but at greater depths they can remain afloat, ballasted so that the lower part of the vessel is in the relatively calm waters

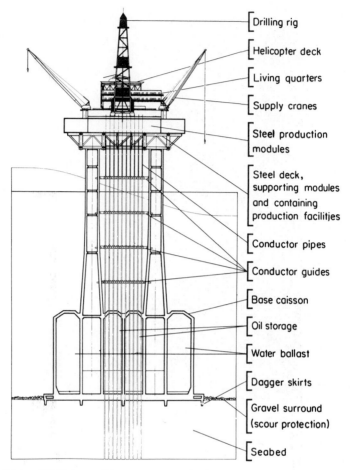

Drilling rig

Helicopter deck

Living quarters

Supply cranes

Steel production modules

Steel deck, supporting modules and containing production facilities

Conductor pipes

Conductor guides

Base caisson

Oil storage

Water ballast

Dagger skirts

Gravel surround (scour protection)

Seabed

Figure 6.5. Submersible concrete platform. (Reproduced by permission of Sir Robert McAlpine and Sons, Ltd., London)

several metres below the surface. Under such conditions anchors are used to keep the vessel on station.

Drill Ships

These are true ships, even to having a ship-shaped hull, and are self-powered, unlike the other types of floating platform just described. They are usually equipped with automatic locating devices in which an ultrasonic transmitter is lowered on to the sea bed, and the ship is kept on station with reference to the transmitter by servo-mechanisms driving a number of propellers, which provide sideways as well as the normal fore-and-aft motion. Since the hull is in the turbulent surface regions of the sea, a drill ship is less stable than a semi-submersible as a drilling platform, and it is necessary to operate in a minimum

Figure 6.6. Sea Quest leaving Belfast Lough on her maiden voyage. (Reproduced by permission of Harland and Wolff, Ltd., Belfast)

water depth of about 30 metres to avoid excessive flexure of the drilling strand. The main use of drill ships at present is in drilling exploratory wells in deep water.

6.3 Extraction and Refining

In a new well, there is usually enough pressure from natural gas, either in solution in the liquid or as a separate phase, to force the oil out of the well and up to the surface. Even if this is not the case, the water associated with the oil reservoir is under pressure and will force the oil out.

As a well continues to produce, the natural drive mechanism may fail, leaving substantial amounts of oil still in the ground (up to 80 per cent of the total in some cases). In order to recover more, artificial drive mechanisms are used. Water or gas is injected under pressure via injection wells, displacing the oil. Pumping is also used, and because of the danger of cavitation in the oil column due to the dissolved gas, it is necessary to locate the pump at the bottom of the well.

On emerging at the surface, the crude oil is separated from gas and water in a series of tanks prior to despatch to the refinery. The gas is processed by cooling under pressure to remove condensables, and to separate the gases which can be liquefied by pressure alone (propane and butane) from those which require refrigeration (methane and ethane). Transport to the refinery is by pipeline or ship, and is dealt with in Chapter 17.

Oil refining is a subject in itself, and a detailed treatment is beyond the scope of this book. However, because of the importance of refining in the production of fuel it is essential to give an outline of the processes involved. The technology of oil fuels is quite different from that of coal. Most coal is burnt with very little pretreatment, and very little of it is subjected to any form of chemical processing before use, while all oil fuels undergo extensive processing before use so as to make the best use of the possibilities of the raw material.

The first stage of refining is distillation, in which the temperature of the crude oil is raised steadily so as to drive off the fractions in order of increasing boiling point. This process separates different products to some extent, but is only the beginning. The fractions resulting from straight distillation are often unsuitable for their intended purpose without further treatment, and the proportions in which they occur do not necessarily coincide with market requirements. For both these reasons further chemical processing is needed. We shall now look at some of the more important processes.

Distillation

The object of distillation is to separate the crude oil into several fractions, known as *cuts*, determined by their boiling point. There is no hard and fast rule for the boiling ranges of the different fractions, for much depends on the composition of the crude and on the requirements of the processes which follow distillation, but a typical split might be as shown in Table 6.6.

Table 6.6 A typical split for the distillation of crude oil

Boiling range °C	Product
− 10 to 0	Propane and butane
25 to 80	Gasoline (petrol)
80 to 180	Naphtha
150 to 250	Kerosene (paraffin)
250 to 350	Gas oil (diesel)
Over 350	Residual oil

The distillation is carried out in towers whose design is illustrated in Figure 6.7. The object of this design is to maximize the separation of the different fractions. The tower is heated by a stream of superheated steam, which creates a temperature gradient. Crude oil is heated and pumped into the tower, where separation of the components occurs according to their boiling points. Liquid

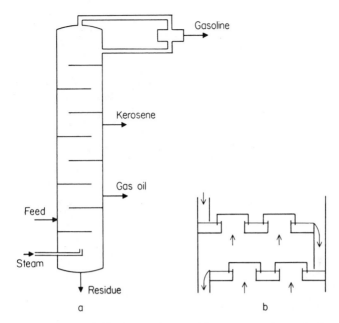

Figure 6.7. Distillation column. The construction of the column is detailed in (b). The gas ascends and bubbles through the descending liquid

fractions are taken off at different levels, and the most volatile components rise up the tower and are collected at the top. The liquid fractions can be subjected to redistillation in smaller columns to improve the separation still further.

Catalytic Cracking

To increase the yield of the commercially important lighter fractions (especially gasoline) and also to improve the octane number of the product, a portion of the heavier fractions is subjected to a catalytic cracking process. The oil is fed in a stream of steam into a reaction vessel (Figure 6.8) containing a fluidized bed of fine particles of fuller's earth (a hydrated alumino–silicate). A reaction occurs in which large molecules are broken down into smaller molecules. Some carbon is deposited on the catalyst, and this leads to a decrease in its effectiveness, so it is continuously removed and regenerated in a second fluidized bed in which the carbon is burnt in a stream of air. The resulting carbon dioxide is released to the atmosphere.

Catalytic Reforming

It is possible to improve the octane rating of the gasoline fraction by passing it over a catalyst of platinum dispersed on high-purity alumina. In this process,

154

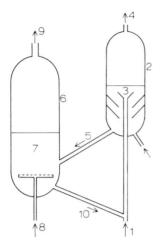

Figure 6.8. Fluidized-bed catalytic cracker. The heated feed (1), together with steam, is fed into the reactor (2), where the catalyst (3) is fluidized. Products of reaction leave at 4. Spent catalyst (5) goes to a regenerator (6) where another fluidized bed (7) is maintained by a stream of air (8). Carbon burns off, and the resulting CO_2 is vented at 9. Regenerated catalyst returns to the feed line via 10. (Modified from Lowson (1970). Reproduced by permission of British Petroleum Company Ltd.)

the straight-chain paraffins can join themselves into a ring, and the resulting cycloparaffins, together with those already present from the distillation, are converted into aromatics.

Sulphur Removal

The presence of sulphur in oil fuels is undesirable because it can cause corrosion of the engines in which it is used, and because when burnt it produces sulphur dioxide which is an atmospheric pollutant. Most of the sulphur compounds in the lightest fractions are mercaptans, which are soluble in sodium hydroxide and can therefore be removed by washing with caustic soda solution. The sulphur compounds in the heavier distillates, especially the gas-oil fraction, are chemically different from those in the light fractions and have to be treated by a different process. The sulphurous oil is made to react with hydrogen at high temperature and pressure in the presence of a catalyst of cobalt and molybdenum oxides dispersed on a base of alumina. The sulphur combines with the hydrogen to produce H_2S. A schematic diagram of the process, called 'hydrofining', is shown in Figure 6.9.

Neither of the processes described above is suitable for residual fuel oils, and this is particularly unfortunate because these usually contain a rather large percentage of sulphur. Residual oil is burnt in large industrial furnaces and power stations which can dispose of the resulting sulphur dioxide via tall chimneys. There is increasing concern (which may not be wholly justified — see Chapter 15) about the effect of this discharge on the environment, and there has been increasing pressure on the oil refiners to reduce the sulphur content of their product. One 'solution' is to use crudes which are low in sulphur, such as those from Nigeria and Libya, but clearly this does not solve the problem. There have been numerous attempts to devise desulphurizing techniques for heavy oils, and several different techniques are now in use.

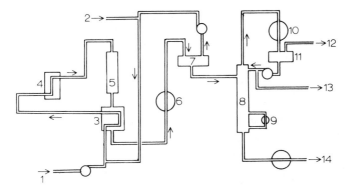

Figure 6.9. Hydrofining process for desulphurization of middle distillates. The feed (1) and a supply of hydrogen (2) are pumped via a heat exchanger (3) and a furnace (4) into a reactor (5) where a catalytic reaction occurs. The reaction products are cooled in the heat exchanger and by a water-cooled condenser (6). Gas and liquid are separated in the separator (7) and the gas is recycled. Liquid proceeds to a distillation column (8) which is heated by a boiler (9) Light fractions leave at the top and are condensed at 10, entering the reflux drum (11) Fuel gas (12) and low-boiling liquid (13) are taken off, as also is the main product, desulphurized middle distillate (14) (Modified from Lowson (1970). Reproduced by permission of British Petroleum Company Ltd.)

The main problem with catalytic sulphur removal from heavy oils is the presence of metal-containing porphyrins which can poison the catalyst. To circumvent this effect, an indirect technique has been devised in which the lighter components of fuel oil are separated, desulphurized and mixed with the heavier components once again. It is clear that such a technique has only a limited effect, and it is found in practice that the lower limit to sulphur content in the mixed output is about 1 per cent. Moreover, the technique relies on the continued existence of appreciable quantities of lighter fractions in the product, which is not, therefore, truly residual oil.

Direct techniques of desulphurizing heavy oils rely either on improved catalysts or on reprocessing catalysts to remove poisons. Among the processes in use are the H-oil process, developed by Hydrocarbon Research, Inc., (Anon., 1969), and the Residfining process developed by Esso Research Ltd. (Blume and coworkers, 1969).

Although refinery desulphurizing is likely to be of importance for the foreseeable future, there is a possibility that large-scale oil-fired plant such as power stations might instead adopt techniques of combustion in which the sulphur is retained rather than released as SO_2. These are considered in more detail later in this chapter.

The characteristics of the different grades of fuel produced from oil determine their applications, and influence equipment design. In the next few sections we shall consider various common fuels and the criteria which are important for their efficient use.

6.4 Gases

The lightest fractions of petroleum are gases at ordinary temperatures. The one with the lowest boiling point is methane, CH_4, which boils at $-162°C$. Methane is the principal constituent of natural gas, and although it is usually a petroleum product, it is the concern of a largely separate industry.

Natural gas associated with oil is first cleansed of condensable matter as mentioned earlier, and then piped (where possible) direct to the consumer. Natural gas sometimes occurs in the absence of oil, as, for example, in the southern North Sea, and in this case no processing is required prior to despatch. In many cases, as in North Africa, the consumer is not immediately to hand, and instead the gas is liquefied by refrigeration and transported by tanker (Chapter 17).

The lightest gases which are usually dealt with by the oil industry itself are propane and butane, which are the familiar bottled gases, known to the industry as liquefied petroleum gas (LPG). Although both propane (C_3H_8) and butane (C_4H_{10}) are gases at ordinary temperatures, they can be liquefied by compression to moderate pressures and can be stored safely and conveniently in steel cylinders, or even, in the case of butane, in specially-designed disposable containers which are little more than 'tin cans'.

The most important characteristic of petroleum gases including natural gas is that they burn with a hot and very clean flame, making them ideal for all applications which demand cleanliness. This includes industrial processes such as firing of ceramics where it is essential that contamination from the gas is minimal, and also domestic cooking where a smoky or smelly flame would be unacceptable to the consumer.

The design of burners for natural and petroleum gases is different from those for coal gas because coal gas has a low calorific value and a high flame speed, while natural gas has a high calorific value and a low flame speed. To burn natural and petroleum gas efficiently, there must be a large admixture of air because of the high calorific value, and the speed of the gas–air mixture at the mouth of the burner must be low otherwise the flame speed will be less than the speed of the gas mixture and the flame will blow out.

A new burner design has been developed recently by the BP Research Centre, in which a honeycomb structure promotes thorough and silent mixing of gas with air. The burner can be modified to burn either coal gas or natural gas (or even LPG) merely by altering the rate at which the gas flows to it (Desty and Whitehead, 1970).

6.5 Gasoline (Petrol)

Gasoline is liquid at ordinary temperatures, but evaporates easily in air

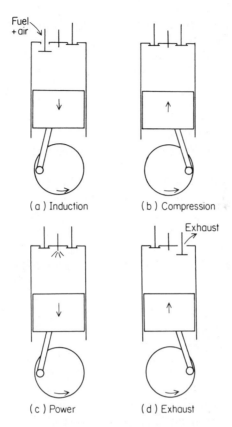

Fuel + air

(a) Induction   (b) Compression

Exhaust

(c) Power   (d) Exhaust

Figure 6.10. Four-stroke engine cycle

to form a highly inflammable mixture. It usually consists mainly of hexane (C_6H_{14}), heptane (C_7H_{16}) and octane (C_8H_{18}). These hydrocarbons can occur in different isomers, and the nature of the isomers has an important effect on the quality of the fuel. To understand the significance of this factor it is first necessary to study the operation of a typical petrol engine.

The essential components of a four-stroke engine are shown in Figure 6.10. A well-fitting piston slides inside a cylinder, and by means of a crank, the reciprocating motion of the piston is transmitted to a rotating shaft. Fuel vapour is mixed with air in the carburettor, and the resulting mixture is drawn into the cylinder through the inlet valve. Burning is initiated by an electric spark, and the products of combustion leave via the exhaust valve.

The cycle of operations is as follows:

(1) The inlet valve opens, and the piston descends, drawing in fuel and air mixture. The inlet valve closes.
(2) With both valves closed, the piston ascends, compressing the mixture.
(3) With both valves still closed, the spark ignites the mixture and rapid

burning occurs, causing a rise in pressure, which pushes the piston down. It is this stroke which provides the power output from the engine.

(4) The exhaust valve opens, and the piston ascends, pushing out the combustion products into the exhaust system. The exhaust valve closes, and the cycle repeats.

The most critical parts of the operation are the carburetion and the ignition, both of which place different demands on the fuel. In the carburettor and inlet system, the fuel must be thoroughly evaporated, otherwise liquid fuel drops will be drawn into the cylinder and combustion may not be complete. This is not usually a problem when the engine is hot because the fuel–air mixture can be heated by waste heat (usually from the exhaust system) before it enters the cylinder. When the engine is cold this is not possible, and to maintain good starting characteristics it is essential that the fuel is readily volatile. This, of course, is why the lighter fractions of petroleum are used in the petrol engine. However, it is important that the fuel does not evaporate before it gets to the carburettor, otherwise the flow of fuel will be interrupted and the engine will stop.

The other critical part of the operation is the ignition and burning of the mixture. The mixture is highly compressed, and there is a tendency for it to ignite spontaneously and to explode rather than to burn smoothly. This pre-ignition is called knocking or pinking and is much more pronounced in engines of high compression ratio. An explosion is both damaging and inefficient, and to avoid it, the nature of the isomers in the fuel has to be controlled. In general, the straight-chain isomers, and especially normal heptane, are poor in this respect, while the aromatics and the branched-chain isomers, especially iso-octane, are good. It is possible to assess the anti-knock properties of a petrol by comparing it with a mixture of pure n-heptane and pure iso-octane in a special single-cylinder test engine with variable compression ratio. The proportion of the two components is varied until the mixture has the same anti-knock properties as the petrol under test, and the percentage of iso-octane in the mixture is quoted as the octane rating of the petrol. Various other octane-measurement techniques have been devised which are more suited to automation; these often involve quantitative chemical analysis of the fuel (Anderson and coworkers 1972).

The octane rating of motor fuels is generally enhanced by the addition of an anti-knock agent such as tetra-ethyl lead, but with the growing public concern about lead pollution from car exhaust, the extent of this practice has been reduced. In these circumstances it becomes much more important to control the octane rating by methods such as catalytic reforming, which unfortunately make the fuel more expensive.

The octane rating of a fuel is of much greater importance in engines with a high compression ratio, and it might appear that the solution to the problem of lead pollution is merely to use engines with a low compression ratio. However, this suggestion raises two problems. The first is that the addition of lead com-

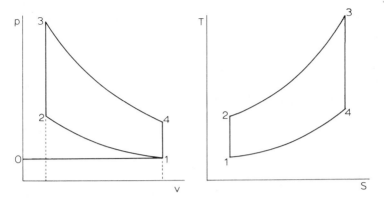

Figure 6.11. Thermodynamic cycle of the petrol engine: 0–1 induction; 1–2 compression; 2–3 ignition and burning; 3–4 power stroke; 4–1 and 1–0 exhaust

pounds to the fuel leads to the deposition of a thin layer of lead on the cylinders and valve gear of the engine, and in present day engine designs the lead layer is an important part of the lubrication system. This problem can be overcome by a change in engine design. The second and more serious problem is that the efficiency of a petrol engine increases with increasing compression ratio, so that a car with a high compression ratio engine has a lower fuel consumption than a similar car driven in a similar manner but equipped with a low compression ratio engine. This can be shown easily by a consideration of the thermodynamics of the petrol engine.

The operation of a petrol engine can be approximated by the thermodynamic cycle shown in Figure 6.11. The efficiency is given by

$$\eta = \frac{Q_{in} - Q_{out}}{Q_{in}} = 1 - \frac{(T_4 - T_1)}{(T_3 - T_2)}$$

Assuming that the compression and expansion are both adiabatic (isentropic), and that the ratios between maximum and minimum volumes are the same in both cases, then from the standard equation for an adiabatic,

$$Tv^{\gamma - 1} = \text{constant}$$

we get

$$\frac{T_1}{T_2} = \left(\frac{v_2}{v_1}\right)^{\gamma - 1} = \left(\frac{v_3}{v_4}\right)^{\gamma - 1} = \frac{T_4}{T_3}$$

Hence we can write the efficiency as

$$\eta = 1 - \frac{1}{(v_1/v_2)^{\gamma - 1}}$$

Here γ is the ratio of the specific heats, which for air has the value 1·4. The ratio of volumes v_1/v_2 is the compression ratio, and it is evident that when this is high, the efficiency is also high. Although this simple theory rather overestimates the efficiency of the petrol engine, the relationship between compression ratio and efficiency is found to hold quite well in practice. Because of the problem of pre-ignition, the compression ratio is usually kept to a maximum of about 10:1 or 11:1, and it is found in practice that engines with a compression ratio of this order have an efficiency of about 25–30 per cent. Consequently, of the heat supplied by the burning fuel, between two-thirds and three-quarters is lost as waste, partly via the cooling system but mostly through the exhaust gases. In the interests of fuel economy it is very important to maximize the thermal efficiency of the engine, and it is for this reason that high compression ratios are favoured.

Apart from the relative inefficiency of the petrol engine, there is a further disadvantage that the exhaust gas contains small quantities of substances which can cause atmospheric pollution. Chief among these are carbon monoxide and hydrocarbons, which arise from incomplete combustion, and oxides of nitrogen, which arise from oxidation of the nitrogen content of the air. The global effects of such emissions are probably negligible, but local effects have caused public disquiet, particularly in America. For example, a rather unusual combination of meteorological circumstances and high traffic density in the Los Angeles area of California has led to the formation of photochemical fogs. The chief offenders in this respect are the unburnt hydrocarbons. To reduce the percentage of such emissions from the exhausts of motor vehicles four distinct approaches have been proposed: catalytic exhaust systems, improved internal combustion systems, alternative fuels and alternative types of engine.

The catalytic exhaust system relies on oxidation of the unburnt hydrocarbons by passing over a hot catalyst after they have left the engine. In a sense this is a confession of failure, since it is not practicable to make good use of the energy released in the catalyst; indeed it can be an embarrassment since it causes very high catalyst temperatures. Nevertheless, such systems have been developed to the point where they are commercially viable.

A more elegant solution is to ensure that combustion within the cylinder is as complete as possible, a prerequisite of which is to ensure that there is an excess of air, or, in other words, a very lean petrol–air mixture. Such a mixture can be reluctant to burn properly, but it can be helped to do so by the presence of a hot surface which in effect catalyses the reaction. A possible way of introducing such a hot surface into the cylinder is to use a 'heat pipe' (a device which uses a molten metal as a high-conductivity heat transfer medium) to conduct heat from the exhaust to the cylinder. Another method of burning a mixture which is on average very lean is to start the combustion in a relatively rich mixture and to allow the flame front to propagate into regions in which the mixture is leaner. This can be achieved by introducing the fuel into the cylinder in an uneven manner, with the richer mixture near the spark plug, producing what is known as a *stratified charge*. Such systems have been developed experimentally, and they may be expected to increase in importance.

A third possibility is the use of fuels other than petrol in an essentially un-modified engine. Methane and LPG (propane or butane) have been used successfully in 'petrol' engines modified slightly by a change in carburettor design, and this has resulted in a considerable reduction in exhaust emissions. Widespread use of gaseous fuels for this purpose is limited to some extent by the shortage of gaseous fractions in the output of oil refineries, although doubt-less this could be overcome by suitable cracking of heavier fractions, and it has been suggested that methane obtained from decomposition of organic waste (see Chapter 15) might be used in a similar way. A more important objection to gaseous fuels is their difficulty of handling compared with liquids. Propane and butane can, it is true, be liquefied by compression and are thus relatively amenable to compact storage, but methane would have to be stored at a low temperature ($-162°C$) or compressed in cylinders, neither of which is likely to appeal to the private motorist. A more unusual fuel for the 'petrol' engine, which again reduces exhaust emissions, is hydrogen, and, again, it is possible to use an engine unmodified apart from the carburettor, although the higher temperature of the combustion reaction compared with hydrocarbons is likely to necessitate some change in materials if a good lifetime is to be achieved. The storage problem is much the same as for methane, although some progress has been made in storage in the form of metastable metal hydrides. The attrac-tion of hydrogen as a fuel for vehicles is not so much its inherent suitability, but rather its integration into a comprehensive energy system known as the 'hydrogen economy', which is discussed further in Chapter 14.

The most drastic solution to the problem of air pollution from vehicle exhausts is to abandon the petrol engine altogether in its present form. Several attempts have been made to develop alternatives. The goal has sometimes been reduction in air pollution, sometimes lower cost, and occasionally impro-ved performance. Most of the proposed designs have failed to match up to conventional engines, but because they offer certain advantages, at least in theory, they will be considered briefly.

The Wankel Engine

This is an attempt to eliminate the reciprocating action of the conventional piston engine, in the interests of smoother running and lower costs. A rotor runs in a specially-shaped case (Figure 6.12), and in so doing three chambers are created which describe a sequence of induction, compression, power stroke and exhaust analogous to the conventional engine. There are no valves or cam shafts, since all the timing is done by correct positioning of the intake and exhaust ports. There are no reciprocating pistons, so the balance of the engine is improved, and there is less vibration. Each cycle of four 'strokes' occupies only one rotation, and there are effectively three cylinders, so the size of the engine, in proportion to its power output, is smaller than for a conventional design. The fuel used is normal low-octane gasoline. The main problem, lubrication of the sliding seal between the rotor tips and the case,

162

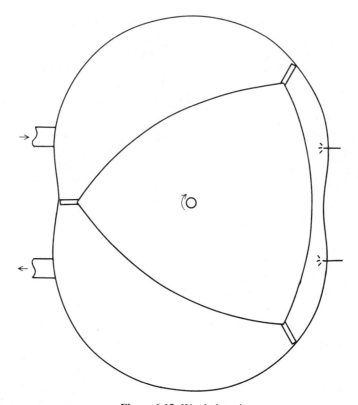

Figure 6.12. Wankel engine

has been partially overcome by suitable choice of materials. At present there are three main disadvantages compared with the piston engine. The fuel consumption tends to be higher; this is being overcome by eliminating leaks past the seals, and by the use of two spark plugs to encourage the fuel to burn faster. There is a higher emission of unburnt hydrocarbons, probably due to leakage past the seals. This has not yet been fully solved, but it is likely to be amenable to improved design and materials. Finally, it is new, and no manufacturer willingly changes his very expensive production lines without more than good reason. The Wankel engine will have to overcome this barrier before it becomes acceptable in large-scale production.

The Stirling Engine

This is an external-combustion engine in which fuel is burnt continuously. The cycle of operations is shown in Figure 6.13. A thermodynamic cycle is performed in which heat is supplied by the combustion reaction at a high temperature to a working fluid (air). Work is done by the fluid in expanding, and waste heat is rejected. The attraction of the system is that as the combustion is continuous, it is exactly like that occurring in an ordinary oil heater or

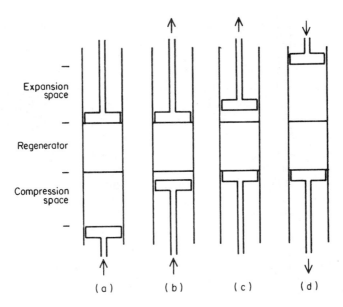

Expansion space

Regenerator

Compression space

(a)   (b)   (c)   (d)

Figure 6.13. Operation of Stirling cycle. The regenerator consists of a medium such as wire wool which is capable of absorbing and emitting heat. Thermodynamics of the process are discussed in Appendix 2

boiler and generates negligible amounts of pollutants other than the acceptable and inevitable CO_2 and water. Moreover, it will tolerate a wide range of fuel, from LPG to heavy fuel oil, and because of the absence of explosions it is quieter than a normal engine. The cycle is already used for low-temperature refrigeration, and Philips, a company which was already engaged in cryogenics, has put considerable efforts into developing the cycle as an engine. Unfortunately, a serious obstacle has appeared. The engine relies on a closed cycle, in which apart from leakage the working fluid is not exhausted to air. Consequently all the waste heat must be extracted by heat exchangers rather than by ejecting it down the exhaust pipe as in a conventional piston engine. It is clear that in a typical engine with a power output of 30 kW and a thermal efficiency of 33 per cent, 60 kW of waste heat must be removed. It has been calculated that in order to achieve the necessary heat transfer, the whole of the bodywork of a vehicle would be needed to act as heat exchanger, and it is not thought likely that this would be popular with the occupants. There appears to be no solution to this problem other than a drastic increase in thermal efficiency, which would reduce the amount of waste heat to be removed for a given power output. This is thermodynamically not possible.

Steam Engines

The advantages of continuous combustion, wide fuel tolerance and low exhaust emissions possessed by the Stirling engine are found also in the various types of steam engine, and for these reasons several groups, both amateur and professional, have studied its possible rediscovery for vehicle propulsion. In the past, the main objection to steam engines has been the long time required to raise steam and also the danger inherent in a boiler at high pressure. Both these objections are overcome by the flash-steam boiler, in which steam is generated (in a superheated state if necessary) by passing water through a hot tube. The warming-up time with this type of boiler is about 30 seconds, and because the amount of water in the boiler at any one time is small, the risk of explosion can be neglected. It would appear, therefore, that steam engines might well prove a useful source of power for road vehicles. However, like the Stirling engine, a steam engine is normally a closed-cycle system, and the condenser must dispose of very large quantities of heat. Alternatively, open-cycle operation might be used, with the waste heat released by venting the steam from the cylinder directly into the atmosphere, as was done in steam-powered railway locomotives. This has two disadvantages. Firstly, the water consumption is high, and it would be necessary to replenish the water supply at frequent intervals, and secondly, the efficiency is inevitably lower in an open cycle because the waste heat is rejected at atmospheric pressure and thus at a temperature of at least 100°C, unlike the closed cycle in which the condenser pressure is lower than atmospheric, and the temperature may be closer to room temperature. Of course, the closed cycle is easily possible in fixed plant such as electricity generators, and also in ships' engines, so the steam turbine is quite suitable in these cases.

6.6 Naphtha

The fraction boiling between about 80 and 180°C, known as naphtha, is extremely versatile, and most of it is converted into lighter components. The products are an important source of gases for petrochemicals, and are also used to supplement the supply of gasoline, especially in the United States, where the demand for gasoline is relatively large, and exceeds the proportion available from crude oil by straight distillation.

The most important process to which naphtha is subjected is a type of catalytic reformation, in which hydrocarbons and steam are reacted in the presence of a catalyst. The process can be applied to most hydrocarbons, and can be illustrated by the reactions which occur with methane:

$$CH_4 + H_2O \xrightleftharpoons{\text{endothermic}} CO + 3\,H_2$$
$$CO + H_2O \rightleftharpoons CO_2 + H_2$$

The product is a mixture of H_2, CO, CO_2 and CH_4, the proportions being controlled by the positions of equilibria of the reactions, and also by the reactivity of the catalyst.

For other hydrocarbons, the reactions involved are somewhat uncertain, but experiments have established that there are three major stages:

(1) Catalytic cracking and dehydrogenation, and, at high temperatures, thermal cracking, leading mainly to olefins of low molecular weight, methane and some hydrogen.

(2) Reaction of these gases with steam, leading to hydrogen and oxides of carbon.

(3) Equilibrium between hydrogen, steam, carbon dioxide, carbon monoxide and methane.

Under normal conditions, methane is the only hydrocarbon which is thermodynamically stable, and consequently all others should be absent in the final products.

The catalysts favoured for the process at present are nickel-based, and a typical commercial version has the composition 21 per cent NiO, 11 per cent CaO, 16 per cent SiO_2, 32 per cent Al_2O_3, 13 per cent MgO, 7 per cent K_2O. (ICI 46–1). Maximum reaction conditions for this catalyst are quoted as 30 bars and 850°C. For more severe conditions up to 1300°C, an alternative catalyst is recommended which has the composition 18 per cent NiO, 15 per cent CaO, 67 per cent Al_2O_3 (Bridger, 1970).

The most important product of this process at present is synthesis gas, hydrogen and nitrogen, for ammonia production. The hydrogen is obtained by raising the temperature and so forcing the chemical reactions to move to the right. This is achieved by the addition of air, which causes partial combustion and also serves to add the required nitrogen to the mixture.

Another interesting product manufactured by the same process but under different conditions is a good-quality town gas, containing methane, hydrogen and carbon monoxide. It is remarkably similar to coal gas (Chapter 5) both in composition and in having a calorific value of about 500 Btu ft^{-3}. The process was used extensively during the 1960's in the United Kingdom for town-gas production, and since the product was very similar to the traditional coal gas it could be burnt in the same equipment without modification, and met with no consumer resistance. When natural gas became available in the United Kingdom, the naphtha process went out of favour except in those areas too remote to benefit from the gas grid, but as the properties of natural gas are quite different from those of coal gas there arose the need for conversion of equipment.

6.7 Kerosene (Paraffin Oil)

Kerosene is obtained by straight distillation in the boiling range 150–250°C. Originally kerosene was a fuel for lamps, but it is now used both as aviation fuel (see next section) and for heating, especially in domestic central heating systems. The quality rating of a kerosene for this purpose is determined by its burning qualities. The *smoke point* is evaluated by burning in a standard

166

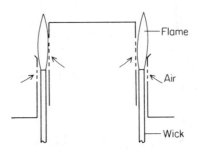

Figure 6.14. Blue flame burner

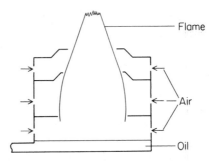

Figure 6.15. Pot burner

wick-fed lamp, and measuring the height of the flame in millimetres at which it starts to smoke. The *char* value is a measure of the tendency to form carbon deposits when burnt. Both these properties can be improved by removing aromatics, which burn with a smoky flame because of their high carbon content. They are removed by solvent extraction with liquid sulphur dioxide.

To burn a liquid fuel it is first necessary to vaporize it. Unlike gasoline, kerosene shows no tendency to evaporate at room temperature, and it must be encouraged to vaporize by heating just before combustion.

In a simple *wick-fed burner* the fuel takes in heat from the flame as it travels up the wick, and it evaporates at the surface of the wick in a zone which can be seen clearly in a candle flame. A simple wick burner produces a luminous and rather smoky flame no matter how good the fuel, but its performance can be improved by supplying excess air to ensure complete combustion. This is done in the blue-flame burner (Figure 6.14) which is familiar in small room heaters. Despite this improvement, it is difficult to run a wick-fed burner at more than about 1kW, and, moreover, the wick needs considerable maintenance. Consequently, wick burners are limited to small portable equipment.

The *pot burner* (Figure 6.15) is a rather more elegant but still simple burner that avoids the problem of the wick. This was common in medium-size portable equipment until a few years ago, when a series of unfortunate accidents led to a decline in popularity. The problem is that if knocked over there is nothing to prevent oil flowing from the tank into the overturned burner. In fact, the

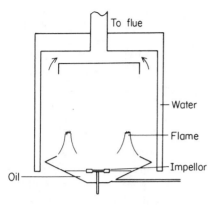

Figure 6.16. Wall-flame burner

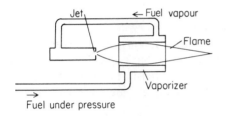

Figure 6.17. Pressure burner

design can be made reasonably safe by incorporating a valve which closes when the burner is not horizontal.

The *wall-flame burner* is the standard type for domestic central heating boilers. Its construction (Figure 6.16) is more complicated than the others described earlier, and it requires an electricity supply for ignition and to drive the spray and air vanes, but it is easily controlled automatically and it is capable of the output (about 15–20 kW) which is required. This type of burner is also suitable for gas oil, which is often used instead of kerosene for domestic central heating.

The *pressure burner* (Figure 6.17) is the standard type for picnic stoves, blowlamps, garden flame guns, etc. The liquid fuel is forced through a vaporizing tube heated by the flame, and its temperature is raised so that it evaporates readily when sprayed through the jet. There are two main problems: the speed of the vapour leaving the jet must not be too high, otherwise it will blow the flame out (see the section on gas burner design), and there is a tendency for the jet to become blocked. The vaporizer tube must be preheated before operation, which means that the equipment cannot be brought into immediate use.

6.8 Aviation Fuel

Piston-engined Aircraft

The fuel burnt by an aircraft piston engine is a high-octane gasoline which is similar to motor fuel but has some special characteristics. The main problem with an aircraft engine is that there are two quite different regimes. When taking off and climbing it operates at extremely high power with a rich fuel–air mixture, and when cruising it operates at a much more economical power level with a lean mixture. The two regimes require different fuel characteristics, and one octane number is no longer sufficient. It is usual to specify two octane ratings, for example 100/130 which indicates a rating of 100 for lean mixtures and 130 for rich mixtures.

The other problem with aircraft is that if the volatility of the fuel is too high, it will evaporate from the tank at high altitude, leading to waste, and it may also evaporate in the fuel system, leading to engine failure. Consequently it is important to control the proportion of low-boiling components in the fuel.

Gas Turbine Engined Aircraft

The gas turbine, in its three different forms, jet, turbofan and turbo-prop, has superseded the piston engine in all but the smallest aircarft. The fuel is basically kerosene, but there are special requirements which are not important in ordinary kerosene. The most important of these is thermal stability, which ensures that the fuel can be heated to relatively high temperatures by air friction without leaving gummy deposits in the fuel system, and, of course, without becoming dangerously inflammable. Fuel meeting the requirements of civil aviation is obtained from distillation in the range 150–250°C, with perhaps some removal of aromatics, and it is referred to as aviation turbine kerosene or Avtur. There is another fuel in common use for aviation turbines, known as aviation turbine gasoline or JP.4, which is obtained by distillation over a wider temperature range, usually 30–260°C. The wider cut enables more fuel to be produced from a given batch of crude oil, and during the recent shortage of oil it became common practice to use JP.4 instead of Avtur. The problem is that the presence of lowboiling fractions increases the risk of fire in an accident, and the use of JP.4 is generally discouraged.

A schematic drawing of a *turbo-jet* engine is shown in Figure 6.18. A multi-stage compresser draws air into the engine and compresses it into a combustion chamber. Fuel is sprayed in and burns, causing an increase in temperature and pressure. The hot gases pass through a turbine, which absorbs only enough power to drive the compressor, and the exhaust gases leave at high speed

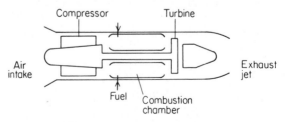

Figure 6.18. Turbo-jet engine

through the exhaust nozzle. The jet engine is an effective means of propelling high-speed aircraft; however, because the mass of the exhaust is relatively small, it is not efficient for large, lower-speed aircraft. The efficiency of the gas turbine for slower aircraft can be improved by modifying it to drive a propeller, thus increasing the mass of the moving gases. The compressor and combustion chambers of this *turbo-prop* engine are similar to those in a pure jet engine, but the turbine has two stages, usually driving two separate concentric shafts. The turbine drives not only the compressor, but also the propeller which supplies most of the thrust, and almost all of the power in the hot gases is absorbed. The turbo-prop is a very satisfactory means of propelling civil aircraft at speeds of up to about 350 knots, and there are very large numbers of such aircraft in service. Its main limitations appear at high speeds, when the airflow over the propeller (which is faster than that over the rest of the aircraft) becomes supersonic.

It is possible to increase the mass of the exhaust stream without the problems associated with the propeller of a turbo-prop, by enlarging the compressor and by-passing some or most of its output around the engine to the exhaust, while still providing enough compressed air to keep the turbine running. This is known as a *by-pass* jet, and it is used in most of the present generation of civil aircraft. The by-pass system can be carried to an extreme in which the by-passed air is propelled by what is really a ducted fan rather than a modified compressor. Such an engine is known as a *fan-jet*, and it is an exceptionally efficient propulsion system for large commercial aircraft.

The gas turbine has proved such a successful engine that it has been adapted for uses other than aviation. As a means of generating electricity, it offers a lower installation cost than steam turbine plant, and also the ability to deliver full power within seconds of starting. It is, therefore, ideal as a standby generator and for dealing with short-term peak loads. The running cost is much greater than steam plant, however, mainly because it is normally run on distillate fuels such as kerosene or gas oil and partly because of the relatively low thermal efficiency (about 25 per cent). Some effort is being made at present to design gas turbines to run on residual fuels, either directly or after chemical processing of the fuel. To achieve better efficiency, it is desirable to use the residual heat in the exhaust gases to preheat the incoming air by means of a heat exchanger. The cost and size of such devices are considerable, and it is not thought desirable at present to use heat exchangers on standby and peak-load plant.

Gas turbines have found applications in high speed marine craft, especially air–sea rescue ships, where their light weight and small size are great advantages. Some vessels of this type are capable of speeds of 50 knots.

There have been many attempts to design gas turbines for motor vehicles. Initially the intention was to produce a smaller, lighter power unit than the conventional piston engine, but lately the emphasis has changed to the reduction of exhaust emissions. The fuel consumption of early designs was excessive but by raising the temperature of operation and by the development of better heat exchangers it has been reduced and is now considered satisfactory. As much as 31 miles per imperial gallon (9 litres per 100 km) has been obtained (Barnard, 1974). The main problem is the high cost of materials in the heat exchanger, and also in the high-temperature parts of the turbine, where the gases may be at 1000–1050°C. The use of ceramics instead of metals may offer a reduction in cost, and may also allow an increase in temperature with consequent improvement in thermal efficiency. It must be borne in mind, however, that even if the materials allow the use of higher temperatures, other factors may impose an upper limit of about 1200°C. For example, there is a rapid increase in nitrogen oxide concentration beyond this temperature.

It is worth noting that a gas turbine vehicle can be fuelled with a wide variety of oils ranging from gasoline to old candle ends, although it is probably happiest with kerosene. There is an inherent improvement in safety in the use of fuels other than gasoline, and this is a very definite advantage of the gas turbine over the petrol engine. Moreover, gasoline is much more expensive than other oil fuels and it has been suggested that because of this a gas turbine would be cheaper to run. However, most of the extra cost of gasoline is accounted for, especially in Europe, by excise duty, and it is unlikely that motorists would be allowed to escape with duty-free kerosene for more than a few months!

6.9 Gas Oil

The fraction boiling between about 250 and 350°C is known as gas oil or diesel oil. The name 'gas oil' derives from its former use in enriching water gas for city supplies; it was sprayed into the reaction vessel and decomposed on hitting the hot coke (see Chapter 5) to yield hydrocarbon gases of high calorific value. With the decline in this method of gas production, gas oil has ceased to be important for this purpose. Nowadays its main uses are as a fuel for diesel engines and as an alternative to kerosene for space heating.

When used as a heating fuel, gas oil is burnt in pressure-jet burners or in wall-flame burners. As a diesel fuel it is used mainly in high-speed engines such as road vehicles, small electricity plants, small boats and civil engineering equipment. Larger, low-speed engines such as marine power units and large electricity plants use a heavier (and thus cheaper) fuel, produced by adding some residual oil to the distillate gas oil. Sometimes these engines are started on light fuel and are changed over to the heavier grade when warm.

The construction of a diesel engine is basically rather similar to that of a petrol engine, but there are considerable differences in the operating cycle, and in the demands made on the fuel. The cycle of operations is as follows:

(1) During the induction stroke, air (not fuel–air mixture) is drawn into the cylinder.

(2) During the compression stroke the air is compressed very much more than in the petrol engine, producing a very high temperature.

(3) There is no spark. Instead, fuel is injected in a fine spray into the hot air in the cylinder and ignites spontaneously. Injection is continued during the power stroke, so the burning proceeds smoothly.

The characteristics of diesel fuel must obviously be quite different from those of gasoline. Volatility is not needed, in fact it may be a nuisance. Spontaneous ignition, far being a disadvantage, is actually a requirement. The performance of a diesel fuel in this respect is measured by its cetane number, which is analogous to the octane number of gasoline in that it is determined by comparison with a two-part mixture, in this case normal cetane and α-methyl naphthalene. The highest cetane numbers are reached by paraffinic components, but these tend to solidify into waxes at low temperatures, and in an unexpectedly hard winter this can lead to an embarrassing number of breakdowns. Diesel fuels intended to be used in cold weather are generally 'winterized' by various methods to prevent this occurrence.

Thermodynamically, the diesel engine can be analysed in a similar way to the petrol engine, and the same conclusion can be reached, that the efficiency is increased by raising the compression ratio. Because air alone, rather than fuel–air mixture, is drawn into the cylinder, there is no difficulty with preignition, and the compression ratio can be made much higher than in a petrol engine, up to 20 to 1. Large diesel engines can achieve practical efficiencies of at least 40 per cent, which compares with the best modern steam turbines. Another advantage of the diesel over the petrol engine is that, perhaps surprisingly, its exhaust is lower in pollutants. Indeed, most road vehicle diesels can already reach the standards required by the United States regulations of the early 1970's. In view of the reputation of the diesel lorry as a major source of nuisance, a word of explanation is perhaps required. There is no doubt that a heavily overloaded, badly-adjusted diesel engine can emit a large amount of unpleasant black smoke. However, if well maintained, especially if the fuel injectors are properly adjusted and if used within its normal power rating, a diesel emits much less carbon monoxide, nitrogen oxides and unburnt hydrocarbons than does a petrol engine of comparable power output. Diesel lorries would be much more socially acceptable if they had larger engines, since there would be less temptation to overload the engine. As a power unit for automobiles, there are several objections to the use of diesel engines. They produce more vibration and noise than comparable petrol engines. They are considerably more expensive, because the higher compression ratio necessitates more robust construction. Although diesel engines are used in certain types of automobiles, including taxis and Land Rovers, they seem unlikely to be used extensively as a replacement for the petrol engine.

6.10 Residual Fuel Oil

After all the distillates have been driven off, the remaining heavy oil is suitable only as a fuel for large-scale furnaces. Its viscosity is so high at ordinary temperatures that it is necessary to use heated tanks and fuel pipes to enable the oil to flow at a reasonable speed. A typical burner is shown in Figure 6.19. The heated oil is atomized into droplets of less than 0·1 mm by blowing under pressure from a nozzle or by means of a blast of air or steam. The droplets are blown

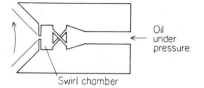

Oil
under
pressure

Figure 6.19. Pressure-jet residual
oil burner

Swirl chamber

into the combustion zone in an air stream and are ignited by means of a pilot flame, sometimes fuelled by propane gas. An alternative type of burner uses a rotating cup as atonizer.

The main use of residual oil is in electricity generation. The plant in an oil-fired power station is essentially similar to that in a coal-burning installation, except that ash-disposal facilities are not needed. The siting of the station is influenced by different criteria, however. The most favourable site is near an oil refinery, and, as the refinery is already a major industrial installation, the addition of a power station stimulates no new environmental objections. Alternatively, power stations can be located at a considerable distance from the nearest refinery because of the fairly low transportation cost of heavy oil.

The most pressing problem in quality control with residual oil is to minimize the sulphur content, which causes sulphur dioxide air pollution. At present the problem is avoided, rather than solved, by blending with oil from low-sulphur crudes. Several techniques have been proposed for removing sulphur from residual oil (see Section 6.3), but they are at an early stage of development. Certain alternative techniques have also been proposed, whose goal is not simply the removal of sulphur but also the conversion of residual oil into other fuels. Some of the products of such processes would be low-quality gases, and because these are expensive to transport, it is considered desirable to site the equipment at the consumer's premises rather than at an oil refinery. An obvious application for such techniques is in power stations.

In one such process, designed by Esso Research Ltd. (Squires, 1972), high-sulphur oil is desulphurized and converted into gas in a fluidized bed of lime-stone. The process incorporates a lime regeneration system, which removes the trapped sulphur as SO_2. This can be sent to a chemical plant. The regeneration process causes some agglomeration of the lime, reducing its reactivity, and it is necessary to replenish with a steady trickle of fresh limestone. The gas produced could be used for a number of purposes, including the fuelling of a gas turbine for electricity generation, when even the exhaust gas could be used to heat a boiler and drive a normal steam turbine plant. This has the interesting advantage that the 'top' temperature of the process is higher than that of a conventional steam turbine plant fired directly with residual oil. It may be possible, therefore, to reach a higher thermal efficiency and to reduce SO_2 emissions at the same time. Unfortunately the capital cost of such a complex is likely to be high.

6.11 Non-fuel Uses of Oil

The most important use of oil products other than as fuel is as a feedstock for the petrochemical industry. The main sources of supply are the naphtha fraction, which is a liquid and thus easy to transport, and olefin gases (principally ethylene) which arise from refinery operations, especially catalytic cracking, and which are most easily used close to the refinery.

The main processes by which petroleum hydrocarbons are converted into raw materials for the chemical industry are as follows:

(1) Removal of hydrogen under mild catalytic conditions to yield olefins. This process produces *ethylene* (for polyethylene, polyesters, polystyrene, vinyl polymers, ethyl alcohol, ethylene glycol and detergents), *propylene* (for polypropylene, acrylics and various chemicals) and *butene* and *butadiene* (for synthetic rubbers, nylon and chemicals).

(2) Removal of hydrogen under more vigorous conditions to yield acetylene. This is used for vinyl polymers, rubbers and chemicals, as well as for oxyacetylene welding.

(3) Removal of hydrogen under catalytic reforming to yield aromatics. These are used in the manufacture of a multitude of organics, including polymers and explosives.

(4) Oxidation, with or without rupture of the carbon chain. This can yield a range of acids, aldehydes and ketones which are used in a great range of chemical processes.

(5) Chlorination. The products of this process include silicones, dry-cleaning fluids, solvents and refrigerant gases.

(6) Decomposition to produce carbon black for tyres, inks, lacquers, etc.

(7) Reaction with steam to yield hydrogen. This is used in the manufacture of ammonia (important for fertilizers), methanol and other alcohols.

The foregoing is ample proof of the importance of oil-based chemicals. It is interesting to note that all the processes mentioned use the lighter fractions that are so much in demand for premium fuels, and there is thus a direct conflict between the demands of the chemical industry and the domestic fuel user. It is worth noting, however, that oil is not the only source of organic chemicals. As we saw in Chapter 6, it is possible to use coal for similar purposes. In view of the threatened shortage of oil, we may yet find ourselves doing just that.

The remaining important non-fuel uses of oil exploit the heaviest components. After distillation of all the lighter fractions, including gas oil, there remains a residue with a boiling point in excess of 350°C. This is further distilled in vacuum. The vacuum distillate is processed into lubricating oil, together with some of the residue, and the remainder is used as bitumen. As can be seen from Tables 6.2 and 6.3, the tonnages involved in these two products are quite large.

Lower-viscosity lubricating oils (such as light instrument oils and engine crankcase lubricants) are obtained from the appropriate distillate fraction and subjected to a dewaxing treatment. Heavier oils are obtained by solvent extraction from the residue. The tailoring of oils for different purposes is a

174

highly specialized task, in which not only viscosity is important, but also viscosity index (the change of viscosity with temperature) and stability against oxidation. The details are beyond the scope of this book. One point of interest to the 'energy crisis' is worth mentioning. In the past the disposal of waste oil from vehicle servicing has been a serious problem, but the increased cost of oil fuel has now made its recovery economic. One procedure which is used at present is to centrifuge the waste to remove solid particles (mostly carbon, with some admixture of steel and lead), and then to blend the cleaned oil with with heavy fuel oil.

Bitumen is used principally for road-making, but it finds application also in roofing felt and coating of steel pipes, including pipelines for the oil industry.

The waxes removed from lubricating oils are used in manufacture of waxed paper cups, polishes and electrical insulation. They are also the basic raw material of candles—a matter of some importance when the more sophisticated modern technologies let us down!

References

Anderson, P. C., J. M. Sharkey and R. P. Walsh, *J. Inst. Petroleum*, **58**, (560), 83–94 (1972).
Anon., *Petrochemical Engineer*, **41**, (5), 39–52 (1969).
Barnard, M. C. S., *Metals and Materials*, **6**, (2), 62–67 (1974).
Blume, J., D. Miller and L. Nicolai, *Hydrocarbon Processing*, **48**, (9), 131–136 (1969).
Bridger, G. W., *Catalyst Handbook*, Imperial Chemical Industries, London, pp. 64–96, 1970.
Desty, D. H. and D. M. Whitehead, *New Scientist*, **45**, (685), 147–149, (1970).
Harvey, H., *New Scientist*, **63**, (904), 5 (1974).
Lowson, M. H., (Ed.), *Our Industry—Petroleum*, British Petroleum, London, 1970.
Squires, A. M., *Scientific American*, **227**, (4), 26–35 (1972).

Further Reading

Berkowitz, D. A. and A. M. Squires, (Eds.), *Power Generation and Environmental Change*, M. I. T. Press, Cambridge, Mass., 1971.
Dunstan, A. E., A. W. Nash, B. T. Brooks, H. Tizard and V. C. Illing, (Eds.), *The Science of Petroleum*, Vols. 1–6, Oxford University Press, London, 1953.
Imperial Chemical Industries, Ltd., *Catalyst Handbook*, World Scientific Books, London, 1970.
Lay, J. E., *Thermodynamics*, Pitman, London, 1964.
Lowson, M. H., (Ed.), *Our Industry—Petroleum*, British Petroleum Co. Ltd., London, 1970.
Nagy, B. and U. Colombo, (Eds.), *Fundamental Aspects of Petroleum Geochemistry*, Elsevier, Amsterdam, 1967.

7 Electricity Generation and Transmission

Electricity is entirely a secondary source of energy, that is it does not occur naturally, but rather is produced from other primary sources. Properly speaking it belongs in Chapter 17, since the transmission of electricity is nothing more than a highly efficient and convenient energy transport system. Nevertheless, electricity is such an important part of modern life that it deserves special treatment. Moreover, this importance is likely to increase in the future, not only because of the apparently inexorable growth in electricity consumption which has occurred ever since a public supply first appeared, but also because many of the new sources of power which have been proposed over the last few decades, such as nuclear fission and fusion, are highly inconvenient in their own right, and must rely on electricity transmission (and possibly thermal transmission, Chapter 15) to convey their energy efficiently and conveniently to the consumer.

As electricity is a secondary source of energy, our first consideration must be the conversion processes by which it is produced. The most important of these is the steam turbine, which is responsible for a very large proportion of the electricity produced at present. Other sources are water turbines (Chapter 13), gas turbines and diesel engines (Chapter 6), and there is some prospect in the future for certain direct energy conversion processes using, for example, magnetohydrodynamics. The importance of these devices is small in comparison with that of the steam turbine generator.

The essential stages involved in a steam turbine electricity plant are outlined in Figure 7.1. Fuel, usually coal or oil, is burnt in an appropriate type of burner, whose design has already been dealt with in earlier chapters. The heat released in combustion is extracted in the boiler, where it is used to raise steam at as high a temperature as possible, so as to maximize the thermodynamic efficiency (Appendix 2). The steam drives the turbine, and then passes to the condenser, is condensed to water and returned to the boiler. The temperature of the condenser is made as low as possible, again to maximize the thermodynamic efficiency. The turbine drives an alternator (AC generator) whose output is transformed to a high voltage and transmitted to the consumer, where the voltage is reduced to a more appropriate value for the consumer's purposes.

Having looked briefly at the overall picture, we now go on to consider the stages in some detail.

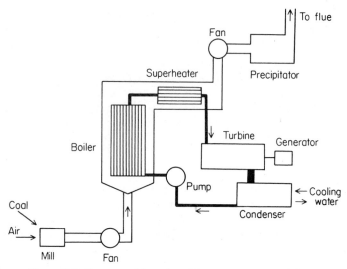

Figure 7.1. Steam turbine electricity generator (coal-fired)

7.1 Fuels and Boilers

The techniques of handling and burning coal and peat have been considered already in Chapter 5, and oil and gas burners have been treated in Chapter 6. The boiler designs are broadly similar for all these fuels, and, although there are important differences in detail, most of the main features of coal-, oil- and gas-fired boilers can be understood by a consideration of Figure 7.2, which shows a boiler fired by oil.

The boiler is constructed of steel tubes placed parallel to each other round the walls of the combustion zone. Water is passed through these tubes, so that the burning fuel is effectively surrounded by a wall of coolant which provides very efficient heat transfer. The water boils, producing steam which is passed through another set of tubes known as the superheater. The object of this is to raise the temperature of the steam as high as possible to ensure that it is well above saturation. Although we have distinguished between water and steam in this discussion, many modern boilers operate in the supercritical region where no distinction can be made. The feed water in the boiler is of very high purity to minimize the deposition of scale and to prevent corrosion. The pressure within the boiler is well above atmospheric, and the boiling point of the water is consequently well above its normal value; typical steam conditions for a modern power station are 160 atmospheres and 560°C.

7.2 Steam Turbines

The simplest possible steam turbine consists of a jet of steam from a simple boiler, issuing through a small hole and impinging on a paddle wheel. This is

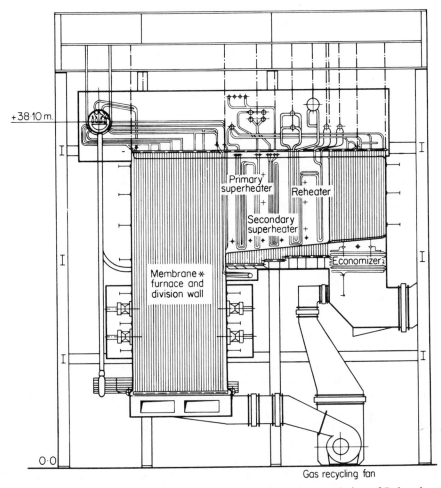

+38·10 m.

Primary superheater

Reheater

Secondary superheater

Economizer

Membrane furnace and division wall

0·0

Gas recycling fan

Figure 7.2. Oil-fired power station boiler. (Reproduced by permission of Babcock and Wilcox (Operations) Ltd., London)

the basic *impulse turbine*. Needless to say, this simple machine is not very efficient. The efficiency of the impulse turbine can be enhanced if optimum use is made of the momentum of the steam, by arranging that the forward speed of the blade is half that of the impinging jet. From Figure 7.3 it is clear that if the incoming speed of the steam (seen by a stationary observer) is c_1, the exhaust velocity is c_2, and the speed of the blade is u, then from the frame of reference of the blade, the incoming velocity is $(c_1 - u)$ and the exhaust velocity is $-(c_2 + u)$. From the viewpoint of the blade's frame of reference these two velocities must be equal and opposite, except for the small slowing-down due to friction, so we can write

$$c_1 - u - c_2 - u = 0$$

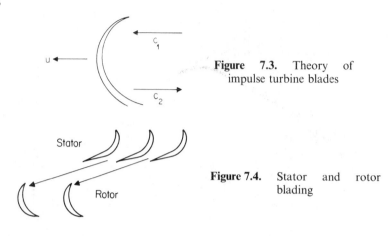

Figure 7.3. Theory of impulse turbine blades

Figure 7.4. Stator and rotor blading

and hence

$$c_1 - c_2 = 2u$$

Returning to the stationary frame of reference, it is clear that if the steam has any appreciable speed when it leaves the blades, a loss of useable kinetic energy has occurred, so the best efficiency must be reached when $c_2 = 0$. Consequently for best efficiency $c_1 = 2u$, i.e. the blade speed is half the speed of the steam impinging upon it.

This simple argument is highly satisfactory for an impulse-type (Pelton) water turbine, as will be seen in Chapter 13, for water is almost completely incompressible. It is not wholly satisfactory for a compressible fluid such as steam, for the steam jet derives its kinetic energy from a loss of pressure as it passes through the jet, and unless the exhaust steam leaving the blade has actually reached atmospheric pressure (or rather, the reduced pressure in the condenser) it still possesses potential energy by virtue of its pressure, even if it is wholly stationary. Accordingly, it can be passed through another jet, gaining kinetic energy and driving another set of blades, and so on until it finally reaches the lowest pressure in the machine. In a practical impulse turbine, a series of multiple jets is formed by a set of inclined stationary blades (known as the *stator*), and the jets impinge on a set of moving blades (the *rotor*) (see Figure 7.4).

An alternative to the impulse turbine is the reaction turbine, which in its simplest form is a rotor arm with tangential jets at its tips, through which steam escapes in the manner of a lawn sprinkler. Once again, this simple design is not very efficient, and a practical reaction turbine employs sets of stator blades which feed steam to sets of rotor blades. In construction this is similar to the blade arrangement in the impulse turbine, but there is an important difference in that the pressure drop is not confined to the stator, but is equally divided between stator and rotor blades (Figure 7.5). Optimum efficiency in this case requires that the incoming steam velocity, c_1 in our previous notation, is approximately equal to the blade velocity u, i.e. the incoming velocity of

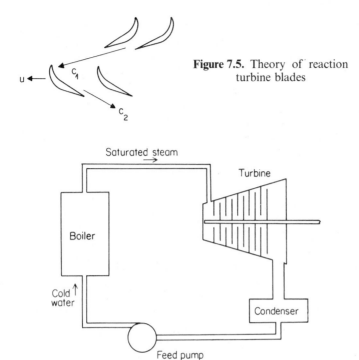

Figure 7.5. Theory of reaction turbine blades

Figure 7.6. Simple turbine

the steam as seen in the blade's frame of reference is almost zero. Clearly this velocity must be positive, otherwise the steam never gets to the blade, but if it is large, there is relative movement between blade and steam which cannot be usefully harnessed to drive the blade forward.

Since for the impulse turbine, $c_1 = 2u$, and for the reaction turbine $c_1 = u$, it is evident that the momentum transferred per stage in the impulse turbine is approximately twice that in the reaction turbine. Thus, a reaction turbine requires more stages of blading than does an impulse turbine of similar capacity and steam pressures.

Most modern steam turbines are a compromise between impulse and reaction types. It is customary to design blading which derives power from both effects, and the percentage contribution from impulse and from reaction may vary from stage to stage through the turbine. It is even possible for the percentage of impulse and reaction to vary along the length of a single blade.

The simplest possible practical turbine is shown in Figure 7.6. Saturated steam from a boiler enters at the boiler pressure and expands through the turbine, giving up energy through both temperature and pressure drops. It is rejected to the condenser where it is condensed to water and returned to the boiler, via a pump which raises the pressure of the water to boiler pressure again.

Such a simple turbine suffers from several problems. First, the difference in pressure between the boiler and the condenser has to be large in order to promote good efficiency, and there is, therefore, a great change in the volume of the steam during its passage through the turbine. The blades at each point along the shaft must be capable of sweeping the appropriate volume of steam and so there must be a great change in the cross-sectional area of the turbine from the small high-pressure end to the large low-pressure end. In practice it is difficult to achieve a sufficiently great change in cross-section in a single unit, and it is customary, therefore, to have separate high-pressure (HP) and low-pressure (LP) turbines, and in large sets, there is often an intermediate-pressure (IP) turbine as well. The low-pressure turbine is usually duplicated as a mirror-image pair so as to double the effective cross-section.

The second problem arises because the steam at the inlet is saturated. During its passage through the turbine it inevitably becomes *wet*, that is, it starts to condense. As a result, the blades can be eroded by high-speed water drops, leading to mechanical failure. The problem can be avoided at least in part by superheating the steam. The amount of superheating, and the pressures and temperatures at the various points in the system, can be chosen so that the steam remains unsaturated right through to the condenser, although it is customary to allow some condensation in the later stages of the LP turbine. In some larger machines the inlet pressure is raised to remarkably high levels (up to 170 bars) in order to achieve greater efficiency. This, unfortunately, tends to increase the degree of condensation at a given temperature, and it is impossible to raise the temperature of the steam to such an extent that saturation at the later stages of the LP turbine is avoided. This is because the materials in the superheater are unable to withstand the very high temperatures involved. A solution to this problem is made possible by a *reheat cycle* in which the steam leaves the superheater at very high pressure and high temperature, expands through the HP turbine, and then returns to the boiler where it is reheated to a temperature close to its previous value but at a lower pressure. It can now pass through the IP and LP turbines in turn without the risk of excessive condensation.

A third problem is the irreversible flow of heat (see Appendix 2) which occurs when cold boiler-feed water is pumped straight into a hot boiler. Such irreversible heat transfer is thermodynamically inefficient, and it is advantageous to increase the temperature of the feed water steadily with the minimum possible temperature difference. This can be achieved by bleeding off steam from various points in the turbine itself, for between the inlet of the HP turbine and the outlet to the condenser there is a supply of steam available at any temperature within the limits of the cycle. Ideally there would be an infinite number of such feed-water heaters, but this is not possible, and in practice about eight separate units with equal temperature increments are used in large installations. This type of cycle is known as a *regenerative cycle*.

These considerations lead to a turbine boiler combination which can be represented schematically as in Figure 7.7. It should be noted that this sketch does not represent any particular turbine, but rather is intended to

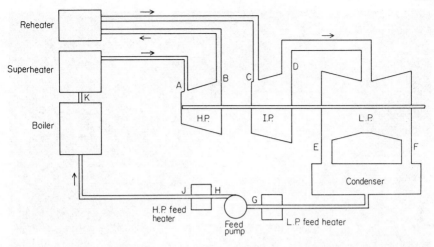

Figure 7.7. Turbine with superheat and reheat. Typical temperatures and pressures might be as follows: A: 540°C, 150 bar; B: 350°C, 10 bar; C: 530°C, 10 bar; D: 250°C, 2 bar; E: 40°C, 0·2 bar; F: 40°C, 0·2 bar; G: 150°C, 0·2 bar; H: 150°C, 160 bar; J: 250°C, 160 bar; K: 380°C, 150 bar

give a general idea of the stages involved in a typical turbine cycle. A photograph of a typical modern installation in shown in Figure 7.8.

A steam turbine is a heat engine, which can be represented to a first approximation by a Carnot cycle (Appendix 2), subject, of course, to the unwarranted assumption that all processes are reversible in the thermodynamic sense. A Carnot cycle for a condensible vapour, such as the water–steam system, is shown in Figure 7.9. A quantity of wet steam at condenser temperature T_C, state 1 in the figure, is compressed at constant entropy to saturation at state 2 where it reaches the boiler temperature T_H. It is vaporized in the boiler at constant temperature T_H to state 3, where it is again saturated, and it then expands at constant entropy through the turbine to state 4. Leaving the turbine, it enters the condenser and is condensed at constant temperature T_C, returning to its original state 1. As usual in a Carnot cycle, the efficiency is given by

$$\eta = \frac{T_H - T_C}{T_H}$$

Putting in typical figures for a modern power station, $T_H = 500$°C (773K) $T_C = 20$°C (293K), which gives an efficiency of 62 per cent.

Unfortunately it is not possible for a practical steam turbine to carry out a Carnot cycle, partly because of the difficulties in compressing wet steam to saturated vapour at constant entropy. An alternative cycle, in which this is not a requirement, was devised by Rankine. In the Rankine cycle without superheat (Figure 7.10a), saturated steam at the condenser temperature T_C at state 1 is pumped at constant entropy into a boiler (state 2) where it is heated at constant pressure, ending up as saturated steam at the boiler temperature

Figure 7.8. A large modern turbine installation. (Reproduced by permission of G.E.C. Turbine Generators Ltd., Manchester)

T_H at state 4. The steam expands through the turbine at constant entropy to state 5, entering the condenser and condensing at constant temperature T_C back to state 1. It is easily shown that the efficiency of such a cycle is given by

$$\eta = \frac{\text{(work obtained from turbine)} - \text{(work put into pump)}}{\text{(heat input from boiler)}}$$

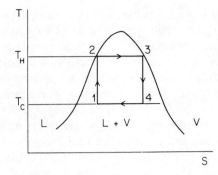

Figure 7.9. Carnot cycle for a condensible vapour. L: liquid; V: vapour

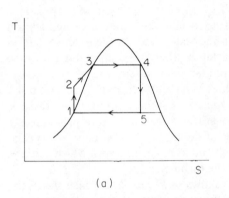

(a)

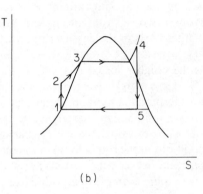

(b)

Figure 7.10. Rankine cycle (a) without superheat, (b) with superheat

which from Figure 7.10a is clearly given by

$$\eta = \frac{(h_4 - h_5) - (h_2 - h_1)}{(h_4 - h_2)}$$

Normally superheated rather than saturated steam is used, and the diagram is then modified to that shown in Figure 7.10b, which has been numbered to correspond with the notation used in Figure 7.10a. Consequently the same equation is valid.

To obtain a value for the efficiency in a typical installation it is necessary to use steam tables to obtain the enthalpies at the various pressures or temperatures. Usually a good modern installation has an efficiency of 40–45 per cent.

The extension of the Rankine cycle to reheat and regenerative cycles is straightforward, and the reader is referred to textbooks of engineering thermodynamics for details.

The speed of rotation of a turbine is of some importance, especially when it is used to drive an alternator. It is shown later in this chapter that an alternator must rotate at a speed which is governed by the frequency of the electricity being generated. A frequency of 50 Hz, which is the European standard, necessitates an alternator rotating at 3000 rpm, 1500 rpm or some other simple fraction of 3000. At 60 Hz, which is the American standard, these speeds are increased to 3600 rpm, 1800 rpm, etc. In order to avoid the necessity of gearing, it is desirable for the turbine shaft to run at the same speed as the alternator, and in order to attain the largest possible output from a given size of machine, it is preferable to run at full speed, that is, 3000 or 3600 rpm. However, as has been noted earlier, the low-pressure section of the turbine is inevitably much larger than the others, and it may be difficult to rotate such a large piece of machinery at high speed because of the large forces involved. The problem can be reduced by separating the shaft of the low-pressure turbine from that of the other stages, and running it at half speed, that is 1500 or 1800 rpm. This involves two separate alternators, and a considerable amount of duplication in other parts of the engineering, which inevitably adds to the expense, but, in very large sets, especially when operating at the American standard which involves higher speeds of rotation, there may be no alternative.

So far we have treated turbines thermodynamically but there are other non-thermodynamic considerations. One of the most important of these is aerodynamic design. Aerodynamic losses in turbines arise from profile losses (that is losses associated with the shape of the blade), losses due to the special conditions at the root and tip of the blade and losses in pipework which conveys the steam between the various parts of the turbine.

A graphic example of a profile loss is shown in Figure 7.11, which shows the discontinuity in flow produced by the finite thickness of the trailing edge of the blade. Other profile losses arise from the boundary layer, in which friction between the steam and the blade surface slows down the relative velocity of the steam with respect to the blade. A similar effect occurs near the tips of the blades, where the turbine casing has a boundary layer, and also near the root of the blades, where the rotor shaft is the offender. Furthermore, the slowing up of the boundary layer at the walls produces a reduction in the transverse acceleration of the steam and in its change of momentum, and, therefore, in the pressure drop across the blade. This gives rise to turbulence and the formation of vortices, the energy of which cannot be extracted usefully by the turbine (Craig and Todd, 1974). Losses in the pipework are most severe where the low-pressure turbine exhausts to the condenser, for it is here that the largest volume of steam must be handled. Optimum efficiency requires a short and unhindered path and an improvement in this respect has been introduced recently by mounting the low-pressure turbine inside the condenser.

A problem related to aerodynamics is erosion of the turbine blades by high-speed water droplets. This is of significance only in the low-pressure section, for it is here that the steam is sufficiently near to saturation to make condensation likely. Since the blade speeds in the large-diameter rotors used in the LP section can be very high, the phenomenon is of considerable practical importance. For example, a blade of 1 m radius rotating at 3000 rpm has a tip speed of 314 m sec^{-1} (700 mph).

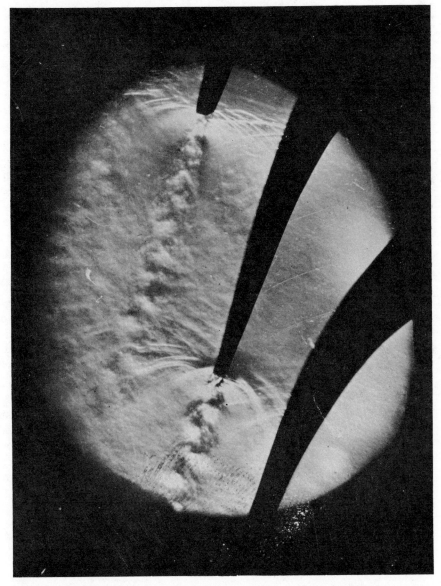

Figure 7.11 Vortices formed at trailing edges of blades. (From Craig and Todd (1974) Reproduced by permission of G.E.C. Turbine Generators Ltd.)

Studies of erosion have been of a semiempirical nature (Bowden and Brunton, 1961; Bowden and Field, 1964; Gardner, 1964; Todd, 1974) and it has been established that the rate of weight loss for a given material is given by

$$\frac{\mathrm{d}W}{\mathrm{d}t} = K(v - v_t)^h$$

Here K is a constant, v is the impact velocity, h is an exponent of the order of 2·5 to 3·0 and v_t is the *threshold velocity*, that is the velocity below which there is no significant erosion. The threshold velocity is a characteristic of the material, and is typically 92 m s^{-1} for mild steel and 305 m s^{-1} for stellite, a cast chromium–cobalt alloy which is frequently used as an erosion shield.

It has been established by internal inspection using introscopes and high-speed photographic techniques (Bowden, 1965) that the most important source of erosion is relatively large drops of water that stream off the trailing edges of the stator blades. These drops have velocities much lower than the steam, so their relative velocity with respect to the rotor blades is high. The leading edges of the rotor blades are thus the most favoured areas for erosion. By contrast, only very small droplets (diameters less than 0·5 μm) are formed by direct nucleation in the wet steam, and these have velocities so close to that of the steam that they do not usually exceed the threshold velocity with respect to the blades; their erosion effect is negligible.

The minimization of erosion is achieved in two ways. The direct approach is to cover the affected region of the blade with an erosion shield made of a material with a high threshold velocity. Predictably these are usually hard metals or intermetallic compounds, which are wrought into the required shape and brazed on to the blade leading edge. It has been found possible to improve the performance of stellite erosion shields by using vacuum-melted material, and although the mechanism of this improvement is not fully understood, it is thought that the amount of non-metallic impurity content is reduced by this process. Various other materials have been tried experimentally, including tungsten carbide and several sintered 'alloys' of titanium carbide with metals, but at present stellite is the most usual choice. A more elegant approach to the problem of erosion is to eliminate the offending water drops as far as possible. This is done by removing the water which has deposited on the stationary blading. A typical method for so doing is shown in Figure 7.12. Water is removed through the passages as indicated, and is exhausted into the steam lines which go to the feed-water heating system, thus usefully using some of the sensible heat contained in the trapped water. An alternative technique which has been tested recently is the use of a slotted trailing edge in the stator blade,

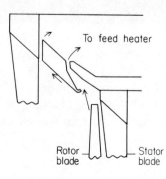

To feed heater

Rotor blade

Stator blade

Figure 7.12 Arrangements for extracting water from LP turbine blades. (From Todd (1974). Reproduced by permission of G.E.C. Turbine Generators Ltd.)

through which the water collecting on the surface of the blade can be sucked away before it is able to run off into the steam flow.

The problems of erosion in low-pressure turbine stages have now been sufficiently well controlled that it has been possible to design *wet-steam turbines* operating on saturated (that is non-superheated) steam. These are of importance in many nuclear power station designs, where it is difficult to arrange for reasonable degrees of superheat, and also for geothermal power plants operating with natural steam, where it would be necessary to burn fuel to raise the steam above saturation. Removal of water is necessary in wet-steam turbines not only to reduce erosion, but also to improve efficiency, for it is an approximate, though reaconably accurate, assumption that the efficiency of a turbine stage is reduced by 1 per cent for each 1 per cent average of inlet and outlet stage wetness (Baumann, 1912). 'Internal' methods of water removal are similar to those described above, and they are usually supplemented by 'external' separators in which the steam is removed from the turbine between stages of blading and passed through a water separator before returning to the turbine. Generally the opportunity is taken to reheat the steam before it returns to the turbine, and it is particularly important to remove 'wet' water before reheating, since otherwise much of the reheating energy is spent on evaporating the wetness, lowering the overall thermodynamic efficiency. External water separators usually depend on inertia of the water drops, and some typical designs are shown in Figure 7.13.

7.3 Condenser

From the low-pressure turbine the steam passes to the condenser where it liquefies and is returned to the boiler. The condenser temperature must be as low as possible in order to maximize the thermodynamic efficiency (Appendix 2), and a very large amount of heat, often equal to about twice the electrical output of the plant, must be disposed of at this low temperature. The condenser is cooled by a large-volume flow of ordinary unpurified water,

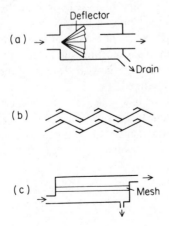

Figure 7.13. External water separators: (a) cyclone; the water droplets fly out centrifugally; (b) wiggle plate; water is collected in channels. (Reproduced by permission of The Peerless Manufacturing Co.) (c) wire mesh filter. (Reproduced by permission of G.E.C. Turbine Generators Ltd.)

obtained where possible from a river or the sea. In the case of an inland power station, where river water may not be available in large enough quantity, an alternative arrangement may be used. The water is circulated to a cooling tower, in which it is sprayed from fine nozzles into a rising stream of air. Some of the water evaporates, cooling the rest, and the cooled water is recirculated to the condensers. In this arrangement the cooling results from a loss of latent heat, about 2000 joules per gram, rather than by a loss of sensible heat which for a temperature rise of 10°C represents 42 joules per gram. From this it would appear that a cooling tower is 50 times more economical in cooling water than is direct river cooling, but, of course, this is an oversimplification since it may be permissible to raise the temperature of the whole river by 10°C but it is certainly not permissible to evaporate its entire flow. Nevertheless, the use of cooling towers does permit a greater concentration of power stations on an existing waterway than does direct cooling. Woodson (1971) has reviewed the different types of cooling tower, and has predicted that they will be used even in coastal power stations in future. Despite their advantages, cooling towers have their own problems, particularly in that they raise the humidity of the surrounding air, causing local mists in unfavourable weather conditions. Dry cooling towers, in which the cooling water does not evaporate but passes through a 'honeycomb' rather like that of a car radiator, have been tried on an experimental basis, for example at Rugeley in the English Midlands, but the cost of construction is inevitably higher than that of conventional wet cooling towers (Chapter 15), and it seems unlikely that they will be used except where unavoidable.

7.4 AC and DC Systems

The first system to be adopted for public electricity supply was *direct current* in which a steady voltage is produced, but this was soon superseded by *alternating current* in which the voltage varies periodically. The main advantage of alternating current (AC) over direct current (DC) is the possibility of changing an AC voltage easily by means of a *transformer*. As we shall see in the section on transmission this is advantageous as very high voltages can be used for long-distance transmission of electricity while relatively low voltages are easily produced for consumers' equipment. There are other advantages in that motor design is simplified, and that simple but highly accurate electric clocks and timing devices can be built, but the main advantage is undoubtedly the ease with which the voltage can be changed.

AC voltages are usually sinusoidal with $V = V_0 \sin \omega t$. Here V_0 is the maximum voltage and ω is the angular frequency. The peak voltage V_0 is not the most suitable parameter to represent AC voltages, and it is easily shown that a more useful choice is the root-mean-square (r.m.s.) voltage $\bar{V}$, which for a sinewave is related to V_0 by

$$\bar{V} = \frac{V_0}{\sqrt{2}}$$

The chief advantage of the r.m.s. value is that it is the equivalent of the steady voltage of a DC system when used for calculations of power in watts.

In both DC and AC circuits it is evidently necessary to have two conductors between the source of power and the consumer to complete the circuit, and one might expect, therefore, that for n circuits between the source and the consumer, there would have to be $2n$ conductors. However, an ingenious saving is possible. Consider three sinewaves which are out of phase with each other by $120°$ $(2\pi/3)$. The equations of these will be

$$V = V_0 \sin \omega t$$

$$V = V_0 \sin (\omega t - 2\pi/3)$$

$$V = V_0 \sin (\omega t - 4\pi/3)$$

These can be represented by three separate curves. An interesting alternative representation, which is widely used in electrical engineering, makes use of the fact that a sinewave can be generated by the projection of a rotating radius onto the diameter of a circle. Since in this case we have three sinewaves with phase differences of zero, $2\pi/3$ and $4\pi/3$, we can represent the situation by three rotating radii, $120°$ apart. The magnitude of the radii can represent either the peak value of the voltage or the r.m.s. value and the latter is more usual.

Now imagine each of the rotating radii to represent an AC current in a separate circuit. Six wires would be required to carry the separate currents, but if the circuits were connected together as in Figure 7.14, only four wires would be needed, since the fourth wire acts as a return conductor for all three circuits. The three wires connected to the 'outer' ends of the circuits are said to carry the three *phases*, and the fourth wire is known as the *neutral*. It is customary, though not essential, to connect the neutral conductor to earth.

By the trick just described, we have saved two conductors, which in a long transmission line is a considerable economy. However, a further economy is possible if the three phases all carry equal currents, for they can then be represented by vectors of equal length at angles of $120°$ to each other, which, as is well known, sum to zero by vector addition. This means that the current in the neutral conductor is *zero*, and the conductor is unnecessary. Even if the three phases are not exactly equal, the neutral conductor is required to carry only

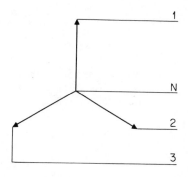

Figure 7.14 Four-wire three-phase connection

the small amount of current necessary to balance the load on the three phases, so the wire can be thinner than the three others.

There is another advantage to poly-phase AC, namely that it can be used to create a rotating magnetic field, which is useful in the design of electric motors. It is well known that two sinewaves of equal amplitude, out of phase by 90°, and acting at right angles to each other give rise to a circular motion, and it requires but a small extension of reasoning to see that three sinewaves of equal amplitude, out of phase by 120°, and acting at 120° to each other can also give rise to a circular motion. If the sinewaves are AC currents passing through coils wound at 120° to each other, a rotating magnetic field is produced.

There may be occasions when the incoming three-phase supply is carried on only three wires, omitting the neutral, but because of imbalance in the loads on the three phases, a neutral is needed. In this case the neutral can be recreated by the connection shown in Figure 7.15, where the incoming wires are connected to the primaries of the sections of a three-phase transformer in a triangular arrangement known as a *delta connection*. The secondaries are connected in another arrangement, known as a *star* or *wye connection*, and at the centre of the star, the neutral appears. Star–delta transformer connection has other advantages as well, particularly in that it eliminates unwanted harmonics in the waveform of the AC supply.

The voltages used for transmission and distribution are generally quoted in terms of the voltage between one phase and another, but for the single-phase consumer it is the voltage between phase and neutral that is important. The two are simply related; the interphase voltage is $\sqrt{3}$ times the phase-to-neutral voltage, as is shown in Figure 7.16.

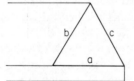 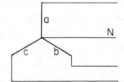

Figure 7.15. Star-delta transformer connection. Each section (a, b, c) functions as a separate single-phase transformer

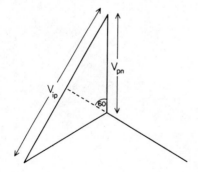

Figure 7.16. Proof that the interphase voltage is $\sqrt{3}$ times the phase-to-neutral voltage

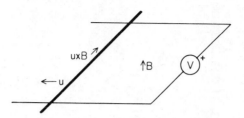

Figure 7.17. Theory of electricity generation

7.5 Electricity Generators

The conversion of mechanical energy into electrical energy relies on a fundamental law of electromagnetism, illustrated in Figure 7.17. If a conductor slides along a set of conducting rails at a velocity u in a magnetic field B, an electromotive force is induced in the conductor. This can be explained by reference to the Lorentz equation for the force on a moving charge q:

$$\mathbf{F} = q(\mathbf{E} + \mathbf{u} \times \mathbf{B})$$

where

$\mathbf{F}$ = force on moving charge (in newtons)
q = charge (in coulombs)
$\mathbf{E}$ = electric field strength (in volts per metre)
$\mathbf{u}$ = velocity (in metres per second)
$\mathbf{B}$ = magnetic flux density (in teslas)

When the conductor is moving with velocity u, the free charged particles within it (which will normally be electrons) will each experience a force $q(\mathbf{u} \times \mathbf{B})$, causing them to move and thus to set up an electrostatic field E. This in turn will operate on the charged particles with a force $q\mathbf{E}$. At equilibrium these forces will be equal and opposite, that is

$$\mathbf{E} = -\mathbf{u} \times \mathbf{B}$$

Because of the field E there is a potential difference between the ends of the conductor. The value of the potential difference e is obtained by line integration along the length l of the conductor:

$$e = \int \mathbf{u} \times \mathbf{B}.ds$$

s being the unit vector along the line of integration. Taking the line integral along the axis of the conductor, and making the simplifying assumptions that **u**, **B** and **l** are mutually orthogonal and that **B** is uniform, we get

$$e = uBl \text{ volts}$$

The nearest practical machine to the theoretical picture just described is

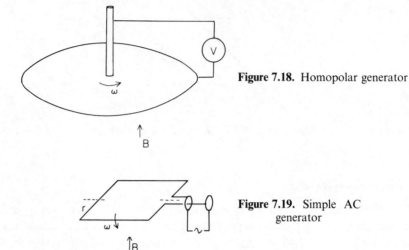

Figure 7.18. Homopolar generator

Figure 7.19. Simple AC generator

the *homopolar* generator, which is illustrated in Figure 7.18. A circular disc rotates in a magnetic field perpendicular to it. Considering an element at radius r (Figure 7.18b) we see that its velocity is ωr perpendicular to the magnetic field, so the induced e.m.f. in the element is $\mathbf{u} \times \mathbf{B}$ directed along the radius. Integrating from the centre of the disc to the periphery, the total voltage e is given by

$$e = \int \mathbf{u} \times \mathbf{B}.\mathbf{ds}$$

$$= \int_0^a B\omega r \, dr = \tfrac{1}{2}B\omega a^2 \text{ volts}$$

Typical numerical values might be $B = 1$ tesla (10^4 gauss), $\omega = 300$ rad s^{-1} (about 3000 rpm), $a = 100$ mm, giving a voltage of 1·5 volts. This is rather low, and consequently the homopolar machine is not useful as a practical generator. However, used as a motor, it offers possibilities in conjuction with superconducting magnets, which can reach higher fields than other types of electromagnet, typically 5 tesla.

An alternative arrangement is shown in Figure 7.19, in which a coil of wire rotates in a uniform magnetic field B at an angular speed ω. The velocity of the conductor is now ωr, and the voltage developed across each side of the coil is

$$e = \omega r B l \sin \omega t$$

For the whole coil, twice this amount is produced. The advantage of this layout is that instead of one single conductor the coil can have any number of turns n, and the output voltage is then

$$e = 2 \, n\omega r B l \sin \omega t$$

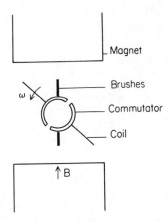

Magnet

Brushes

Commutator

Coil

Figure 7.20. Schematic of a commutator and brushes for generating DC

↑B

Thus the output voltage can be as high as is required, merely by increasing n. Note that the output is an alternating current, which as we have seen is generally the most useful for power transmission.

This simple alternating theory can be extended to three phase AC generation. If a second and third coil are added to the existing one, but placed at angles 120° (2 π/3) and 240° (4 π/3) to the first, with separate pairs of *slip-rings* (sliding contacts), three separate sinusoidal voltages are developed, whose equations are

$$e_1 = 2n\omega rBl \sin \omega t$$

$$e_2 = 2n\omega rBl \sin (\omega t - 2\pi/3)$$

$$e_3 = 2n\omega rBl \sin (\omega t - 4\pi/3)$$

Connections to the coil in Figure 7.19 are made by means of brushes and slip-rings. Suppose instead that the connection is made by brushes and a divided ring or commutator, as in Figure 7.20. The output is then unidirectional, that is direct current, although it is not very smooth. By increasing the number of coils and commutator segments (Figure 7.20), it can be arranged that the coil connected to the brushes is always the one in the most favourable position for generation, that is travelling at right angles to the field. In this situation the average DC voltage appearing at the brushes is close to the maximum value, which as can be deduced easily from the foregoing theory, has a value

$$e = 2n\omega rBl$$

Armed with this simple theory, we can now go on to consider practical AC and DC generators.

DC Generators

In the simple machine illustrated in Figure 7.20, the magnetic field is not as great as it might be, because of the large gap between the magnet poles and because the coil has an air core with low permeability. In a practical DC

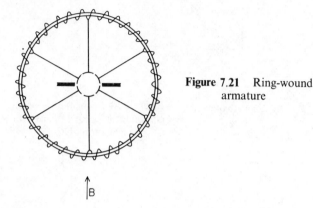

Figure 7.21   Ring-wound armature

generator the poles of the field magnet are shaped to fit closely around the moving coils, which are wound on an iron core of high permeability producing an assembly known as an *armature*. An additional fault in the simple design above is the inefficient use which is made of the coils because only one is connected to the external circuit at any given time. It is possible to improve on this by the arrangement of Figure 7.21, in which *all* the coils are connected together in series, so that all of them can contribute to some extent to the generation of power. This connection is known as a ring-wound armature, and although for practical reasons it is not used in this form, a variant which is very similar to it in general principle is in common use in practical DC generators.

The field in a DC generator can be provided most conveniently by an electromagnet, since this offers a larger field than most permanent magnets and also allows control of the output voltage by variation of the field strength **B**. The power required to drive the electromagnet can be obtained from an external source of DC (such as another DC generator), producing a *separately-excited* machine, or from the DC produced by the machine itself, producing a *self-excited* machine. The advantage of separate excitation is that the field is easily controlled by light-current circuitry and the large armature current is controlled by varying the small field current. Self-excited machines can be controlled in a similar way, but clearly there must be some interaction between the field current and the armature current, making control more complicated. There are various techniques for making a self-excited DC generator control its own output in ways appropriate to its function, such as series or shunt connection of the field coil and armature coils, and the reader is referred to textbooks of electrical engineering for details.

Alternators (AC Generators)

The Lorentz equation does not necessarily require the magnetic field to be stationary and the conductor to move; the opposite is just as effective in generating an e.m.f., and for practical reasons it is more convenient. Consequently it is usual to use a stationary coil (known as the *stator*) and a rotating field magnet (known as the *rotor*). Such an arrangement obviates the need for slip-rings. Further, the large air gap in the magnetic field implied in Figure 7.19 is again not desirable, so it is reduced as far as possible, by curving the poles. This presents a problem in that the current generated is no longer sinusoidal, and it is desirable for several reasons to correct this as far as possible, by suitable shaping of the poles of the stator or rotor and by careful spacing of the coil windings. Details of how this is achieved are beyond the scope of this book.

The rotor field in an alternator is generally provided by an electromagnet for the same reasons as in the DC generator. Since the output is AC, and the electromagnet requires DC, it is not possible to employ self-excitation unless some form of rectification is used. Consequently the exciter for an alternator is usually a small DC generator attached to the same shaft as the alternator rotor. Power has to be transmitted to the rotor magnet via a pair of slip rings, but this is not a serious problem since the excitation current is usually quite small. Control of the excitation is required, as in the DC machine, in order to regulate the output voltage, and it is achieved either by direct control of the rotor current or, more usually, the field current in the exciter generator. This has the advantage of minimizing the current-handling capacity of the control circuitry. Some recent alternators have been built with AC exciter generators, the output of which is rectified electronically.

Single-phase and poly-phase alternators differ only in the number of separate windings on the stator. Like multi-pole DC machines, poly-phase AC generators make better use of space and materials and operate more smoothly than their single-phase counterparts, so they are preferred for all but the smallest plants. The advantages of three-phase AC transmission provide a further justification for their use.

Alternators are extremely efficient, often better than 95 per cent, but there is some loss of power from three main sources.

(1) Windage loss, arising from aerodynamic drag on the rotor. Careful design of the rotor surface and of the way in which the copper conductors are slotted into the iron can reduce the losses from this cause, as can the use of hydrogen instead of air as coolant. This point will be discussed below.

(2) Ohmic losses in the copper conductors of the stator and, to a lesser extent, the rotor. These can be reduced by using lower currents and higher voltages, but high voltage involves greater insulation thickness, and since insulation is useful neither for conducting current nor for magnetic purposes, it is in effect a waste of space. For this reason it is undesirable to use excessively high voltages. The designer is forced into a compromise, which in the case of typical 500 MW sets is around 25 kV and 20 kA.

(3) Hysteresis and eddy current losses in the iron cores. These are minimized by choice of suitable magnetic material and by lamination to reduce circulating currents.

All these losses give rise to heat which must be removed. Small alternators are air-cooled, but in larger machines of 60 MW and above it is usual to use hydrogen at a pressure of about four atmospheres, since this not only reduces the windage loss (see above) but also offers a higher thermal capacity. Further, because of the lower density, the power required to propel the coolant through the machine is reduced. It is evident from Table 7.1, however, that a liquid is a more effective cooling medium, and the latest types of alternators are water cooled.

Table 7.1 Thermal capacities of different coolants for alternators

Fluid	Thermal capacity (arbitrary units)
Air	1
Hydrogen (4 bar)	4
Oil	1750
Water	4000

From D. R. Treece in *Turbine Generator Engineering*, G.E.C. Turbine Generators Ltd., Manchester, 1974. (Reproduced by permission of G.E.C. Turbine Generators Ltd.)

Alternators are synchronous machines, that is they rotate at an angular speed which is exactly related to the angular frequency of the alternating current which they generate. A simple two-pole generator of the type described earlier rotates at an angular speed equal to that of its AC output, which will be 3000 rpm (50 Hz) in Europe and 3600 rpm (60 Hz) in America. Such alternators are ideal for single-shaft turbines, but as we have noted in Section 7.2 it is sometimes necessary to run the LP stage of a large turbine at reduced speed — and, therefore, on a separate shaft. In this case the alternator on the slower shaft is not geared up as might be expected, but rather is designed with extra rotor poles. It should be clear from the theory discussed earlier that a four-pole rotor (one with two pole-pairs) will give rise to an AC which has an angular frequency equal to twice the mechanical angular speed of rotation. Thus it is convenient to run the LP stage at half the speed of the other stages. This can be extended to machines with any number of pole-pairs, and if the number is large, the speed of rotation can be very low. Such low-speed alternators are used with water turbines (Chapter 13) where the speed of rotation is often between 150 rpm (20 pole pairs) and 300 rpm (10 pole pairs). It is also necessary to use low-speed alternators with the large diesel engines which sometimes provide power in remote communities, although there has been a tendency for modern diesel engines to be designed for higher speeds of rotation so as to reduce the physical size of the generator and motor.

The fact that alternators must rotate at a fixed speed leads to the interesting problem of how a generator, starting initially from rest, can be connected to an existing AC network. Such a problem occurs many times daily in any public electricity supply, and it is overcome by *synchronization* of the 'new' machine with the network before connection. Synchronization is simply a matter of matching the amplitude, frequency and phase of the generator voltage to those of the system voltage. When this is so the generator 'sees' the same conditions before and after connection, and there is no problem of transient surges at the moment of switching.

Once connected, the generator is completely matched to the system in phase and voltage, and thus is supplying no power. Now, a method has to be found to force power from the generator into the system. By analogy with DC this might be expected to be simply a matter of raising the voltage of the generator above that of the system, but, while power will indeed be transferred, this approach is not acceptable because the amount varies during the cycle. Instead, the waveform of the generator is made to *lead* that of the system. In fact, this lead arises naturally if an attempt is made to run the turbo-alternator faster than synchronous speed. If the output of the generator is small compared with the power in the system which can, therefore, be viewed as being infinite (the *infinite bus bars* concept), then, instead of running faster, a phase lead is developed, and power flows from the machine into the system. In contrast, if the alternator is allowed to *lag* behind the system by shutting off the steam to the turbine, it withdraws power but continues to rotate at synchronous speed, becoming a synchronous motor.

The infinite bus bars concept is usually a very good approximation. However, no system is truly infinite, and its frequency tends to increase or decrease if the power input exceeds or falls short of demand for more than a few seconds. This is exactly analogous to mechanical inertia, and indeed it arises largely from the inertia possessed by the rotating machinery (both generators and motors) in the system. Electricity supply authorities strive to maintain the constancy of the supply frequency and because of this it is possible to use simple synchronous electric clocks which have good accuracy.

The synchronous alternator is by far the most important type of electricity generator, but there are others which, although not of importance at present, may be useful in the future and which are worth reviewing briefly.

Induction Generators

An induction machine is mechanically rather similar to an alternator, but it differs in that the rotor is not magnetized from an 'outside' source. Instead the magnetic field is produced by currents *induced* in the rotor coils by transformer action from the stator. Such a machine generates no power when the rotor is running at synchronous speed, acts as a motor when its speed is less than synchronous, and as a generator when its speed is greater than synchronous. Consequently the frequency at which such a machine generates is determined not by its own speed but entirely by the frequency of the system to which it is connected. It is thus suitable for primary power sources where the speed cannot be easily controlled or where the available power is likely to fluctuate.

198

The obvious case is wind-power (Chapter 13). It must be borne in mind that an induction generator requires to be connected to an existing system (preferably one where the infinite bus-bar assumption is justified), as it cannot generate at all if it has no source of AC to excite its windings. Thus it is not suitable for small-scale windmills, but in schemes involving the connection of wind plants to the public electricity supply, an induction generator is a suitable choice, provided, of course, that the plant is disconnected when the speed falls below synchronous (when the plant would become a fan rather than a source of power!). It is worth noting that an induction generator is self-synchronizing. This avoids one of the main problems in wind power, namely the necessity of repeated resynchronizing with the system on reconnection after a calm period.

Magnetohydrodynamics

At the beginning of this chapter we referred to the homopolar generator, in which a conducting disc rotated between the poles of a magnet and produced DC. A variant of this device, in which the moving conductor is a stream of hot ionized gas, offers attractive possibilities for improving the efficiency of thermal power stations when used as a 'topper' (Chapter 15). This is known as magnetohydrodynamic generation, thankfully abbreviated to MHD. We shall now look briefly at such a device. (see also chapter 15.)

Consider an ion in a gas flowing through the apparatus sketched in Figure 7.22. Initially the electric field **E** is zero. From the Lorentz equation the force developed on the ion, $q(\mathbf{u} \times \mathbf{B})$, will cause it to be displaced, and when enough ions have been displaced in the same direction an opposing electric field will be set up. Eventually the force due to this field will be equal and opposite to that due to the electromagnetic force, so that

$$q\mathbf{E} = q(\mathbf{u} \times \mathbf{B})$$

which, with the geometry shown in Figure 7.22, is simplified to

$$qE = quB$$

In this situation no movement of charged particles occurs. The voltage developed between the parallel plate electrodes spaced a distance d apart is $V = Ed$. We have, therefore, a source of DC with an open-circuit voltage given by $V_0 = Bud$. Typical values might be $B = 1$ tesla, $u = 1000\,\mathrm{ms}^{-1}$, $d = 1$ m, resulting in a V_0 of 1000 volts.

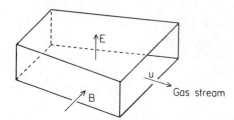

Figure 7.22. Principle of MHD generator

Such a generator is very simple in concept, and requires no moving parts. It is thought that it could make use of the very high temperatures which are theoretically available in the burning of fuels (and in nuclear reactors) without the serious materials problems which arise when conventional boilers or turbines are subjected to these high temperatures. It is intended that an MHD plant would be used as a 'topper' (Chapter 15), feeding its waste gas into a conventional boiler installation, and thus improving the thermodynamic efficiency.

Unfortunately, as in so many cases, there are severe problems. It will be noted that in the above theory we calculated the open-circuit voltage only, and there is a good reason for this, for when an appreciable current is drawn the output voltage drops drastically owing to the low conductivity of a hot gas. To improve the performance it is necessary to seed the gas with ions of low ionization potential. Caesium with its low ionization potential of 3.89 V is preferable, but it is expensive, and potassium with a slightly higher ionization potential of 4·34 V is the usual choice. Even though potassium, in the form of salts such as potassium sulphate, is cheap, such large quantities are needed that the seed must be recovered after use in order to keep costs within bounds. Recovery is also important because of the problems of boiler corrosion that would arise if potassium-rich gases were passed on to conventional plant. The removal of seed has presented a major problem in the design of experimental MHD apparatus, and it is likely to present an even more serious problem in large-scale application.

The materials used to construct the duct and the electrodes must be selected with care, for although there is no rotating machinery, the temperature is very high and there is likely to be considerable erosion. Ceramics are an obvious choice, but have been less successful than expected owing to abrasion and thermal spalling. Water-cooled metal walls, as used in rocket motors, have been found more satisfactory, but as the walls of an MHD apparatus must be non-conducting, it is necessary to alternate metal with ceramic insulating layers. The coolant carries off some useful heat, and in principle it is possible to use this in the steam plant which follows the MHD generator, but whether the conflicting requirements of good cooling (low coolant temperatures) and thermodynamic efficiency (high coolant temperatures) can be resolved is a matter that has yet to be examined.

A final problem with MHD generation is that the output consists of a high-current, low-voltage DC which must be inverted to AC before it can be sent out. Methods of inversion are well established, but they are expensive, and they will add considerably to the cost of adding an MHD topper to a conventional power station.

Non-heat-engine Generation

All the techniques of electricity generation so far discussed suffer from one major drawback—they all involve a heat engine, whose efficiency is limited by the Carnot cycle to a value much less than 100 per cent. There is considerable interest, therefore, in methods of electricity generation which do not require a heat engine. These include fuel cells (Chapter 14), thermionic and thermoelectric generation (Chapter 12), photoelectric generation (Chapter 12) and also the natural sources of wind and water power (Chapter 13). The reader is referred to the appropriate sections for information about these sources of electricity.

7.6 Electricity Transmission

We have already mentioned earlier in this chapter that the main advantage of AC over DC is that the voltage can be readily changed by means of a transformer. For transmission over long distances there is a great advantage to be gained in the use of high voltages since the transmission capacity of a conductor is limited by resistive heating which is proportional to the square of the current being carried. By raising the voltage it is possible to reduce the current, and it is easily shown that for a given amount of resistive heating, the quantity of power that can be carried on a given conductor is proportional to the square of the voltage. High voltage thus makes it possible to transmit large amounts of power without excessively thick conductors. Moreover, because the resistive loss is dependent only on the current, it is evident that the percentage of power lost in transmission is inversely proportional to the square of the voltage, that is, high-voltage transmission is more efficient.

The foregoing advantages are not without penalty, for the higher voltage raises problems of insulation. Overhead lines are usually bare, so air acts as the insulator. The breakdown voltage of air is approximately 3 MV m^{-1}, but for several practical reasons the conductors in an overhead line are usually much further apart than implied by this figure. As a result the insulating properties of the air are rarely a limiting factor. A more serious problem arises at the towers from which the cables hang, for these are at earth potential and an insulating suspension is needed, the length of which increases roughly linearly with voltage. The surface of an insulator is easily contaminated with water, salt and other deposits which rapidly degrade its performance. There is always some discharge at insulators, especially in wet weather, but the power lost in these discharges is negligibly small by comparison with resistive losses in the conductors.

Another problem which arises at high voltages is corona discharge from the ionization of the air surrounding the conductors. Very little power is lost by this mechanism, but the discharge gives rise to radio interference and it must be minimized. Corona can be prevented by reducing the field at the surface of the conductor, and this can be achieved in principle by increasing the radius of curvature. However, this would imply the use of thicker wires (or hollow tubes) which is not an ideal solution from an engineering point of view. Fortunately it has been found that groups of conductors at the same potential, separated by spacers, can reduce corona discharge by behaving as if the radius of the conductor were equal to the overall radius of the group. This produces the effect of a large-diameter conductor without the engineering problems which would otherwise result.

The problems associated with transmission at very high voltages can be overcome, but they are sufficiently severe that it is not economic to use excessively high voltages, and a compromise must be achieved between the desire for efficient transmission and the problems of high voltage. In general the further the distance over which the power is to be transmitted, the higher is the optimum voltage. In the United Kingdom the most common transmission voltages are 400, 275 and 132 kV, while in the United States, where the distances involved are usually longer, higher voltages are used, up to 765 kV.

Transmission lines are liable to mechanical damage and to lighting strikes, and to guard against these and other hazards they are protected by sophisticated electronic systems which perform the same functions as the fuse and the earth-fault relay, both of which are familiar in domestic installations. The unlikely event of breakage of one or more of the three conductors in a three-phase system leads to an imbalance which is detected by the protection gear as a *phase fault*, while the earthing of one or more conductors is detected as an *earth fault*. In either of these cases the line is switched off. The protection equipment reacts sufficiently quickly that if a conductor breaks, a phase fault is detected and the line isolated long before it reaches the ground (and becomes an earth fault!). Falling conductors do not, therefore, constitute a hazard. A much more common occurrence is a lightning strike on or near the line, leading to ionization of the air in the vicinity and the creation of a conduction path to earth. This creates an earth fault, which will continue to exist as long as the current from the line continues to flow to earth. When the protection gear has operated the discharge ceases, the ions disperse and there is no longer any fault. To deal with this situation, automatically-reclosing protection systems have been developed, which after switching off the line under fault conditions, check to see if the fault is still there, and if it is not, reconnect the supply. Such devices have considerably reduced the impact of unplanned power cuts, and have saved the transmission engineers many outings in thunderstorms. Another lightning-control device is to run an earthed wire along the top of the transmission towers to intercept the lightning flash before it hits the conductors.

The conductors in overhead lines are usually of steel-cored aluminium; the steel provides the tensile strength while the aluminium is the conductor. Insulators are usually glass or ceramic. Towers are usually galvanized steel, which is not painted, partly to save labour and partly because the natural matt grey colour blends into the sky.

A three-phase circuit requires three conductors. Single-circuit lines are common in the United States but double-circuit lines, with six conductors, are often found in the more congested United Kingdom. The two circuits can be treated independently, and it is possible to do maintenance work on one of the circuits while the other is live. The Japanese electricity system has taken the multi-circuit line to extremes, with up to six circuits (18 conductors) per line.

It is not always possible or desirable to use overhead transmission lines, and underwater or underground cables may be necessary, especially in heavily built-up areas, or in areas of great natural beauty. It is usual to have a separate cable for each phase, except at lower voltages where insulation problems are less severe.

The construction of a typical cable requires an outer protective sheath, a layer of insulation and a conductor. Conductors are normally aluminium, although copper is still used to some extent. The insulator is sometimes a liquid, usually oil, and sometimes a gas, usually sulphur hexafluoride. In both these cases the insulating fluid can be circulated to improve the transfer of heat from the conductor to the outer sheath, where it is lost into the surrounding soil or water. This heat mainly arises through ohmic losses in the conductor, but there is an appreciable dielectric loss in the insulator. In addition to the load current, the cable is also required to carry the charging current needed to charge its own capacitance, which leads to further conductor losses. The removal of the heat generated is not as easy as with an overhead line, so the power handling is less for a given conductor cross-section. Further, the charging current is proportional to the length of the cable, assuming a constant capacitance per unit length, and there is thus a limit to the maximum length of cable, which in typical cases may be as short as 20 miles. But these technical objections are often of secondary importance in comparison with the great expense involved in laying an underground cable. Nevertheless, there are many circumstances in which the additional cost is justified, and the use of underground cable is likely to become of increasing importance in future.

Despite the great advantages of AC transmission, it suffers from certain problems, which can be solved by the use of high-voltage DC. In particular, DC is better suited than AC to cables, and with the likely increased use of cables in preference to overhead lines, it may be expected to grow in importance. Moreover, DC is better suited to superconducting cables, which are thought in some quarters to offer the best long-term means of electricity transmission.

As we saw earlier, the peak voltage of a sinewave AC is $\sqrt{2}$ times the r.m.s. voltage. The effect of this is that an AC circuit must be insulated to withstand a voltage 1·4 times greater than the 'useful' voltage. This means that a given cable has a greater power-transmission capacity with DC than with AC. Moreover, the problem of dielectric heating in cables does not arise with DC as the dielectric is under a steady stress.

A further problem with cables is the current required to charge the capacitance of the cable. In an AC system this charging current has to flow into and out of the cable capacitance every cycle, but with DC it is required only when the cable is switched. The effect of this is not only to increase the power handling of the cable, but also to allow the use of longer cables.

It is evident that since more power can be transmitted along a given cable with DC than with AC, a DC cable can be cheaper. This is true also for overhead lines, though to a lesser extent. However, since electricity is generated and distributed as AC, it must be rectified to DC and inverted to AC at the ends of the DC cable. Rectifiers and inverters using large mercury-arc tubes or solid-state devices are well developed and are quite acceptable items of electrical engineering, but they are expensive, and their installation cost must be added to the cost of the cable. Of course, only two such installations are needed regardless of the length of the line, so the economics of DC cable and overhead line installation becomes more favourable as the distance increases.

There is another technical reason for DC transmission under certain circumstances, which may lead to its use over short distances where it would normally be regarded as uneconomic. As we noted earlier in this chapter all the machinery in an AC system is synchronized completely. If a generator is out of synchronization at the moment of connection to the system, massive power flows can occur which can lead to severe damage. The same is true of two nominally separate AC transmission systems, belonging for example to different nations or companies, which are linked by an interconnection; unless the two systems are accurately in synchrony when the connection is made, massive power flow occurs along the interconnection, and the line can be damaged, or at the very least, the protection equipment will disconnect the circuit. Furthermore, it is possible that during normal operation the systems will move out of phase to an extent which necessitates the transfer of power along the interconnector, and unless the interconnection is of reasonably large power capacity in comparison with the sizes of the systems, the line may be overloaded. These problems are avoided completely if the interconnector is DC, for then the two systems can run asynchronously.

The uses of DC lines can best be illustrated by a few examples.

(1) A 750 kV overhead line ($\pm$ 375 kV relative to earth potential) connects the Los Angeles area to Celilo on the Washington–Oregon border, a distance of about 850 miles. This is a very large installation, with a capacity of 1500 MW, and takes advantage of the lower installation cost of a long distance DC line.

(2) A 200 kV ($\pm$ 100 kV) line connects the electricity systems of England and France under the English Channel. Clearly DC is a likely choice for an underwater cable, but as the underwater section is only about 20 miles long, the cost of an AC system would be comparable to the DC system actually chosen. The decisive reason for the choice of DC was the possibility of very large oscillations when the two systems were connected by a relatively slender 160 MW link.

(3) An interconnection which must qualify as the shortest DC line in the World, only 30 feet long, connects two power systems at New Brunswick, Canada. The sole reason for DC in this case is the possibility of asynchronous operation.

The electrical resistivity of metals falls as the temperature is lowered, and for very pure metals with low dislocation densities the resistivity can become very low indeed. Consequently, if an electric cable is refrigerated to a low enough

204

temperature, the losses due to resistance can be reduced considerably. A cable of this type is termed *cryoresistive*. A typical choice of refrigerant might be liquid nitrogen (78 K), and for aluminium the reduction in resistance on cooling to this temperature from 300 K is about tenfold. This might appear to promise well, but the refrigeration plant consumes power, and when due allowance is made for a realistic coefficient of performance, it is found that the power consumed by the refrigerator is almost equal to the power saved by lower resistance. Consequently no power is saved by the use of cryoresistive cables. However, there is a benefit in that the power dissipation is removed from the cable itself and transferred to the refrigeration plant, where it is easier to deal with. It is found that the current-carrying capacity of a cable of given cross-section can be roughly doubled by refrigeration, and this can lead to savings in capital cost. At present there are no such cables in use, but they offer some promise for the future.

When the temperature of certain metals, alloys and compounds is reduced to near 0 K, the electrical resistivity falls sharply to zero, and they become superconducting. At the same time, all magnetic flux is expelled from the material, which becomes perfectly diamagnetic. The temperature, magnetic field and electric current to which the specimen is subjected all affect the transition to the superconducting state, as shown in Figure 7.23; the highest temperature at which it can be superconducting is known as the critical temperature, T_c. The magnetic effects are rather more complicated, for there are two types of superconductor, type I in which the magnetization characteristic is reversible, and type II in which hysteresis can occur, with the dissipation of energy. There is great attractiveness in the possibility of using superconducting cables for power transmission, for despite the large amount of power required to refrigerate the conductors, there is a considerable net saving. AC is inherently less satisfactory than DC because of the hysteresis effects mentioned earlier, but designers have developed experimental cables for both AC and DC operation. Superconducting cables have not yet been used commercially, because at present there is no material which will carry reasonable currents and remain superconducting unless liquid helium (4 K) is used as the refrigerant. There is little likelihood of such cables entering general use until a material can be

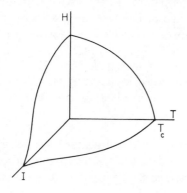

Figure 7.23. Limits of superconductivity as a function of magnetic field *H*, current *I* and temperature *T*. T_c is the critical temperature

found which will remain superconducting at high current with the much cheaper liquid hydrogen (21 K) as refrigerant. If such a material can be developed, there is the intriguing possibility of transmitting both electricity and fuel gas along the same pipe!

References

Baumann, K., *J. Institution of Electrical Engineers*, **48**, 830, 1912.
Bowden, F. P., (Ed.), Royal Society Joint Meeting with C.E.G.B., *Phil. Trans. Roy. Soc.*, **A260**, (1110) (1965).
Bowden, F. P. and J. H. Brunton, *Proc. Roy. Soc.*, **A263**, (1315), 433–450 (1961).
Bowden, F. P. and J. E. Field, *Proc. Roy. Soc.*, **A282**, (1390), 331–352, 1964.
Craig, H. R. M. and K. W. Todd, in *Turbine-generator Engineering*, G.E.C. Turbine Generators Ltd., Manchester, pp. 47–52, 1974.
Gardner, G. C., *Proc. Institution of Mechanical Engineers*, **178**, 593 (1964).
Todd, K. W., in *Turbine-generator Engineering*, G.E.C. Turbine Generators Ltd., Manchester, pp. 38–46. 1974.
Woodson, R. D., *Sci. American*, **224**, (5), 70–78 (1971).

Further Reading

Edwards, J. D., *Electrical Machines*, International Textbook Co. Ltd., Aylesbury, England, 1973.
Forsyth, E. B., (Ed.), *Underground Power Transmission by Superconducting Cable*, Brookhaven National Laboratory, New York, 1972.
G.E.C., *Turbine-generator Engineering*, G.E.C. Turbine Generators, Ltd., Manchester, 1974.
Lay, J. E., *Thermodynamics*, Pitman, London, 1964.

8 Nuclear Physics

The production of electricity through the controlled use of nuclear fission is now an accepted fact, and nuclear power production is an important part of all national energy planning. It is the most recent large-scale development in electric-power production, the World's first commercial nuclear power station having opened in the United Kingdom at Calder Hall in 1956. However, the first experimental results that led to this development appeared in a flurry at the end of the nineteenth century.

In 1895, Roentgen discovered X-rays. In 1896, Becquerel discovered radio-activity in uranium salts while following up Poincare's suggestion that there might be a relationship between these newly discovered X-rays and optical fluorescence. In 1898, Mme Curie isolated polonium and radium from the ores from which the uranium salts were extracted, having observed that the radioactivity from the ores was greater than that from the salts alone. In so doing, she established the fundamental processes of radiochemistry.

A further shock came when it was found that these radioactive materials changed their chemical identity with time. Intensive study led Rutherford and Soddy, in 1903, and von Scheidler, in 1905, to the theory of radioactive decay. In this theory, radioactive atoms disintegrate spontaneously, the number disintegrating in a given time being proportional to the number of radioactive atoms present. In the disintegration, an atom emits a small particle and changes its chemical character as a result. The investigation of the rays emitted during these disintegrations led to their classification into three types; alpha particles, beta particles and gamma rays.

Alpha particles are strongly ionizing and are absorbed in a few centimetres of air. When their deflections in electric and magnetic fields were studied, it became apparent that they were helium atoms with a doubly positive charge.

Beta particles were found to be much more penetrating than alpha particles and required several millimetres of aluminium to stop them. They were negatively charged, and electric and magnetic deflection experiments in 1900 showed them to be identical with the electrons discovered by J. J. Thomson in 1899.

Gamma rays were found to penetrate several centimetres of lead, to be undeflected by electric or magnetic fields, and hence to be highly energetic electromagnetic radiation identical in character with X-rays.

It must be emphasized that, at that time, no real physical model for the structure of the atom existed. The discovery that atoms could change their chemical character by emitting either negative or positive particles led J. J.

206

Thomson to propose a model in which the atom was assumed to be a sphere of positive electricity of uniform density, through which the negative electrons were distributed in such numbers that the electrical charge of the atom became zero. This became known as the *plum pudding model*. Unfortunately, this model could not account for the results of experiments performed by Rutherford, in 1911, which involved firing alpha particles at atoms, and examining the direction in which they scattered. In 1911, Rutherford proposed the basis of our present model. In order to account for the observed scattering of the alpha particles, he had to suppose that there was a very intense electric field near the centre of the atom. In order to achieve this, he suggested that the atom consists of a *nucleus*, which is very small, massive and carries the entire positive atomic charge, and that this is surrounded by a large sphere over which the negative electronic charge is distributed. The diameter of this sphere is comparable to that of the atom at about 10^{-8} cm. In 1913, Geiger and Marsden published detailed experimental results of alpha particle scattering which showed a remarkable agreement with Rutherford's theory, and confirmed that the positive nucleus of the atom must be minute. Further experiments using higher energy alpha particles confirmed the Rutherford model of atomic structure and established the radius of the nucleus at about 10^{-12} cm.

In 1919, Rutherford succeeded in breaking up the nitrogen nucleus by alpha particle bombardment, and showed that the particles emitted were hydrogen nuclei. These were called *protons* from the Greek $\pi\rho\omega\tau o\nu$ meaning first.

Because protons often appeared in nuclear disintegrations, and some nuclei were known to emit electrons, a model of the nucleus was proposed in which a nucleus of mass number A and atomic number Z was composed of A protons and $A - Z$ electrons. This model quickly encountered difficulties arising from the magnetic properties of electrons and nuclei, and had to be abandoned in favour of one involving the proton and the *neutron*, a particle proposed by Rutherford in 1920, and discovered by Chadwick in 1932, which is electrically neutral and has a mass slightly greater than that of the proton.

In 1934, I. Joliot-Curie and her husband F. Joliot discovered that many stable elements could be converted into other, radioactive, elements by bombardment with alpha particles. Thus *artificial radioactivity* was discovered. Several workers, including Fermi and Segré, showed that neutrons could be slowed down to thermal speeds and then became very efficient at disintegrating other nuclei. This discovery was followed by that of fission, by Hahn and Strasseman in 1939. Fission is a process in which a neutron strikes a heavy nucleus and is absorbed by it to form a compound nucleus. This compound nucleus is unstable and may break into two roughly equal parts liberating considerable amounts of energy in the process.

The discovery of fission was followed by the first atomic bomb in 1945, by the first electricity generation at Arco, Idaho, on December 20th, 1951, and the first commercial power station at Calder Hall, England, in 1956. It is perhaps ironic that the Arco reactor, EBR-I (Experimental Breeder Reactor No. 1) was a breeder reactor, and that these have not yet been fully developed for commercial power production.

In order to understand the reason why energy can be released during the fission process, we must go back again to the basic Rutherford model of atomic structure, and examine the properties of the miniscule nucleus.

8.1 Properties of the Nucleus

By 1932, the existence of both protons and neutrons was accepted. This led to the assumption that every nucleus was made up of combinations of different numbers of these two elementary particles, a hypothesis which, in that same year, was to form the basis of a detailed theory of the nucleus formulated by Heisenberg. Under this theory, the mass of the nucleus is almost equal to that of the total number of individual elementary particles (protons and neutrons). The number of protons, Z, is the *atomic number*, or positive charge of the nucleus. The total number of elementary particles, A, is known as the *mass number*, because one would expect of mass of the nucleus, M, to be equal to the sum of the masses of its constituents. In practice, this is not quite true, and the mass of the nucleus is slightly less than the total of its isolated components.

Einstein's theory of relativity tells us that mass and energy are interchangeable and related by the well-known expression $E = mc^2$. As a consequence, we can presume that this difference between the mass of the nucleus and the total of the independent particles forming it, the *mass defect*, has been released either as electromagnetic radiation or as kinetic energy of the nucleus when it was formed. This difference then represents the energy that must be supplied to the nucleus to split it into its component parts again.

The mass of the 'elementary particle' corresponding to unit mass number must be chosen. One logical choice would be that of the proton. Another would be the mass of the neutron. Neither of these two would be ideal as the nucleus is made up from combinations of the two. Instead, in 1961, the International Union of Physics and Chemistry set the value as one-twelfth of the mass of that isotope of carbon which has twelve nucleons (six protons and six neutrons). On this basis

$$1 \text{ atomic mass unit (amu)} = 1 \cdot 660420 \times 10^{-27} \text{ kg} = 931 \cdot 478 \text{ MeV}$$

Table 8.1 gives the masses of a few of the more important nuclei in both atomic mass units and MeV.

It was found at quite an early date that the same chemical element could exist as several different *isotopes*, differing only in the number of neutrons in the nucleus. For example, oxygen, with an atomic number of eight, can exist with mass numbers of 16, 17 and 18 as stable nuclei, while those isotopes with mass numbers of 14, 15, 19 and 20 are also known but are unstable.

It is also possible for different chemical elements to have the same mass number. This means that the number of neutrons in the nuclei are such as to

Table 8.1 Masses of some important nuclei

	amu	MeV
Electron	5.48597×10^{-4}	0·511006
Proton	1·0072766	938·256
Neutron	1·0086654	939·550
Deuterium	2·01410	1876·090
Helium	4·00260	3728·33
Carbon-12	12·00000	11177·736
Uranium-235	235·0439	$21·8938 \times 10^4$
Uranium-236	236·0457	$21·9871 \times 10^4$
Molybdenum-95	94·9046	$8·8402 \times 10^4$
Lanthanum-139	138·9061	$12·9388 \times 10^4$

compensate for the different numbers of protons. For example, at $A = 50$, both vanadium, with $Z = 23$, and titanium, with $Z = 22$, have stable isotopes. Such nuclei are called *isobars*.

In order to identify fully any specific nucleus, both A and Z must be given. The notation that has developed is one in which the chemical symbol is printed with Z and A as subscript and superscript, respectively. Thus carbon-12, with $Z = 6$ and $A = 12$, is written as $_6C^{12}$ or $^{12}_6C$.

Another important quantity is the *binding energy per nucleon*. This, given by the mass defect divided by the number of nucleons, and expressed in MeV, is a measure of the amount of energy released or absorbed if a nucleon is either added to or subtracted from the nucleus. For example, suppose a proton and a neutron combine to form a deuteron.

$$_1H^1 + _0n^1 \longrightarrow _1H^2 + \gamma \tag{8.1}$$

Table 8.1 gives the masses of the proton and the neutron as 1·0072766 and 1·0086654 amu, respectively. Adding these gives a hypothetical mass of 2·015942 amu for the deuteron. The experimental value of the mass of the deuteron is given by the mass of deuterium, minus the binding energy of the electron, as 2·013551 amu. This is some 0·00239 amu lighter, corresponding to a binding energy of 2·226 MeV. This is the energy that will be released when the deuteron is formed, or the energy that must be supplied if it is to split up. For the deuteron, the binding energy per nucleon is 1·113 MeV.

If the binding energy per nucleon is plotted as a function of mass number for various nuclides, a clear picture emerges. For very light nuclei, with mass number less than about 55, it increases rapidly from 1·113 MeV at $A = 2$ to 8·5 MeV at $A = 32$. This is shown in Figure 8.1. A striking feature of this figure is the very high stability of the $_2He^4$ nucleus—the alpha particle. This suggests that the alpha particle may exist as a subunit of the nucleus, a supposition that is reinforced by the high binding energies of $_4Be^8$, $_6C^{12}$, $_8O^{16}$ and $_{10}Ne^{20}$. It would appear that the nucleus is more stable when it is composed of an integral number of alpha particles.

Above about $A = 20$, the curve varies fairly smoothly and passes through

210

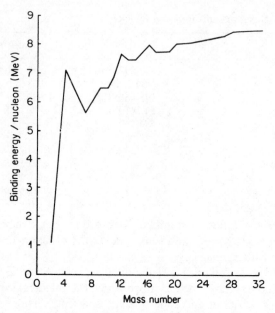

Figure 8.1. Binding energies of light nuclei

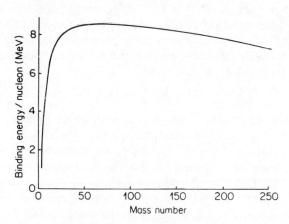

Figure 8.2 Binding energies of nuclei

a broad maximum at about $A = 60$. The iron nuclide Fe^{56} has one of the highest binding energies at 8·79 MeV/nucleon. This makes iron one of the most stable nuclei, and accounts for its high abundance, both in the Earth's crust and as a component of extra-terrestrial matter. Above $A = 60$, the binding energy per nucleon begins to fall off again until, by U^{238}, it has reached 7·58 MeV/nucleon.

Thus the overall picture of the binding energies of nuclei is as shown in Figure 8.2. This curve is often confusing at first sight because the most stable nuclei lie near the top. If the usual procedure of atomic physics is adopted,

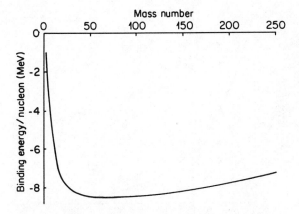

Figure 8.3. Binding energies of nuclei referred to free
nucleons as zero of potential energy

and the free nucleon is represented as being at zero potential energy, stable
nuclei will appear with negative energies and this diagram can be reflected
through the A axis to produce Figure 8.3, which shows the stability of the
middle mass nuclei in a more usual form.

Two facts emerge from Figures 8.2 and 8.3. If a heavy nucleus can be induced
to break into two small components, the sum of the binding energies of the
components will be greater than the original binding energy of the parent. As a
result, energy will be released and will appear as kinetic energy of the product
nuclei, and as gamma rays. This is nuclear *fission*. If two light nuclei can be
induced to coalesce (*fusion*), the binding energy of the resultant nucleus will
be greater than the sum of the components and, once again, energy will be
released. This is the possible alternative process for the release of nuclear energy
for commercial exploitation.

8.2 Radioactivity

The study of nuclear physics began with the discovery of radioactivity.
Becquerel's original discovery involved uranium salts, but in 1900 Crookes
made a further development. He found that, if a uranium salt was precipitated
from solution by the addition of ammonium carbonate and then redissolved
in excess reagent, a small residue remained. This residue was found to be highly
radioactive. In addition, the product obtained by evaporating the solution had
very little radioactivity. This meant that most of the activity of the uranium
salts was not due to uranium at all, but to some other material. The active
substance contained in the residue was called uranium X (UX) to distinguish
it from uranium. Becquerel then found that, if the two samples of uranium and
uranium X were allowed to stand for some time, the activity of the uranium X
decreased and that of the uranium fraction increased. Similar results were
obtained for thorium salts by Rutherford and Soddy in 1902. Exactly as with

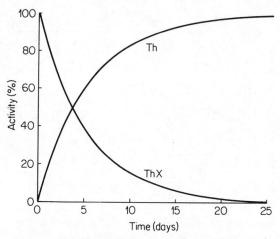

Figure 8.4. Radioactive decay of ThX—Rutherford
and Soddy's results

the uranium salts, an active material, thorium X, could be separated from the thorium salts which were then inactive. After some time, the activity of the thorium X sample decreased while that of the thorium increased. Rutherford and Soddy studied this behaviour quantitatively and produced the results shown in Figure 8.4. The experimental decay curve for the thorium X sample was exponential in character, $A_x(t) = A_x(0) \exp(-\lambda t)$, and the recovery curve for the thorium sample was found to fit the curve

$$A(t) = A(0)\,(1 - \exp(-\lambda t)) \tag{8.2}$$

where λ, the constant in the exponential term, is called the *disintegration constant* and is the same in both cases. The curves for uranium and uranium X are equivalent to these except that the time scale is longer. Thorium X loses half of its activity in 3·64 days while uranium X takes 24·1 days.

These observations led Rutherford and Soddy to a theory of radioactive decay. They suggested that the atoms of a radioactive material undergo spontaneous disintegration with the emission of alpha particles or beta particles and the formation of a new element. If this is true then the intensity of the radioactivity, or *activity*, is proportional to the number of atoms undergoing change per unit time. The activities in the exponential expressions can, therefore, be replaced by the number of atoms and the expression for the radioactive decay of the X sample can be written as

$$N(t) = N_0 \exp(-\lambda t) \tag{8.3}$$

Differentiation of this expression shows that the rate of change of N with time is proportional to N, the constant of proportionality being λ. This is the fundamental equation of radioactive decay. With this equation and with two other assumptions, it was possible to account for the growth of activity in the thorium or uranium sample from which the ThX or UX had been extracted. These two assumptions are:

(1) There is a constant rate of production of a new radioactive substance (e.g. UX) from the parent radioactive substance (uranium).
(2) The new radioactive substance itself decays according to equation (8.3).

Under these assumptions, if M atoms of UX are produced per second from the parent uranium, and if these decay exponentially with a disintegration constant λ, then we can write

$$\frac{dN}{dt} = M - \lambda N$$

where N is the number of UX atoms present at time t. If this expression is multiplied through by $\exp(\lambda t)$, regrouped and integrated with respect to time, we find, after imposing the boundary condition $N = 0$ at $t = 0$, that

$$N = \frac{M}{\lambda}(1 - \exp(-\lambda t))$$

which shows its clear relationship to equation (8.2) if $N_0 = M/\lambda$ and $A = N$.

The law of exponential decay, equation (8.3), reflects the statistical nature of radioactivity. If there is a fixed probability that an atom will disintegrate in a given time, then the number of disintegrations per unit time will be proportional to both this probability and to the number of atoms present in the sample, that is to some such expression as λN. This is a statement of equation (8.3). The experimental measurements on radioactive decay .confirm the statistical nature of the process. Thus, if measurements are made on a very short time scale, so that few disintegrations are recorded in each measurement, statistical fluctuations about the expected values will be observed. As a consequence, the theory of probability and statistical considerations must be taken into account in the design and interpretation of radioactivity experiments.

A further quantity associated with exponential decay laws which is of great importance is the *half-life*, T. This is the time taken for the population of the radioactive species to fall to one half of its initial value. It is a constant, characteristic of the radioactive material, and is given by

$$T = 0 \cdot 693/\lambda$$

After each half-life, the fraction remaining is one half of what was present at the beginning of that period. After n half-lives the fraction remaining is $(\frac{1}{2})^n$.

We can now explain the success of Rutherford and Soddy's explanation of the observed performance of Th and ThX. Their first assumption that the rate of production of ThX from the parent Th is constant obviously cannot be true. However, if the half-life of thorium is extremely long compared to the duration of the experiment, then the exponential character of the decay of

214

thorium will not be apparent and to all intents and purposes, the rate of production of thorium X will indeed be constant. In fact, the half-life of thorium is $1·39 \times 10^{10}$ years, so that the assumption of a constant decay rate over a of a month or so was very good. Incidentally, it is only because the half-lives of these heavy radioactive nuclei are so long—comparable to the age of the Earth—that they still exist in fairly large numbers. If they were shorter lived they would have decayed away long ago.

It was also found that not only did thorium produce thorium X, but that thorium X produced a daughter product which produced another daughter product and so on, producing a *natural radioactive series*. Three such series were found, each starting with a very long lived radioactive isotope. These three series were called the Uranium series, the Actinium series and the Thorium series. They were all originally classified with suffixes after the name of an identifiable element as with thorium and thorium X, but we can now properly ascribe the chemical name and appropriate A and Z to each element in the series. It was the discovery and identification of the elements in the series that led Soddy, in 1913, to the discovery of isotopes. The three series are presented with their original nomenclature in Figures 8.5, 8.6 and 8.7, while the full lists with type of disintegration, half-life and nuclide identification are shown in Tables 8.2, 8.3 and 8.4.

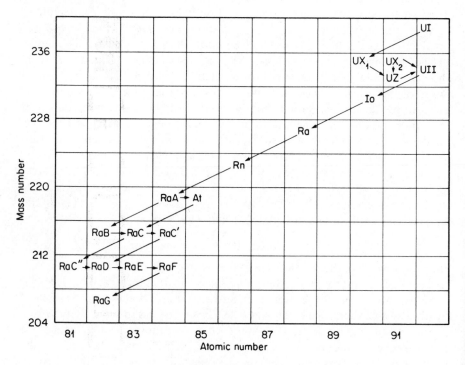

Figure 8.5. Uranium series

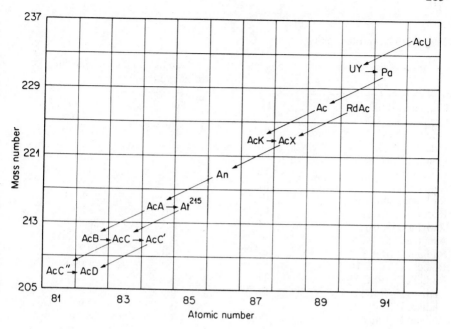

Figure 8.6. Actinium series

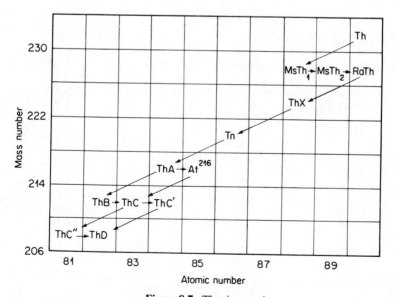

Figure 8.7. Thorium series

Table 8.2 The uranium series

Radioactive species	Nuclide	Type of disintegration	Half-life
Uranium I (UI)	$_{92}U^{238}$	α	$4\cdot50 \times 10^9$ y
Uranium X$_1$ (UX$_1$)	$_{90}$Th234	β	$24\cdot1$ d
Uranium X$_2$ (UX$_2$)	$_{91}Pa^{234}$	β	$1\cdot18$ m
Uranium Z (UZ)	$_{91}Pa^{234}$	β	$6\cdot7$ h
Uranium II (UII)	$_{92}U^{234}$	α	$2\cdot50 \times 10^5$ y
Ionium (Io)	$_{90}Th^{230}$	α	$8\cdot0 \times 10^4$ y
Radium (Ra)	$_{88}Ra^{226}$	α	1620 y
Ra Emanation (Rn)	$_{86}Em^{222}$	α	$3\cdot82$ d
Radium A (RaA)	$_{84}Po^{218}$	α,β	$3\cdot05$ m
Radium B (RaB)	$_{82}Pb^{214}$	β	$26\cdot8$ m
Astatine-218 (At^{218})	$_{85}At^{218}$	α	$1\cdot5$–$2\cdot0$ s
Radium C (RaC)	$_{83}Bi^{214}$	α,β	$19\cdot7$ m
Radium C′ (RaC′)	$_{84}Po^{214}$	α	$1\cdot64 \times 10^{-4}$ s
Radium C″ (RaC″)	$_{81}Tl^{210}$	β	$1\cdot32$ m
Radium D (RaD)	$_{82}Pb^{210}$	β	22 y
Radium E (RaE)	$_{83}Bi^{210}$	β	$5\cdot0$ d
Radium F (RaF)	$_{84}Po^{210}$	α	$138\cdot3$ d
Thallium-206 (Tl^{206})	$_{81}Tl^{206}$	β	$4\cdot2$ m
Radium G (RaG)	$_{82}Pb^{206}$	Stable	

Table 8.3 The Actinium series

Radioactive species	Nuclide	Type of disintegration	Half-life
Actinouranium (AcU)	$_{92}U^{235}$	α	$7\cdot10 \times 10^8$ y
Uranium Y (UY)	$_{90}Th^{231}$	β	$24\cdot6$ h
Protoactinium (Pa)	$_{91}Pa^{231}$	α	$3\cdot43 \times 10^4$ y
Actinium (Ac)	$_{89}Ac^{227}$	α,β	$22\cdot0$ y
Radioactinium (RdAc)	$_{90}Th^{227}$	α	$18\cdot6$ d
Actinium K (AcK)	$_{87}Fr^{223}$	β	21 m
Actinium X (AcX)	$_{88}Ra^{223}$	α	$11\cdot2$ d
Ac Emanation (An)	$_{86}Em^{219}$	α	$3\cdot92$ s
Actinium A (AcA)	$_{84}Po^{215}$	α,β	$1\cdot83 \times 10^{-3}$ s
Actinium B (AcB)	$_{82}Pb^{211}$	β	$36\cdot1$ m
Astatine-215 (At^{215})	$_{85}At^{215}$	α	10^{-4} s
Actinium C (AcC)	$_{83}Bi^{211}$	α,β	$2\cdot16$ m
Actinium C′ (AcC′)	$_{84}Po^{211}$	α	$0\cdot52$ s
Actinium C″ (AcC″)	$_{81}Tl^{207}$	β	$4\cdot79$ m
Actinium D (AcD)	$_{82}Pb^{207}$	Stable	

Table 8.4 The thorium series

Radioactive species	Nuclide	Type of disintegration	Half-life
Thorium (Th)	$_{90}Th^{232}$	α	$1\cdot39 \times 10^{10}$ y
Mesothorium 1 (MsTh1)	$_{88}Ra^{228}$	β	$6\cdot7$ y
Mesothorium 2 (MsTh2)	$_{89}Ac^{228}$	β	$6\cdot13$ h
Radiothorium (RdTh)	$_{90}Th^{228}$	α	$1\cdot90$ y
Thorium X (ThX)	$_{88}Ra^{224}$	α	$3\cdot64$ d
Th Emanation (Tn)	$_{86}Em^{220}$	α	$54\cdot5$ s
Thorium A (ThA)	$_{84}Po^{216}$	α,β	$0\cdot16$ s
Thorium B (ThB)	$_{82}Pb^{212}$	β	$10\cdot6$ h
Astatine-216 (At^{216})	$_{85}At^{216}$	α	3×10^{-4} s
Thorium C (ThC)	$_{83}Bi^{212}$	α,β	47 m
Thorium C' (ThC')	$_{84}Po^{212}$	α	$3\cdot0 \times 10^{-7}$ s
Thorium C" (ThC")	$_{81}Tl^{208}$	β	$2\cdot1$ m
Thorium D (ThD)	$_{82}Pb^{208}$	Stable	

8.3 Artificial Nuclear Disintegration

It was a logical step to consider the possibilities of inducing nuclear disintegrations by bombarding nuclei with sub-atomic particles. Alpha particles were used because they are massive and energetic. The first disintegration based on these ideas was caused by Rutherford in 1919. He bombarded nitrogen with alpha particles from radium C (bismuth-214) and observed scintillations on a screen placed 40 centimetres away. Since it was known that the alpha particles could not penetrate 40 centimetres of air, it was apparent that some reaction had been induced. Magnetic deflection measurements showed the particles causing the scintillations to be protons and Rutherford was able to rule out the possibility that they might have come from hydrogen impurity in the nitrogen gas. The only possible conclusion was that the alpha particles had caused the nitrogen atoms to disintegrate. This work was extended by Rutherford and Chadwick to other light elements, and positive results were found for all those between boron and potassium with the exception of carbon and oxygen.

The actual process involved in the disintegration was not identified until the work of Blackett, in 1925, who succeeded in taking cloud-chamber photographs showing an alpha particle track abruptly stopping and being replaced by two other tracks, one corresponding to the proton and the other to an ion of the daughter atom. It was, therefore, proved that the alpha particle was absorbed by the nucleus to form a *compound nucleus* which was unstable and which disintegrated by emitting a proton. It had previously been thought possible that the proton might have been ejected by some process involving the alpha particle's passing close to the nucleus but not colliding with it.

The reaction can be written in a manner analogous to a chemical reaction as

$$_7N^{14} + _2He^4 \longrightarrow [_9F^{18}] \longrightarrow _8O^{17} + _1H^1$$

or in shorthand form as $N^{14}(\alpha,p)O^{17}$. In these reaction expressions the mass and charge numbers must balance, and for convenience are often referred to by the incident and emitted particles. Thus this is an example of an *alpha–proton* reaction.

Other sub-atomic particles were found to be capable of causing artificial radioactivity, and in particular the neutron proved to be extremely effective. The reason why is not difficult to find. The neutron is electrically neutral and, as a consequence, experiences no coulomb force as it approaches the charged nucleus. It is, therefore, much more likely to penetrate the nucleus to induce a nuclear reaction. Because of this property, more disintegrations have been produced with neutrons than with any other particle.

The reaction between a neutron and a nucleus gives rise, in most cases, to either an alpha particle, a proton, a gamma ray or two neutrons. These are, in the shorthand notation, (n,α), (n,p), (n,γ) and $(n,2n)$ reactions, respectively. The commonest is the (n,γ) reaction. This is called *radiative capture* because the neutron is captured and electromagnetic radiation is emitted. The simplest such reaction is the hydrogen–deuterium reaction, $_1H^1(n,\gamma)_1H^2$ or $p(n,\gamma)d$, of equation (8.1).

With the development and study of these induced reactions, many of which produced new isotopes which were not known in nature and which were radioactive, interest developed in the problems of nuclear stability and in the theory of the structure of the nucleus. If a plot is made of the number of neutrons versus the number of protons in nuclei, an interesting picture appears. This is shown in Figure 8.8 where both stable nuclei and the naturally radioactive nuclei are included. It is apparent that, in order for a nucleus to be stable, two criteria must be satisfied: the nucleus must not be too large, as evidenced by the fact that there are no stable nuclei near the top of the chart, and the ratio of neutrons to protons appears to be critical. For a given nuclear mass, either an excess or a deficit of neutrons leads the nucleus into instability. Any theory of nuclear structure must account for these features.

One experimental fact remains to be discussed before looking at nuclear models. Until 1930 it was believed that, with the exception of some rare long range alpha particles, all the alpha particles from a given nucleus were emitted with the same energy. In that year, Rosenblum demontrated by very careful magnetic deflection experiments, that the normal alpha particles emitted by some nuclei, for example ThC (Bi^{212}), fall into several closely spaced velocity groups. This range of well-defined energies is called a *spectrum* and has led to much knowledge about nuclear structure.

The discovery of this alpha particle spectrum led to the suggestion that, in view of the success of the interpretation of atomic spectra in terms of discrete energy levels, it was reasonable to try to account for the new phenomenon in terms of discrete nuclear energy levels. A suggestion was made that for these disintegrations with several alpha particle energies, the emission of the particle leaves the nucleus in an excited state, from which it decays to the ground state by the emission of a gamma ray. When the energies of the emitted gamma rays

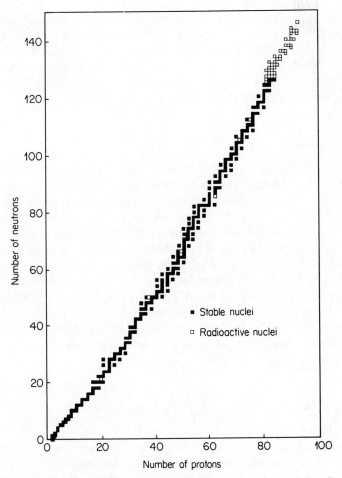

Figure 8.8. Numbers of neutrons and protons in the naturally occurring nuclei

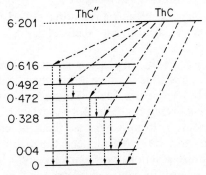

Figure 8.9. Alpha decay of ThC, showing possible combinations of α particles and γ rays

were measured, and corrections were made to the experimental alpha particle energies to allow for the recoil of the nucleus, these could indeed be fitted together to support this theory. This is illustrated in Figure 8.9 where the data for the decay of ThC to ThC″ is given. Notice how the maximum alpha particle energy, 6·201 MeV, can be made up in several other ways. One example is the 6·161 MeV alpha and the 0·040 MeV gamma ray.

8.4 Nuclear Models

In the absence of any detailed theory of nuclear structure, attempts have been made to correlate nuclear data in terms of rough pictures or models of the nucleus. Each of these is useful in its way, but each suffers from certain limitations. It has been impossible to form a detailed nuclear theory because of the lack of knowledge of the forces acting within the nucleus. Because of the nature of the problem, all our information on internal nuclear structure is derived from observations on the behaviour of emitted particles and radiation. As a result, the information is inferential rather than direct. Each of the proposed nuclear models correlates some experimental results, usually within a narrow range, and fails once it is applied outside this range. Only two models will be discussed briefly here; the *shell model*, because of its importance in the development of nuclear theory, and the *liquid drop model*, because of its usefulness in correlating the data of interest in the fission process.

Nuclear Shell Model

This model derives from the empirical correlations of nuclear data, and the fact that many nuclear properties seem to vary periodically in a way similar to that shown in the periodic table of the elements. In particular, many nuclear properties seem to show discontinuities near certain values of proton and neutron number. Especially stable nuclei appear when either Z or N (or both) is equal to one of the so-called *magic numbers* 2, 8, 20, 50, 82, 126. These magic numbers have been interpreted as representing closed shells of protons or neutrons by analogy with the electronic shells of 2, 10, 18,… electrons shown in the periodic table.

Some of the evidence for this analogy is striking. The binding energy for each of the 3rd, 9th,… nucleons is much lower than that of their immediate predecessors, usually by one or two MeV. The magic character of the number 126 for neutrons is especially noticeable in alpha particle decay. If alpha emission separates the 126th neutron from a nucleus, then the energy of the particle is much lower than in other cases. This reflects the high binding energy of this neutron. Neutron capture reactions are relatively unlikely for nuclei with 50, 82 or 126 neutrons, because the captured neutron will be only weakly bound, and is, therefore, more likely to escape again in the reaction.

The shell model is, therefore, rather similar to the electron model of atomic physics, in that the protons and neutrons occupy defined orbits and move independently of each other. The orbit of each nucleon is determined by the potential well in which it finds itself. This, of itself, poses one of the problems facing the model as it is not at all clear where the centre is for the potential well. The simplifying assumption that the nucleons find themselves in a rectangular potential well enables the wave equation to be solved for the energy levels of the nucleus. Application of the Pauli exclusion principle to both protons and neutrons, and a specific allowance for the two different types of nucleons, leads to the appearance of independent shells. Unfortunately they appear with the wrong numbers of nucleons in the closed shells, at 2, 8, 20, 40, 70, 112, 168. The experimental result can be attained, however, if, again as in atomic physics, the concept of spin is introduced and it is assumed that there is a strong energy dependence according to whether the spin angular momentum of the nucleon is parallel or anti-parallel to its orbital angular momentum.

The shell model has been applied successfully to a range of nuclear problems. For example, it has been possible to predict the total angular momenta of nuclei with good agreement between theory and experiment. It has, therefore, been possible to assign values to nuclei for which it has not been measured, in particular for beta-radioactive nuclei and the assignment has been very useful in the study of beta decay. There is also a correlation between the magic numbers and the number of long lived excited states of nuclei (isomers). This occurs because the neighbouring energy levels have a large difference in total angular momentum making the transition highly forbidden. The shell model can be used to predict the angular momenta of low lying excited levels and shows that the conditions for isomerism exist just below the magic numbers 50, 82 and 126, but not above them. The model also predicts when isomerism should exist for unfilled shells.

Liquid Drop Model

Much importance is attached to the mass and binding energy of nuclei. Consequently, a formula that would allow the calculation of these would be very useful. Such a formula has been developed within the framework of the liquid drop model of the nucleus, and is called the *semiempirical mass formula*.

We know from Figure 8.2 that over much of the range of mass numbers the binding energy per nucleon is approximately constant. This means that the binding energy of the nucleus is approximately proportional to the number of nucleons within it. The volume of the nucleus is also approximately proportional to the number of nucleons. Further, nuclear forces are of short range, and they appear to be saturated, that is they act only between a limited number of nucleons. These characteristics are analogous to the forces that hold a liquid drop together and so it is possible to regard the nucleus as a drop of incompressible fluid of very high density ($\approx 10^{17}$ kg m^{-3}). The 'liquid' is assumed to be composed of two types of particles, protons with their positive charge, and the neutral neutrons. Electrostatic forces act between the protons tending to force the nucleus apart, but these are resisted by the surface tension of the drop. The semiempirical mass formula is derived from these ideas by considering the various factors that could affect the nuclear binding energy and weighting them with constants derived from theory where possible and from experimental results in cases where theory cannot predict them.

The main contribution to the nuclear binding energy comes from the term proportional to the mass number A. Since the nuclear volume is also proportional to A, this term can be regarded as a 'volume energy' and written as

$$E_1 = a_1 A \qquad (8.4)$$

Tending to drive the protons apart, and decrease the binding energy is the coulomb interaction between the positive charges. If the protons are assumed to be uniformly distributed throughout the nucleus it can be shown that the total coulomb energy for a nucleus of charge Z and radius R is given by

$$E_2 = - a_2' \frac{3}{5} \frac{Z(Z-1)}{R} e^2$$

If the relationship between A and the nuclear volume is used to replace R by $A^{1/3}$ then for the larger values of Z this can be approximated by

$$E_2 = - a_2 Z^2 A^{-1/3} \qquad (8.5)$$

The nucleus also has a surface and a nucleon on the surface interacts with only half as many particles as a nucleon in the bulk. It is necessary, therefore, to subtract a term from the binding energy to allow for this *surface energy*. This term will be proportional to the surface area and can, therefore, be written as

$$E_3 = - a_3 A^{2/3} \qquad (8.6)$$

The binding energy must also contain a term to allow for the so-called *symmetry effect*. This is the effect already discussed by which too many or too few neutrons in a nucleus render it unstable. For a given value of A there is a particular value of Z for which the nucleus is most stable. Figure 8.8 shows that, for the light nuclei, this value is $Z = A/2$. As A becomes larger, the number of neutrons in the stable nuclei exceeds the number of protons, and so a term proportional to some power of $(A - 2Z)$, the excess of neutrons over protons, will provide a measure of the magnitude of this effect. In fact $(A - 2Z)^2$ is used as this has the useful property that both it and its second derivative are zero for $Z = A/2$. It has also been observed that the symmetry effect falls off with increasing A so that its contribution to the binding energy can be written as

$$E_4 = - a_4 (A - 2Z)^2 / A \qquad (8.7)$$

One further term must be added. It has been observed that the nucleus is more or less stable depending on whether the numbers of protons and neutrons are even or odd. Nuclei with odd numbers of both protons and neutrons are the least stable, while those with even numbers of both nucleons are the most stable. This *odd–even* effect can be represented by a term E_5 whose value depends on the numbers of both protons and neutrons, and which falls off as A increases. Thus E_5 is given by

Z	N	E_5
even	even	δ
even	odd	0
odd	even	0
odd	odd	$-\delta$

where N is the number of neutrons in the nucleus.

By bringing together equations (8.4) to (8.7) and E_5, the binding energy (B.E.) of the nucleus can now be written as

$$\text{B.E.} = a_1 A - a_2 Z^2 A^{-1/3} - a_3 A^{2/3} - a_4 (A - 2Z)^2 A^{-1} + E_5$$

Since the mass of the nucleus is given by the sum of the masses of the A nucleons less the binding energy expressed in mass units, we can write the *semiempirical mass formula* as

$$M = Z M_p + (A - Z) M_r - \frac{1}{c^2} (a_1 A - a_2 Z^2 A^{-1/3} - a_3 A^{2/3} - a_4 (A - 2Z)^2 A^{-1} + E_5)$$

Numerical values can be inserted for the constants, and one such set was produced by Green (1958) in which

$$a_1 = 0.01671, a_2 = 0.00075, a_3 = 0.0185, a_4 = 0.1, \delta = -0.036 A^{-3/4}$$

These values are expressed in atomic mass units and, therefore, represent a_n/c^2. The resulting expression for M becomes

$$M = 0.9819554 A - 0.0013888 Z + 0.0075 Z^2 A^{-1/3} + 0.0185 A^{2/3}$$
$$+ 0.1 (A - 2Z)^2 A^{-1} - 0.036 A^{-3/4} b \qquad (8.8)$$

where b is zero for zero E_5, and is ± 1, as appropriate, otherwise.

Several different formulae have been produced corresponding to equation (8.8). The values in these expressions depend largely on the effort put into their determination. These formulae reproduce quite accurately the nuclear masses for many nuclei, in this case to within 2 MeV throughout the range, with the exception of those nuclei in the regions of strong shell effects. In particular, equation (8.8) is sufficient for our application in the discussion of nuclear fission.

While the shell model and the liquid drop model each has its range of validity, accounting accurately for a group of nuclear properties, it is unfortunate that they have so little in common. The shell model assumes basically that the nucleons are non-interacting while the liquid drop model assumes a tumultuous mass of nucleons bouncing around within the nucleus. Obviously both models cannot be true, and both models offer some insight into different aspects of the overall problem. For our purposes the liquid drop model will be adequate.

8.5 Nuclear Reaction Cross-sections

Two particular nuclear reactions have already been discussed; the neutron–proton reaction to form a deuteron, and the alpha-proton reaction on nitrogen. One aspect of this problem that was not discussed was the relative probability of a reaction proceeding. Some reactions are very likely to occur, and some are extremely unlikely.

Even intuitively, the probability that a nuclear reaction will occur is directly related to the probability that the incoming particle will strike the nucleus. In simple terms this can be visualized as a straightforward projectile problem in which a ball is fired at a target. The probability that the ball will strike the target is related to the cross-sectional area of the target. An extension of this analogy leads to the concept of a *reaction cross-section*. Any given nuclear reaction is regarded as a ball and target problem, and the probability of its proceeding is related to the *effective* cross-section of the target for the incoming particle for the particular reaction being considered. This word *effective* is important because if a neutron, say, can be absorbed by a certain nucleus to form a compound nucleus, then this gives rise to a *total* cross-section for neutron absorption. If the compound nucleus can then decay by different competing processes—for example, by proton emission, gamma emission and two-neutron emission—then each of these processes gives rise to a *specific* cross-section. Thus there are cross-sections $\sigma(n,p)$, $\sigma(n,\gamma)$ and $\sigma(n,2n)$, and the sum of these must equal the total cross-section σ.

The dimensions of reaction cross-section are those of area, and the unit is the *barn* (1 barn = 10^{-24} cm^2). This name arose through a remark that when a reaction cross-section was as large as that it was 'as big as a barn'. The term then came into international usage. In fact, it is a reasonable unit because it is approximately the same size as the geometric cross-sectional area of the nucleus.

The reaction cross-section reflects the probability that a nuclear reaction will occur. In this it reflects the difficulties that the bombarding particle experiences in reaching the nucleus. The cross-sections for reactions induced by alpha particles must be less than those for slow neutron induced reactions because the alpha particle experiences a strong repulsive coulomb force as it approaches the nucleus, while the neutron does not. Cross-sections as low as 10^{-20} barns have been measured for reactions involving neutrinos, while, for (n,γ) reactions using slow neutrons, the value may be as high as 1000 barns.

8.6 Neutron Physics

So far, the neutron has been considered only with reference to its role in understanding the properties of the nucleus. Since its discovery in 1932, however, its importance has grown to such an extent that an area of study called neutron physics has developed. This is a wide field, having applications throughout fundamental physics and chemistry and into engineering. The applications range from nuclear power reactors and the nuclear bomb, to monitors for ensuring that unscrupulous oil tanker captains do not pump water with the oil when the tanker is unloading its cargo. The nuclear reactions in which the neutron plays a part help to produce information about the nucleus. When slow neutrons are scattered from a crystal lattice, information is gained about the dynamics of the lattice, and hence about the forces between atoms and molecules. However, the best known area of importance for neutrons is that of the chain reactions involving fissile materials.

Some aspects of neutron physics have already been covered, for example the discovery of the neutron and its part in producing new nuclear species. Here we will be concerned primarily with the interactions of the neutron with bulk matter.

Nuclear reactions are the only source of neutrons and there are several different approaches that have been successful in producing them for experimental work. There are, of course, the (γ,n) reactions on light elements that led to the discovery of the neutron originally. Photoneutron sources are popular because the neutrons produced here are almost monoenergetic; two such reactions are Be^9 $(\gamma,n)Be^8$ and $H^2(\gamma,n)H^1$. The binding energy of the neutron is very low in Be^9 and H^2 and, therefore, the thresholds for these reactions are low at 1·67 and 2·23 MeV, respectively. Very intense gamma emitters can be manufactured cheaply in nuclear reactors and so quite intense neutron sources can be conveniently made. Neutrons can also be produced in particle accelerators, and the bombardment of heavy ice or deuterated paraffin with deuterons accelerated in a van de Graaff generator provides an efficient source. The energy range of the neutrons can be tailored by choosing the particular target material and accelerated nuclei from a list of possible choices.

Neutron Moderation

The neutrons produced from any of these sources have intermediate or high energies. If slow neutrons are needed, the faster ones must be slowed down or *moderated*, by allowing them to pass through a block of suitable material called a *moderator*, in which the neutrons slow down by collisions. There are certain characteristics that a good moderator must have, and the principal two are that the energy lost per collision should be as large as possible, and that the cross-section for neutron absorption should be as low as possible for all possible neutron energies.

The interactions of neutrons with matter are very different from those of charged particles. All of these cause ionization of the target material and so are slowed down basically by electrostatic processes. Neutrons, on the other hand, are uncharged, do not induce ionization in the target material, and must be slowed down by what are essentially billiard ball collisions. One of the interesting features of neutron-induced reactions is that in general they are more efficient when the neutrons are travelling slowly. Moderation is, therefore, an important aspect of all neutron physics. The complete theory of neutron moderation is complex but all the underlying principles and some of the most useful results

226

can be shown quite simply. At energies above 1 MeV, the inelastic scattering of neutrons by intermediate or heavy nuclei can be important as a moderating mechanism, but below 1 MeV this becomes negligible and by far the most important process is the elastic scattering from light nuclei. For simplicity only this process will be considered.

It is a straightforward problem to calculate the amount of energy lost by the neutron in each collision. It is only necessary to write down and solve the equations for conservation of energy and momentum in the collision. It is useful to formulate the problem in the centre of mass frame of reference thus making use of the symmetry of the billiard ball problem in this reference frame. Neither quantum mechanics nor relativity need be introduced because the classical billiard ball model is sufficient, and the speed of a 1 MeV neutron, at 1.38×10^7 ms^{-1} is only one-twentieth of the speed of light.

If we give the neutron mass m, and speed v_0, and the target nucleus mass M and zero initial velocity, then we can write the speed of the centre of mass as

$$v_c = v_0 \frac{m}{M + m}$$

The representation of the collision in the centre of mass frame is shown in Figure 8.10. Before the collision the neutron is moving from left to right with speed

$$v_n = v_0 - v_c = v_0 \frac{M}{M + m}$$

and the nucleus is moving from right to left with speed v_c. After the collision, the total momentum must still be zero, and so the directions of motion of the neutron and the nucleus must still be opposite to each other. In addition, their speeds are unchanged because of the condition of conservation of energy, and so they depart in opposite directions along a line making some angle ϕ with the original line of approach. Thus, in the centre of mass system, the total effect of the collision is to change the direction of motion of the particles, but not their speeds.

In the laboratory frame of reference, however, Figure 8.11 shows that the

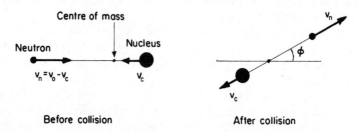

Before collision   After collision

Figure 8.10. Collision between neutron and moderator nucleus in centre-of-mass frame of reference

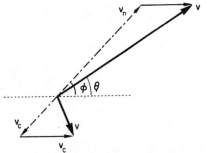

Figure 8.11. Relation between centre-of-mass and laboratory frames of reference.

speeds of the particles do change and also gives the vector relationship between the two frames of reference after the collision.

There are two cases of interest for which the energy lost by the neutron can be calculated very simply. The first is the case of a glancing collision in which $\phi = 0$. In this collision the velocity of the neutron is unaffected and so it loses no energy in the collision. The other case is that of a head-on collision. In this case $\phi = 180°$, and the final neutron speed is

$$v = v_0 \frac{M}{M + m} - v_0 \frac{m}{M + m} = v_0 \frac{M - m}{M + m}$$

This can be expressed in terms of the fraction of the original energy now carried by the neutron as

$$\frac{E}{E_0} = (\tfrac{1}{2}mv^2)/(\tfrac{1}{2}mv_0^2) = \left(\frac{M - m}{M + m}\right)^2 = \left(\frac{A - 1}{A + 1}\right)^2 \tag{8.9}$$

since M/m is approximately equal to the mass number of the target nucleus because m is almost unity and M is almost an integer. Not surprisingly, the neutron loses most energy in a head-on collision, and least in a glancing collision, or miss.

From this simple analysis it can easily be seen that the best moderator for neutrons must be hydrogen, with a mass almost equal to that of the neutron. In this case, a head-on collision will almost exactly stop the neutron and give the neutron's original energy to the proton, or hydrogen nucleus. If the moderator is graphite, with $A = 12$, the maximum fractional energy loss is $1 - E/E_0$ or 28·4 per cent.

Most collisions, however, are not head on, and it is necessary to calculate the average energy lost in each. In general the neutron will be deflected through some angle ϕ in centre of mass coordinates. As a result, the cosine law can be applied to Figure 8.11 to show that

$$v^2 = v_0^2 \left(\frac{M}{M + m}\right)^2 + v_0^2 \left(\frac{m}{M + m}\right)^2 + 2v_0 \left(\frac{M}{M + m}\right)\left(\frac{m}{M + m}\right) \cos \phi$$

Consequently, the fractional energy change is now

$$E/E_0 = v^2/v_0^2 = \frac{M^2 + m^2 + 2mM \cos \phi}{(M + m)^2} = \frac{A^2 + 1 + 2A \cos \phi}{(A + 1)^2}$$

It is convenient to express this in terms of the ratio $r = ((A - 1)/(A + 1))^2$ which was introduced in equation (8.9) and represents the maximum possible energy loss in the collision. The fractional energy, therefore, becomes

$$E/E_0 = \frac{1}{2}(1 + r) + \frac{1}{2}(1 - r) \cos \phi \qquad (8.10)$$

Before calculating the average value of E/E_0, let us look at the angular distribution of scattered neutrons. Figure 8.11 shows that, in the laboratory frame of reference, the scattering angle θ is given by

$$\cot \theta = (\cos \phi + 1/A)/\sin \phi$$

Hence

$$\cos \theta = \frac{1 + A \cos \phi}{(1 + A^2 + 2A \cos \phi)^{1/2}}$$

If we can determine $<\cos \theta>$, the average value of $\cos \theta$, we will obtain a picture of the dependence of the scattering of neutrons on the mass of the moderator nuclei. Unfortunately, the determination depends on the probability that a neutron will be scattered through an angle ϕ in the centre of mass system. This is not known *a priori*. Experimental and rigorous theoretical evidence suggest, however, that this scattering is spherically symmetric for neutrons with energy less than 10 MeV. Since this condition is satisfied for most of the neutrons involved in the fission process, we can assume this result, and calculate $<\cos \theta>$ by integrating $\cos \theta$ over the solid angle $2\pi \sin \phi \, d\phi$ resulting in

$$<\cos \theta> = \frac{1}{4\pi}\int_0^\pi \cos \theta / 2\pi \sin \phi \, d\phi = \frac{1}{2}\int_{-1}^{+1} \frac{1 + A \cos \phi}{(1 + A^2 + 2A \cos \phi)^{1/2}} \, d(\cos \phi)$$

Hence

$$<\cos \theta> = \frac{2}{3A} \qquad (8.11)$$

This result shows that when A is large, $<\cos \theta>$ is small, and the scattering in the laboratory system is almost isotropic. Neutrons will, therefore, be scattered forward as frequently as they are scattered backward. When A is small, however, $<\cos \theta>$ is large and more neutrons are scattered forward than back.

We can now determine the energy lost per collision in a large number of collisions. Since we are dealing with a random number of collisions, we can expect the processes to obey an exponential law of some sort. Consequently we can introduce ξ, the average decrease in the logarithm of the neutron energy at each collision. Since E/E_0 is a linear function of $\cos \phi$, as shown in equation (8.10), and since all values of $\cos \phi$ are equally probable in the centre of mass frame of reference, it follows that all values of E/E_0 are equally likely. The probability PdE that a neutron of initial energy E_0 will have an energy between E and $E + dE$ after a collision is, therefore,

$$PdE = \frac{dE}{E_0(1 - r)}$$

This is so because $E_0 (1 - r)$ represents the entire range of energies that the neutron can have after one collision. ξ is the average decrease in the logarithm of the neutron energy at each collision, and so

$$\xi = <\ln E_0 - \ln E> = <\ln(E_0/E)>$$

is given by

$$\xi = \int_{rE_0}^{E_0} \ln(E_0/E)P dE = \int_{rE_0}^{E_0} \ln(E_0/E) \frac{dE}{E_0(1 - r)}$$

$$= 1 + \frac{r}{1 - r} \ln r$$

which, on substitution for r, becomes

$$\xi = 1 - \frac{(A - 1)^2}{2A} \ln\left(\frac{A + 1}{A - 1}\right) \tag{8.12}$$

For $A > 10$, a convenient expression for ξ is

$$\xi = \frac{2}{(A + \frac{2}{3})} \tag{8.13}$$

This is accurate to about one per cent and shows that for graphite the average energy lost per collision is represented by $\xi = 0.158$. This implies that E/E_0 is 0.854 and represents a loss of 14.6 per cent of the neutron energy in the average collision. For a target nucleus with $A = 200$, the corresponding energy loss is 0.99 per cent showing that heavy nuclei make less efficient moderators than light ones.

Equation (8.12) for ξ breaks down when $A = 1$ since the right-hand side becomes indeterminate. Taking limits as $A \to 1$, however, does produce $\xi = 1$, which shows that the neutron energy decreases on average by the factor $e = 2.718$ in each collision. Hydrogen is, therefore, a very efficient moderator.

Once ξ is known, the average number of collisions to produce a given change in energy can easily be evaluated. One case of great interest for nuclear reactor calculations is a neutron of initial energy 2 MeV (the average energy of a fission neutron) slowing to 0.025 eV (the energy of a neutron at room temperature). The number of collisions needed can be determined as the ratio of $\ln(E_1/E_2)$ to ξ, where $E_1 = 2$ MeV and $E_2 = 0.025$ eV. Thus the number of collisions is

$$n = \frac{1}{\xi} \ln\left(\frac{2 \times 10^6}{0.025}\right) = \frac{18.2}{\xi} \tag{8.14}$$

The values of ξ and the number of collisions needed for slowing 2 MeV neutrons to thermal speeds are shown for a range of substances in Table 8.5.

As a result of this slowing down process, the neutrons eventually reach a state of thermal equilibrium with the atoms of the moderator lattice. In a given collision thereafter a neutron may gain or lose a little energy, but on average

Table 8.5 Moderator efficiencies for 2 MeV neutrons

Moderator	A	ξ	n = number of collisions from 2 MeV to 0·025 eV
H	1	1	18·2
D	2	0·7253	25·1
C	12	0·157	115·9
N	14	0·1364	133·5
Au^{197}	197	0·0101	1798·8
U^{238}	238	0·00838	2171·9

energy is neither gained nor lost. The behaviour of the neutrons is then similar to that of a gas of atoms, and they can be described quite accurately by kinetic theory. Thus the neutron speeds obey the Maxwell distribution; the most probable neutron speed v_0 is $(2/mkT)^{1/2}$ where k is Boltzmann's constant, m is the mass of the neutron and T is the absolute temperature; the average speed is $1·1284v_0$, and one can consider the neutrons as having a mean free path between collisions.

There are several competing processes involving the collisions of neutrons in the moderator. This means that there are several different mean free paths to be considered. There is the *scattering mean free path*, λ_s, or the average distance between collisions in which the neutron is scattered from the colliding nucleus. This is given by

$$\lambda_s = (1/N \sigma_s) \tag{8.15}$$

where N is the density of nuclei, and σ_s is the scattering cross-section per nucleus. There is the *absorption mean free path*, λ_a, given by

$$\lambda_a = (1/N \sigma_a) \tag{8.16}$$

where σ_a is the absorption cross-section. This represents the average distance travelled before the neutron is absorbed by a nucleus of the moderator. Obviously this is a very important parameter since any material which has a high absorption cross-section will be of limited value as a moderator because of the number of collisions involved in the slowing-down process. Finally there is the *transport mean free path*, λ_t, which measures the effect of forward scattering in the moderator. It is defined by

$$\lambda_t = (1/N \sigma_t) = \frac{1}{N \sigma_s(1 - <\cos \theta >)} \tag{8.17}$$

where $<\cos \theta >$, given by equation (8.11), represents the average value of the scattering angle in the laboratory reference frame. Since the scattering is always predominately forward for real moderator materials, λ_t is greater than λ_s, and reflects the fact that a neutron will travel further forward in a given number of collisions than if there was no preferred direction of scatter. It is a measure of the rate at which a neutron loses its forward momentum, or its memory of the direction in which it is travelling.

Neutron Diffusion

Once a neutron has been slowed down to thermal speeds, it continues to diffuse through the moderator. This process is characterized by the *diffusion coefficient, D*, which is a parameter of the particular moderator material. The diffusion coefficient must be controlled by the various mean free paths introduced above and it is apparent that the important one must be λ_t, in which case it can be written as

$$D = \frac{1}{3}\lambda_t v \qquad (8.18)$$

where v is the speed of the neutron. This must appear in the diffusion coefficient since it is intuitively apparent that faster neutrons will diffuse further through a material than will slow ones.

Once D is known, the current density for neutrons flowing through a slab of moderator in the x direction can be written as

$$J = -D\frac{dn}{dx} = -\frac{1}{3}\lambda_t v \frac{dn}{dx}$$

where n, the neutron density, is a function of x.

If we consider a neutron source producing J neutrons per unit area per second incident onto an infinite slab of moderator, then, if the element of slab between x and $x + dx$ is considered, we can say that the number of neutrons lost to motion in the x direction is denoted by $L_x dx$ and is given by

$$L_x dx = J(x) - J(x + dx)$$

$$= \frac{1}{3}\lambda_t v \left(\frac{dn}{dx}\bigg|_{x+dx} - \frac{dn}{dx}\bigg|_x \right)$$

$$= \frac{1}{3}\lambda_t v \frac{d^2 n}{dx^2} dx$$

At equilibrium, the number of neutrons leaking into the element of slab must be equal to the number being absorbed in it, which is $nv\,\sigma_a N dx$, and so

$$\frac{1}{3}\lambda_t v \frac{d^2 n}{dx^2} = nv\,\sigma_a N$$

which reduces to

$$\frac{d^2 n}{dx^2} = \frac{1}{L^2} n \qquad (8.19)$$

where L is called the *thermal diffusion length* and is given by

$$L^2 = \frac{1}{3}\lambda_t \lambda_a \qquad (8.20)$$

It is clear from equation (8.19), that there is an exponential decay in the neutron density with distance travelled in the moderator, and the diffusion length L represents the average distance travelled by a neutron in the x direction before absorption. This is distinct from λ_a which represents the average total distance travelled in all directions before absorption takes place.

Table 8.6 Constants for some moderating materials

	Density	Diffusion length (cm)	Migration length (cm)	λ_t (cm)	λ_s (cm)	λ_a (cm)	Slowing down power	Moderating ratio
H_2O	1·00	2·67	6·25	0·48	0·43	51·8	1·28	58
D_2O	1·10	123·0	116·0	2·65	2·40	13400	0·18	21000
Be	1·85	20·8	23·0	1·47	2·10	2480	0·16	130
BeO	3·0	29·0	30·5	1·29				
Graphite (C)	1·67	49·0	56	2·39	2·7	705	0·065	200

The various constants for some moderators are shown in Table 8.6 where it can be seen that, while water, with its high hydrogen content, is an excellent moderator for neutrons, it is also, unfortunately, a high absorber of them as well. This restricts its usefulness as a moderator for nuclear reactors.

Superficially, since hydrogen has a high absorption cross-section for neutrons, deuterium should be the best moderating material. This is apparent because of the high energy loss per collision as shown in Table 8.5. Unfortunately deuterium is a gas and so a large volume would be necessary to ensure that the neutrons will undergo the necessary number of collisions. A more dense material would be a more efficient moderator, because the mean free path between collisions would be reduced. This effect can be taken into account by introducing the *slowing down power* which is defined as

$$\text{Slowing down power} = \xi N\sigma_s = \xi\sum_s = (N_0\rho\xi\sigma_s)/A \qquad (8.21)$$

where N_0 is Avogadro's number and ρ is the density of the material.

The quantity $\sum_s = N\sigma_s$ is called the *macroscopic scattering cross-section*, and is the probability that a neutron will be scattered in unit distance travelled in the moderator. Since ξ is the average loss in $\ln E$ per collision, the slowing down power represents the average loss of energy per unit length of neutron travel. It should obviously be large for a good moderator.

It is possible to represent the performance of different moderator materials in terms of their relative efficiency at slowing down neutrons and their absorption effects by introducing a quantity called the *moderating ratio*. This is defined as the ratio of the slowing down power to the absorption cross-section. It is, therefore, a measure of the relative slowing down power and absorbing power of a moderator. It is defined as

$$\text{Moderating ratio} = \xi\sum_s /\sum_a = \xi\sigma_s/\sigma_a \qquad (8.22)$$

Obviously this should be high for a good moderator.

Table 8.6 includes values of the slowing down power and moderating ratio for some potential moderators. According to these results, D_2O is the best of the moderators listed, and H_2O the worst. Light water comes off so badly because, despite its extremely high slowing down power, its moderating ratio is very low, reflecting the fact that neutrons are lost to the moderator in unduly large numbers. It is interesting that graphite has a moderating ratio some 2·5 times larger than that of light water despite the much lower slowing down power.

This discussion of moderation is, of course, extremely limited, as has been the whole treatment of nuclear physics. There are other important considerations such as the energy distribution of the neutrons during moderation, and the distance travelled by neutrons during the process. A detailed treatment of these problems can be found in the further reading list.

References

Green, A. E. S., *Rev. Mod. Phys.*, **30**, 569 (1958).
Hahn, O. and F. Strasseman, *Naturwiss.*, **27**, 11, 89 (1939) (in German); Hahn's account of this work in his Nobel Lecture is in English translation in: Hahn, O., *New Atoms*, W. Coage, (Ed.), Elsevier Publishing Co. Inc., New York, 1950.

Further Reading

Bennet, D. J., *The Elements of Nuclear Power*, Longmans, London, 1972.
Glasstone, S., *Source Book on Atomic Energy*, D. van Nostrand Co. Inc., New York, 1958.
Heckman, H. H. and P. W. Starring, *Nuclear Physics and the Fundamental Particles*, Holt Rinehart and Winston Inc., New York, 1963.
Kaplan, I., *Nuclear Physics*, Addison Wesley Publishing Co. Inc., Reading Mass., 1955.
Segré, E., *Nuclei and Particles*, W. A. Benjamin Inc., New York, 1963.
Soodak, H., (Ed.), *Reactor Handbook*, Vol. IIIa (Physics), Wiley Interscience, New York, 1962.

9 Nuclear Fission

Nuclear fission was discovered during a series of attempts, using neutron bombardment, to produce the heavy transuranic elements which have atomic numbers greater than 92. The intended processes were (n,γ) reactions followed by beta decay of the product nucleus. The interpretation of the early experiments was difficult and the results could only be understood eventually in terms of fission.

A radioelement formed by nuclear reaction was only available in very small quantities, often as small as 10^{-12} g. In order to separate these radionuclides, *carriers* were used. These were stable compounds with chemical properties similar to those of the radionuclide of interest. They were added to the solution containing the radionuclide and precipitated as an insoluble salt. The radioelement was precipitated with the carrier to be separated later by other chemical reactions.

Barium was used for the separation of radium from the other elements of the uranium series. Radium and barium are both in Group IIA and form insoluble sulphates. Their chlorides can also be precipitated from concentrated hydrochloric acid, after which they can be separated by repeated fractional crystallization from hydrochloric acid. In a similar way, lanthanum can act as a carrier for actinium, and stable iodine can be used for the radioactive isotope of iodine.

In the early experiments, uranium was bombared with neutrons and several different beta activities were detected. Carrier techniques were tried to separate the elements, but the results could not be made to fit into any scheme consistent with the known properties of the heavy nuclides. In particular there were four activities assumed to be due to isotopes of radium because they were precipitated using barium as the carrier, which seemed to produce actinium decay products because these could be precipitated using lanthanum. These possibilities raised some interesting questions. For example, how likely was the $(n,2\alpha)$ reaction involved in producing radium from uranium—especially at low neutron energies? Also, why was it that the two sets of activities could not be separated from their respective carriers? In 1939, Hahn and Strasseman showed that the activities were, in fact, due to isotopes of barium and lanthanum, and they were even able to identify one known isotope of each element in the activities. Thus the existence of nuclear fission was proven. It was possible, by bombarding uranium with slow neutrons, to cause its nucleus to split into smaller components. From the way in which the fission nuclei appeared, it was postulated that there were two fission fragments occurring with each fission of a uranium nucleus, and that several neutrons and gamma rays were also emitted in the process. Evidence supporting fission built up over a short period, and the whole range of intermediate mass nuclides were found among the fission fragments.

Along with the pairs of fission fragments, neutrons and gamma rays, the emission of light elements with masses greater than four and less than twelve is a fairly common event, occurring about once in 80 fissions; long range alpha particles are emitted about once in every 400 fissions; and although the emission of three fragments of comparable size is very rare it has been observed. It is important to remember that fission is a statistical phenomenon and is subject to the usual range of fluctuations to be expected from such phenomena.

9.1 Aspects of the Fission Process

The probability of fission as compared with the probability of other reactions is of critical importance to the nuclear power industry. For example, the U^{235} nucleus may absorb a neutron to form U^{236} or the neutron may be scattered without inducing a nuclear reaction at all. Once the compound nucleus of U^{236} is formed, it may undergo fission, or it may emit gamma rays and decay to its ground state. Thus there are at least three competing processes, and their relative importance is reflected in the reaction cross-sections. The cross-sections for these three processes for the important fissile materials are shown in Table 9.1. It is notable that the cross-sections for radiative capture are a sizable fraction

Table 9.1 Reaction cross-sections for common fissile materials (thermal neutrons)

	Fission (barns)	Radiative capture (barns)	Absorption (barns)	Scattering (barns)	η
U^{233}	527	54	581		2·27
U^{235}	580	107	687	10	2·06
U^{238}	$<0.5 \times 10^{-3}$	2·71	2·71	8·3	
Pu^{239}	746	280	1026	9·6	2·10
U (nat.)	4·18	3·5	7·68	8·3	1·33

of those for fission for both U^{235} and Pu^{239}. It is also notable that since the fissile isotope of uranium, U^{235}, forms only 0·7 per cent of naturally occurring uranium, the fission and radiative capture cross-sections for natural uranium are substantially different from those of U^{235}, even to the extent that scattering becomes more likely than absorption of thermal neutrons to cause either fission or radiative capture.

The products of these fission reactions provided some insight into the mechanisms of the fission process. The first relevant observation has already been mentioned in Chapter 8, namely that fission must be exothermic for the heavy nuclides. This is apparent from the binding-energy curves in Figures 8.2 and

8.3. It is further apparent from Figure 8.8 that, if a heavy nucleus such as U^{236} splits into two roughly equal parts, each of the fission fragments will have too many neutrons to be stable. For example, $_{92}U^{236}$ has a neutron–proton ratio of 1·57; the range of ratios for the stable isotopes of some typical fission products—krypton, iodine, xenon and caesium—is 1·17 to 1·52. As a consequence, the fission fragment nucleus is unstable and decays either by beta decay or, if the excitation energy is high enough, by neutron decay. We have, therefore, a simple and straightforward prediction that the fission products will be radioactive. That this is true is evidenced by the fact that it was this radioactivity that led to the original discovery of fission.

The fission fragments typically occur in pairs such that their atomic numbers add up to 92—that of uranium—and their mass numbers total something rather less than 236. The most probable fission pair, occurring in about 6 per cent of all cases, yields products with mass numbers 95 and 139, giving a sum of 234. The lowest mass number found in the common two-product processes is 72, and the highest is 158. Between these limits there lie 87 different nuclides which can, therefore, be taken as defining the range of nuclides formed as direct fission fragments in the fission of uranium-236. In fact, over 60 different fragments have been identified giving at least 30 independent fission schemes. Each of these fragments, being radioactive and highly excited, decays in its search for stability, and starts a small radioactive series involving the emission of beta particles. These series are called *fission decay chains* and each chain has three members on average. Longer and shorter chains are quite common, of course, and the problem of determining the fission products and their decay schemes is a difficult one. Nevertheless, well over 60 chains have been identified and over 200 radio-nuclides assigned to them.

An example of a long chain, of importance because it contains two of the nuclides involved in the original discovery of fission, is

$$_{54}Xe^{140} \xrightarrow[16s]{\beta^-} {}_{55}Cs^{140} \xrightarrow[66s]{\beta^-} {}_{56}Ba^{140} \xrightarrow[12\cdot8d]{\beta^-} {}_{57}La^{140} \xrightarrow[40h]{\beta^-} {}_{58}Ce^{140}$$

An interesting short chain, which contains the element *promethium*, hitherto unknown, is

$$_{60}Nd^{147} \xrightarrow[11d]{\beta^-} {}_{61}Pm^{147} \xrightarrow[4y]{\beta^-} {}_{62}Sm^{147} \text{ (about } 10^{11}y)$$

Promethium was unknown in nature because it existed only in fission chains, and having a four year half-life, it has decayed out of all natural fission chains long ago.

We now have a picture of fission as a process in which the nucleus, in an excited state due to neutron absorption, breaks into two components. These components are produced by chance, and a range of pairs is possible. Is there any way of assuring ourselves of the validity of this model? Can we find a way of ascertaining that these fission fragment pairs actually exist? There are two things that can be done. The number of times a given mass number is identified

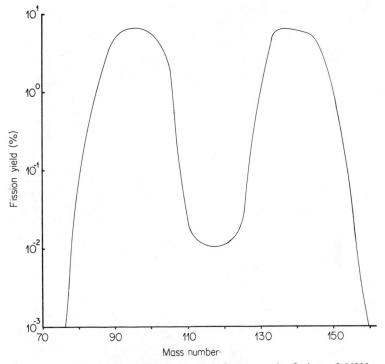

Figure 9.1. Distribution of mass among fragments in fission of U^{235} by thermal neutrons

in fission processes can be plotted against mass number. If the fission fragments do occur in pairs, then corresponding mass numbers should be equally probable. When this is done for U^{235} and Pu^{239}, the results are as shown in Figures 9.1 and 9.2. The two curves show two distinct peaks, representing the two most probable masses, and also, the probabilities of occurrence of the two masses are almost identical. The clear inference is that the two nuclides occur in pairs. There is a more convincing argument, however. It can be assumed with safety that, just before fission occurs, the fissile nucleus is at rest. It then absorbs a neutron and undergoes fission. To a good approximation, since momentum is conserved, the two fission fragments should leave the parent nucleus in opposite directions and with speeds related by

$$M_1 V_1 = M_2 V_2$$

We know therefore the ratio of the speeds of the two nuclei and can determine the ratio of their energies as

$$E_1/E_2 = (\tfrac{1}{2}M_1 V_1^2)/(\tfrac{1}{2}M_2 V_2^2) = M_2/M_1$$

Thus, the energies are inversely related to the masses. The measured energy distribution for the fission fragments from U^{236} is shown in Figure 9.3 and it can be seen that the ratio of high energy to low energy for the two peaks,

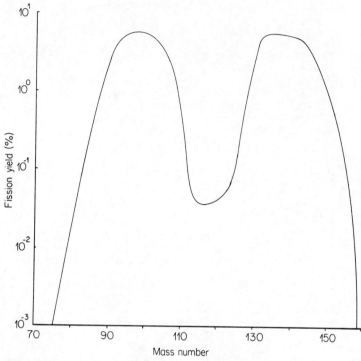

Figure 9.2. Distribution of mass among fragments in fission of Pu²³⁹ by thermal neutrons

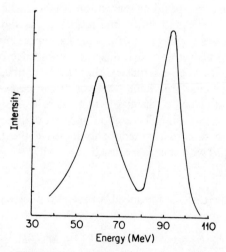

Figure 9.3. Distribution of fragment energies in fission of U²³⁵ by thermal neutrons

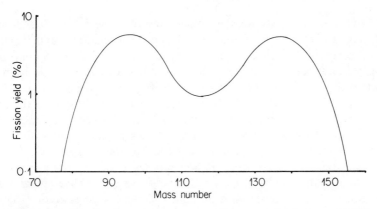

Figure 9.4. Distribution of mass among fragments in fission of U^{235} by 14 MeV neutrons

at 1·48, is very close to that for the ratio of high mass to low mass from Figure 9.1, namely 1·46. Again the implication is clear, the two masses occurred as a pair and emanated from the same fission reaction.

It is very interesting that fission is non-symmetric, that is the two fission fragments are on average of different sizes. This is clearly shown in Figure 9.1 and 9.2 and is unexpected. It is also known that as the energy of the bombarding neutron increases, the depth of the well between the two peaks decreases and the fission becomes more symmetric. This is illustrated in Figure 9.4 where the mass number distribution for the fission of U^{235} by 14 MeV neutrons is shown. If the liquid drop model of the nucleus is used to account for the fission process, then it predicts that the fission should be symmetric. For this reason it cannot represent a true model of the nucleus. Nevertheless, despite this limitation, the model does provide a sufficiently good qualitative picture of fission for most purposes and also gives reasonable quantitative values for that data for which it was designed, namely the energy released on fission.

Before looking at the model for fission, there are two other features that should be discussed: the neutron emission that occurs during the fission process and the energy released. When the neutron to proton ratio for some of the fission fragments was compared with that of the unstable excited U^{236} nucleus, it was suggested that some neutrons might be emitted during fission. This was tested by placing a neutron source at the centre of a large vessel containing a uranium solution and surrounding it with detectors at various distances. The vessel was first filled with ammonium nitrate solution for comparison purposes. The average neutron density was found to be higher with the uranium solution and showed in a rough way that neutrons were being produced. The average number of neutrons produced per fission is of great interest and for the fission of U^{235} by thermal neutrons, it is 2·47.

This discussion applies to the so-called *prompt* neutrons, the neutrons emitted with the fission process itself. There are also other *delayed* neutrons which derive from the decay of the fission products having lives ranging from 0·05 to 55·6

240

seconds. These obviously appear some time after the fission has occurred, and, though their total contribution to the neutron flux is small at 0·76 per cent, their influence on the time-dependent behaviour of a chain reacting pile is noticeable, and they play an important part in reactor control. One example of the origin of the delayed neutrons is the 22·0 second group. I^{137} is a fission product and a beta emitter with a half-life of 22·0 seconds. In some of the decays, the daughter Xe^{137} is formed in a sufficiently excited state that it can decay to Xe^{136} by neutron emission. This is a very quick reaction, and so the appearance of these neutrons is controlled by the half-life of the decay of I^{137}.

The neutrons produced in the fission process, either by the prompt or by any of the delayed mechanisms, are fast neutrons. They have energies which are in general much higher than thermal and the lowest is about 0·05 MeV. The upper limit is about 17 MeV and the average energy is about 2·0 MeV. The spectrum for the fission of U^{235} is shown in Figure 9.5.

The energy released in fission is striking, about 200 MeV. This must be compared with the several MeV of other nuclear reactions, and the few electron volts of chemical reactions. The energy to be released in fission can be estimated from the binding-energy curve (Figure 8.2) of the last chapter, where we see that in the range of mass numbers from 80 to 150, where most of the fission

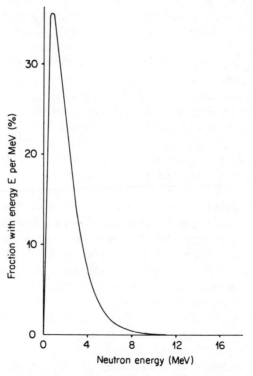

Figure 9.5.   Energy distribution of neutrons released in fission of U^{235}

fragments lie, the binding energy per nucleon is about 8·4 MeV. In contrast, this is reduced to about 7·5 MeV in the neighbourhood of uranium. Thus in the fission process about 0·9 MeV must be released for each nucleon. The total energy released should be approximately equal to the number of particles (236) multiplied by the energy released per nucleon, and so is approximately 200 MeV.

Alternatively, the energy released can be calculated from the mass data. The most probable fission pairs are about $A = 95$ and 139, and obviously one must really consider the stable end products of the fission chains. Such chains could end in the pair Mo^{95} and La^{139}. This pair has a total mass number of 234 and so, apparently, two neutrons are also produced in the process. Summing the masses of these four particles, using the data in Table 8.1, we arrive at a total of 235·918 amu. The mass of the U^{236} nucleus is 236·133 amu and so apparently 0·215 amu has been converted from mass into energy. Since 1 amu is 931·47 MeV, this is equivalent to 200·26 MeV.

The two estimates agree surprisingly well and so we can have some confidence that approximately this amount of energy is released. How is it distributed among the various possible processes? At the moment of fission, the energy stored in the part of the mass of the uranium nucleus that is to disappear is suddenly released in a variety of different manifestations. Some of it appears in the kinetic energy of the two fission fragments as they fly apart under the influence of the repulsive coulomb forces; part of it appears in the kinetic energy of the neutrons released at the same time; part of it appears in the gamma rays that are released promptly as the excited nuclei lose excess energy, and the rest is temporarily stored in the nuclear mass of the fission fragments to be released in the kinetic energy of the beta particles and their associated neutrons, gamma rays and the delayed neutrons. By far the greatest proportion is released in the kinetic energy of the fission fragments, and an approximate energy budget for the process is shown in Table 9.2 where it can be seen that about 80 per cent of the energy is released in this way.

Table 9.2 Energy budget in fission

Kinetic energy of fission fragments	168 MeV
Kinetic energy of neutrons	5 MeV
Fission gamma radiation	5 MeV
Betas emitted by fission products	7 MeV
Gammas emitted by fission products	6 MeV
Neutrinos (not available as heat source)	11 MeV
Total	202 MeV

9.2 Liquid Drop Model of Fission

By far the easiest way of visualizing the mechanism of fission is to use the liquid drop model introduced in the last chapter. If the fissile nucleus is viewed

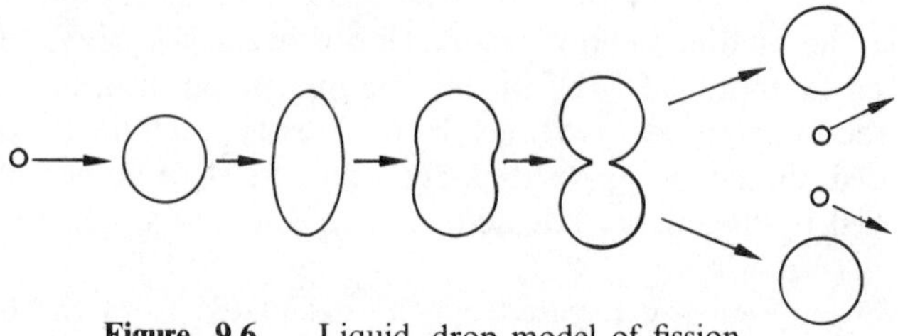

Figure 9.6. Liquid drop model of fission

as being a liquid drop, and the incoming neutron as a little droplet, then, provided that the incoming droplet has sufficient kinetic energy to penetrate the surface of the drop, we can visualize the situation where the addition of this droplet will cause the larger drop to oscillate in shape. Once these oscillations start there are only two possibilities; either they die down and the drop stabilizes, or they gradually build up, as shown in Figure 9.6, until a neck develops and the surface tension of the liquid causes the radius of the neck to shrink, eventually causing the drop to split into two – perhaps with a few small droplets formed at the moment of separation. Our two possibilities for the drop correspond to the possibilities of radiative capture and fission. The relative probabilities of the two events are reflected in their cross-sections as given in Table 9.1, 107 barns for radiative capture of thermal neutrons by U^{235} and 580 barns for its fission. In the liquid drop model of the nucleus the repulsive forces arise through the coulombic repulsion of the protons, and the surface tension forces holding the drop together are allowed for by a surface energy term. The stability of the nucleus to one of these oscillations in shape is tested by assuming that initially the nucleus is spherical and that the arrival of the neutron causes a distortion such that it becomes ellipsoidal. If the potential energy of the ellipsoidal nucleus is greater than that of the spherical one, then the nucleus will relax back to the stable lower potential energy state, whereas, if the potential energy of the distorted nucleus is less than that of the spherical one, the distortion is permanent and will develop until eventually the nucleus falls apart in the same way as the drop did in the model.

The basis for our calculation of the stability of the nucleus lies in equations (8.4) to (8.8), but appropriate modifications must be made to allow for distortion. If we consider equation (8.4), we can set up our distortion in such a way that the volume remains constant. This is a simplifying assumption and eliminates the necessity of considering changes in the volume energy. However, the surface area of the nucleus must change with the distortion, and this is reflected in a change in the value of the surface energy and the repulsive coulomb energy. Consequently, there will be changes in equations (8.5) and (8.6). The contributions to the binding energy from the symmetry effect and the odd–even effect will be unchanged since the number of nucleons and the ratio of neutrons to protons is unaffected by the distortion. The change in binding energy of the nucleus due to the distortion is, therefore, given by

$$\Delta E = \Delta E_2 + \Delta E_3 = -a_2 Z^2 A^{-1/3} - E_2' - a_3 A^{2/3} - E_3' \qquad (9.1)$$

where E'_2 and E'_3 represent the contributions to the binding energy of the nucleus, corresponding to equations (8.5) and (8.6), after the distortion is taken into account.

Assume that the nucleus is distorted into an ellipsoid of revolution of semi-major axis a and semiminor axis b. To ensure the constancy of the volume let a and b be related by $a = R(1 + \varepsilon)$, $b = R(1 + \varepsilon)^{-1/2}$, when the volume $V = 4\pi ab^2/3$ is constant. The surface area of this ellipsoid is

$$S = 2\pi b^2 + 2\pi \frac{ab}{e} \sin^{-1} e$$

where $e = (1 - (b/a)^2)^{1/2}$. Expanding this equation in terms of R and ε, and limiting the expansion to second order in ε, we find that

$$S = 4\pi R^2 (1 + \tfrac{2}{5}\varepsilon^2 + \ldots)$$

which is a factor of

$$(1 + \tfrac{2}{5}\varepsilon^2)$$

greater than the surface area of the sphere. The surface energy of the distorted nucleus is, therefore, given by

$$E'_3 = -a_3 A^{2/3}(1 + \tfrac{2}{5}\varepsilon^2) \tag{9.2}$$

There is a similar problem for the repulsive coulomb energy. The electrostatic energy is given by integrating the term for the interaction between two volume elements dv_1 and dv_2, separated by a distance r_{12}, over the entire volume of the nucleus. Thus

$$E_c = -\tfrac{1}{2}\rho^2 \int r_{12}^{-1} dv_1 dv_2$$

where ρ is the charge density, assumed uniform. While this reduces to E_2 in the case of a spherical distribution, for the ellipsoid it becomes

$$E_c = -\frac{3}{10} \frac{Z^2 e^2}{(a^2 - b^2)^{1/2}} \ln \left\{ \frac{a + (a^2 - b^2)^{1/2}}{a - (a^2 - b^2)^{1/2}} \right\}$$

Once again, expanding this and restricting the expansion to second order in ε, and using the notation leading to equation (8.5), we find

$$E'_2 = -a_2 Z^2 A^{-1/3}(1 - \varepsilon^2/5) \tag{9.3}$$

Combining equations (9.1), (9.2) and (9.3) leads to the expression for the change in energy due to a deformation of the nucleus

$$\Delta E = \frac{1}{5}(2a_3 A^{2/3} - a_2 Z^2 A^{-1/3})\varepsilon^2 \tag{9.4}$$

244

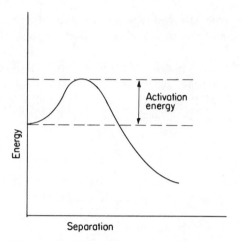

Figure 9.7. Activation energy and potential
barrier opposing spontaneous fission

If ΔE is positive, then the nucleus is stable, and will relax back to its spherical state. If ΔE is negative, then the distorted nucleus has a lower energy than the spherical one, and distortion will, therefore, increase, making the nucleus unstable. Substitution of the values of a_2 and a_3 from Chapter 8 ($a_2 = 0.00075$, $a_3 = 0.0185$) shows that for $Z^2/A > 49$, the perturbed nucleus is unstable.

Even if this calculation shows the nucleus to be stable, it is still true that for a heavy nucleus, the splitting into two smaller parts is energetically advantageous. This means that the stability predicted by this calculation is true only for small ellipsoidal disturbances. It also means that the potential energy minimum in which the nucleus sits is a false minimum and that the potential energy curve for the separation of the two component nuclei which would be involved in the fission is as shown in Figure 9.7. This figure shows that there is an energy barrier opposing fission for such nuclei and that if the nucleus is given energy greater than the height of this barrier, fission will become a probable process. The height of this barrier is called the *activation energy*. Even if sufficient energy is not supplied to the nucleus, it is still possible that fission will occur spontaneously through the quantum mechanical process of *tunnelling* or *barrier penetration*. In general this can be taken as a somewhat improbable process for most of the materials of interest to the nuclear power industry. In the case of U^{238}, for example, there are about 25 spontaneous fissions per gram per hour, and the half-life is about 10^{17} years.

The activation energies for a range of heavy nuclides are shown in Table 9.3, together with the excitation energies caused when the last neutron was added. This can be calculated by evaluating the binding energy of the last neutron from the mass data. For example, if a neutron is added to U^{235}, then the final mass will be that of the stable U^{236} nucleus. This is 236.0457 amu. The mass of the U^{235} nucleus is 235.0439 amu, and that of the neutron is 1.0086654 amu, so that their sum is 236.0525654 amu. The mass difference is

Table 9.3 Activation energies for heavy nuclides

	Activation (MeV)	Neutron binding energy (MeV)
Th^{232}	5·9	5·1
U^{233}	5·5	6·7
U^{235}	5·8	6·4
U^{238}	5·9	4·8
Pu^{239}	5·5	6·4

0·00687 amu which corresponds to an energy difference of about 6·4 MeV. We know, therefore, that, if a thermal neutron (one with almost zero kinetic energy) combines with a U^{235} nucleus, the resultant U^{236} nucleus will be formed in an excited state some 6·4 MeV above its ground state. The activation energy for U^{236} is shown in Table 9.3 to be somewhat less than this, and so the nucleus is free to deform and undergo fission; fission becomes a probable process.

The values of the excitation energies shown in Table 9.3 clearly reflect the importance of the odd–even term in the binding energy. Those nuclei with an odd number of neutrons—U^{233}, U^{235} and Pu^{239}—have a much greater excitation energy than those with an even number of neutrons. Fission with thermal neutrons occurs much more often with nuclei possessing an odd number of neutrons, and this is reflected in Table 9.1 which showed the fission cross-sections. Specifically, the comparison between U^{235} and U^{238} is important. The excitation energy of U^{235} is greater than the activation energy of U^{236}. As a result, U^{235} readily undergoes fission with thermal neutrons. By contrast, the excitation energy of U^{238} is less than the activation energy of U^{239} by about 0·6 MeV. Consequently, U^{238} does not undergo fission when bombarded by thermal neutrons. It will undergo fission, however, when bombarded by fast neutrons of energy greater than 0·6 MeV.

This discussion illustrates the way in which many of the important features of fission can be accounted for by the liquid drop model of the nucleus. There is a good qualitative picture of the process and sometimes the predictions are quantitative as well. Unfortunately there are some serious defects to the theory. It does not predict the correct photofission thresholds ((γ, fission) reactions), or spontaneous fission rates, and even more damaging, it predicts that the most likely mode of fission is the symmetric one in which both fragments have the same mass. Nonetheless, its superb pictorial representation of the process and general qualitative agreement with experiment make it the most popular model for general use.

9.3 Power from Nuclear Fission

It was the enormous release of energy in the fission process that led to both the idea that it might be used for electricity generation, and that it could be used to build a bomb. If it were possible to induce many nuclei to undergo fission within a very short time, then one would have the conditions necessary for an explosion. If one could induce a much smaller number to undergo fission each second, but ensure that the number doing so could be maintained and controlled, then it might be possible to extract power on a continuous basis. In this way nuclear power could be used for peaceful purposes.

Fortunately, the emission of further neutrons as part of the fission process provides a way of achieving these aims. For the bomb, the neutrons from the first fission are encouraged to cause as many further fissions as possible as quickly as possible, so producing energy and further neutrons. These in turn are encouraged to cause fission and the chain reaction so produced avalanches, producing greater and greater amounts of energy at each step. The time between fissions is short, and the build up of kinetic energy proceeds exponentially with a time constant of about 10^{-8} seconds. As a result, a large fraction of the available energy is released within a microsecond. From this stage, the nuclear and chemical explosive bombs behave in the same way, the explosion resulting from the failure of the bomb casing under the effects of excessive temperature and pressure. The radiation effects are due to the prompt emission of radiation and to the fission fragments and a few other neutron induced reactions. They are, to a certain extent, divorced from the actual explosion.

The same chain reaction can be used to maintain a steady rate of consumption of the fissile material, and to produce power, provided that steps are taken to limit the number of neutrons available for inducing fission. This is done by introducing suitable absorbers into the core of the reactor so as to remove some of the neutrons and leave only a sufficient number for the reaction to proceed at a uniform rate.

As soon as one starts to consider this type of control of nuclear fission, many factors become important. Which is the best fuel to use? How best would one extract the energy? How critical are the control problems? These and many other questions must be answered before a nuclear power plant can become a reality. The question about the best fuel, for example, is very interesting in itself. The readily fissile isotope of uranium is U^{235}. U^{238} will undergo fission by fast neutrons, but under the conditions pertaining in a nuclear reactor, this is one of the less likely possibilities. U^{238} is much more likely to absorb neutrons as they slow down by collisions with the nuclei in the core, to form Pu^{239} which is itself a fissile material. Because of the existence of several excited states of the U^{238} nucleus, there is a high probability that a neutron will be absorbed by resonant absorption when its energy is close to that of one of these levels. This means that as a fission fuel U^{238} is useless. Unfortunately U^{235} forms only 0·7 per cent of naturally occurring uranium. It is necessary, therefore, that either the 235 isotope be separated from the 238 or that some way be found of ensuring that the controlled fission process can continue despite the existence of a large component of U^{238} in the core. In particular, it would be useful if a reactor could be made to operate using natural uranium as fuel.

What are the various factors that influence such operation?

The dominant consideration for this operation is the high resonant absorption cross-section of U^{238} as the neutrons slow down in the core. The only solution to this problem is to ensure that they slow down as quickly as possible so that they spend as little time as possible with energies near the resonant values. The discussion of moderation in Chapter 8 showed that the best moderator for achieving this was water. Unfortunately Table 8.6 also showed that the high neutron absorption cross-section of hydrogen makes water less useful and that neutrons will be lost to the moderator if it is used. This can be compensated for by using *enriched* fuel, that is fuel in which the ratio of U^{235} to U^{238} has been artificially increased. Alternatively, one can accept somewhat less efficient moderation with higher losses to the U^{238} The two most common moderators of this type, both of which allow operation of the core with natural uranium as fuel, are heavy water and graphite. Of these heavy water is by far the better moderator, but it is very expensive compared to graphite.

We can now set up a rough model of the processes involved in sustaining a controlled nuclear fission reactor. This is shown in Figure 9.8. Suppose that initially a nucleus undergoes fission producing n fast neutrons. Of these, some will directly induce fission in U^{238} nuclei to produce some extra energy and an extra fraction of neutrons known as the *fast fission fraction* ε. There are now $n\varepsilon$ fast neutrons in the core of the reactor. These neutrons diffuse through the pile, slowing down by collisions with the nuclei of the moderator, the uranium fuel and the supporting structure. A fraction l_f will escape from the reactor before they are slowed down leaving $n\varepsilon(1 - l_f)$ neutrons which can maintain the reaction. Some of these will be absorbed by resonant absorption in U^{238} to produce the reaction chain

$$U^{238} + n^1 \longrightarrow U^{239} \xrightarrow{\beta^-} Np^{239} \xrightarrow[4\cdot2\ \text{day}]{\beta^-} Pu^{239}$$

Since Pu^{239} is itself fissile this does not represent a loss to the amount of nuclear energy available from the fuel, and in fact it constitutes a slight gain. It does, however, represent an immediate loss to the neutron balance resulting from our initial fission. There are now $n\varepsilon(1 - l_f)p$ neutrons available if a fraction p escape this fate. These are the neutrons that reach thermal energies. Some of them will diffuse through the structure and escape, leaving a fraction $(1 - l_t)$, while a further fraction $(1 - f)$ will be lost by absorption in the structure, the moderator and the coolant. This leaves a number of neutrons $n\varepsilon(1 - l_f)p(1 - l_t)f$ that may induce fissions. The number of fissions induced is limited by the relative probabilities of fission and absorption of thermal neutrons by the fuel itself. This is given by $\eta = n\sigma_f(U)/\sigma_a(U)$ which represents the number of fast fission neutrons produced per thermal neutron *absorbed* in uranium (that is the number of neutrons produced per fission multiplied by the relative probability of fission to absorption). The number of second generation fissions caused by one first generation neutron is called the *reproduction factor* or *multiplication factor*, and is, therefore, given by

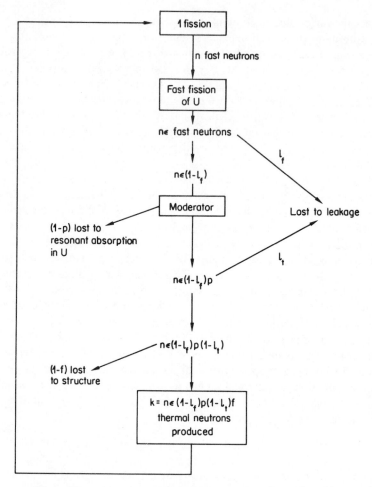

Figure 9.8. Processes controlling neutron population in a nuclear reactor

$$k = \eta \varepsilon p f (1 - l_f)(1 - l_t)$$

If $k > 1$, then the fission chain will avalanche and the pile will eventually melt. If $k < 1$, there will be insufficient neutrons to sustain the chain reaction. If $k = 1$, the reaction is in a steady state, and is said to be *critical*.

If we assume that the reactor core is infinitely large, then there can be no leakage of neutrons from the system, making l_f and l_t both zero. The equation for the multiplication factor then becomes

$$k_\infty = \eta \varepsilon p f \qquad (9.5)$$

This formula is called the *four-factor formula* and is important in the practical design of reactors. η depends on the nuclear properties of the fuel. ε depends on both the fuel and the size and shape of the reactor core. The other two parameters p and f depend on the fuel and also on the nuclear properties and geometric arrangement of moderator, coolant, structural materials and any other material that may be present in the core. One of the basic design considerations is to optimize these various factors so as to yield the largest possible value of pf. For a given fuel, this produces the largest value of the infinite multiplication factor, k_∞.

9.4 Reactor Criticality

Equation (8.19) can be used to give an appreciation of the concept of criticality in a nuclear reactor. If this equation is extended to three dimensions, it can immediately be written as

$$\nabla^2 n - n/L^2 = 0 \qquad (9.6)$$

We can make the simplifying assumption that our reactor has a neutron flux which is entirely thermal—there are no fast neutrons. Under this assumption the effect of the multiplication factor k can be introduced by rewriting equation (9.6) as

$$\nabla^2 n + (k - 1)n/L^2 = 0 \qquad (9.7)$$

This reflects the loss of n neutrons to the fission process, and the subsequent creation of kn more. The boundary condition can be taken as $n = 0$ on the edge of the pile, so producing the conditions of the four-factor formula with no neutron leakage—again a simplifying assumption.

Equation (9.7) now represents a simple eigenvalue problem, with the lowest eigenvalue corresponding to the criticality condition. If we consider the core as being a cube of side a, the solutions are given by

$$n = C \sin\left(\frac{\pi x}{a}\right) \sin\left(\frac{\pi y}{a}\right) \sin\left(\frac{\pi z}{a}\right)$$

where

$$\frac{3\pi^2}{a^2} = (k - 1)/L^2 \qquad (9.8)$$

The quantity $(\nabla^2 n)/n$ is known as the *material buckling* of the region of the reactor, and the term *buckling* arises because the function represents the curvature of n in space. Corresponding to the material buckling is the *geometric buckling*, which relates the material buckling to the geometry of the particular reactor being considered. In this case it is given by $3\pi^2/a^2$. One representation of the critical equation is that the material buckling of the reactor system should equal the geometric buckling of the particular shape in which it is built. Geometric bucklings and flux distributions for some reactor geometries are shown in Table 9.4.

Equation (9.8) gives an immediate estimate of the critical size of a reactor with a given multiplication factor k. For example, if $k = 1.1$, and $L^2 = 350 \text{ cm}^2$,

Table 9.4 Geometric bucklings for some possible reactor shapes

	Flux distribution	Geometric buckling
Sphere Radius R	$\dfrac{1}{r}\sin\dfrac{\pi r}{R}$	$\left(\dfrac{\pi}{R}\right)^2$
Cylinder Radius R Height H	$\dfrac{2\cdot405}{R}\,r\cos\dfrac{\pi z}{H}$	$\left(\dfrac{2\cdot405}{R}\right)^2+\left(\dfrac{\pi}{H}\right)^2$
Rectangular block Sides a, b, c.	$\cos\left(\dfrac{\pi x}{a}\right)\cos\left(\dfrac{\pi y}{b}\right)\cos\left(\dfrac{\pi z}{c}\right)$	$\pi^2\left(\dfrac{1}{a^2}+\dfrac{1}{b^2}+\dfrac{1}{c^2}\right)$

this result gives an estimate of the minimum size for our cubic reactor to be critical as 3·22 m. Unfortunately the assumption that all the neutrons are thermal is unrealistic, and must be replaced.

A rather better model can be formed by reformulating the problem slightly. At any point in the pile the increase in the thermal neutron density per unit volume per second will be controlled by three effects, diffusion, the production of new thermal neutrons by the slowing down process and absorption. The assumption is again made that no neutrons are lost through leakage from the reactor core. Since the diffusion of neutrons into a volume in unit time is represented by $D\nabla^2 n$, the slowing down contribution can be represented by a source, q, the loss through absorption as $-n/T$ where $T = \lambda_a/v$ is the mean lifetime of a thermal neutron, and the expression for the increase in thermal neutron density with time becomes

$$\frac{\partial n}{\partial t} = D\nabla^2 n + q - n/T \tag{9.9}$$

It is now necessary to find some form for q. This depends entirely on the model chosen to represent the moderation process, and a common choice is the so-called *age-diffusion method*, which assumes that the neutrons are slowed down in a continuous process. If we assume that no neutrons are lost to the moderating process by absorption, then the change in the slowing down density over an energy interval dE at a given point in the medium, may be set at the rate at which neutrons of energy E diffuse away from that point. Thus

$$\frac{\partial q(E)}{\partial E}\,dE = -D(E)\nabla^2 n(E)\,dE$$

or

$$\frac{\partial q(E)}{\partial E} = -D(E)\nabla^2 n(E) \tag{9.10}$$

where $n(E)$ represents the neutron density with energy E. Now in the steady state, the neutron balance requires that the rate of scattering out of the interval dE shall be the same as the rate of scattering into it, neglecting absorption; hence

$$\xi \Sigma_s(E) \, n(E) \, dE = \frac{q(E)}{E} \, dE$$

following the discussion of moderation in Chapter 8. If this is substituted into equation (9.10), it appears that

$$\nabla^2 q = -\frac{\xi \Sigma_s E}{D(E)} \frac{\partial q}{\partial E} \qquad (9.11)$$

If a new variable $\tau(E)$ is introduced, defined by

$$\tau(E) = \int_{E_0}^{E} \frac{D(E)}{\xi \Sigma_s E} \, dE$$

where E_0 is the energy of the source neutrons, equation (9.11) can be rewritten as

$$\nabla^2 q = \frac{\partial q}{\partial \tau} \qquad (9.12)$$

This is known as the *Fermi-age equation*, and $\tau(E)$ is called the *Fermi age*. It is not a unit of time, but rather a length squared. Although τ is not a unit of time it nevertheless relates to the chronological age of the neutrons as the time between the creation of the neutron and its reaching the energy E. As the neutron slows down, its age increases.

A solution of immediate interest for the Fermi-age equation is that for a monoenergetic point source producing fast neutrons which undergo continuous slowing down in a non-absorbing medium. In this case, q is a function of space coordinates and age and is given by

$$q(r,\tau) = \frac{\exp(-r^2/4\tau)}{(4\pi\tau)^{3/2}}$$

where $q(r,\tau)$ is the slowing down density for neutrons at age τ at a distance r from a point source emitting one neutron per second. This can be used as the distribution function in calculating the mean square slowing down distance, $<r_s^2>$, about a point source, as

$$<r_s^2> = \frac{\int_0^\infty r^2 (4\pi r^2 q(r,\tau)) \, dr}{\int_0^\infty 4\pi r^2 q(r,\tau) \, dr}$$

which becomes

$$<r_s^2> = \frac{\int_0^\infty r^4 \exp(-r^2/4\tau) \, dr}{\int_0^\infty r^2 \exp(-r^2/4\tau) \, dr} = 6\tau$$

Thus the Fermi age is one-sixth of the mean square distance between the point of origin of the neutron, where its age is zero, and the point at which its age is τ.

Equation (9.12) can now be used to determine the q term in equation (9.9). It is separable into two terms, one depending on the geometry of the reactor core, and the boundary conditions, and the other determined by the age. Thus with the assumption of zero leakage of neutrons from our cubic reactor core, equation (9.12) has a solution

$$q(x,y,z,t) = Q\Theta(\tau) \sin\left(\frac{\pi x}{a}\right) \sin\left(\frac{\pi y}{a}\right) \sin\left(\frac{\pi z}{a}\right) \tag{9.13}$$

where Q is a constant ensuring that at $\tau = 0$, q becomes the number of fission neutrons generated in the pile. Substituting into equation (9.12) and setting $\tau = 0$, when Θ must take the value 1, leads to

$$\frac{d\Theta}{d\tau} = -(3\pi^2/a^2)\Theta$$

which has the solution

$$\Theta(\tau) = \exp(-B^2\tau)$$

where $B^2 = 3\pi^2/a^2$.

It is also possible to determine Q from the known density of thermal neutrons using the four-factor formula. If there are n/T thermal neutrons absorbed, then nf/T of these are absorbed by the fuel. These will lead to $nf\eta\varepsilon/T$ fission neutrons. Since the infinite multiplication factor is $k_\infty = \eta\varepsilon pf$, where p is the resonance escape probability (assumed here to be unity), it follows that k_∞/p fission neutrons are produced for each thermal neutron absorbed. Therefore, $Q = k_\infty n/(pT)$.

Equation (9.9) can now be written as

$$\frac{\partial n}{\partial t} = D\nabla^2 n - \frac{n}{T} + \frac{k_\infty n}{pT} \exp(-B^2\tau) \tag{9.14}$$

In the steady state, $\partial n/\partial t = 0$, and for a critical reactor we can write $B^2 = B_c^2$. If, in addition, we substitute for T (given above), D and L (given in equations (8.18) and (8.20)), then equation (9.14) becomes

$$\nabla^2 n + B_c^2 n = 0$$

where

$$B_c^2 = \frac{k_\infty \exp(-B_c^2\tau) - 1}{L^2}$$

and τ is the Fermi age of thermal neutrons. This expression can be rewritten as

$$\frac{k_\infty \exp(-B_c^2\tau)}{1 + L^2 B_c^2} = 1 \tag{9.15}$$

which is the age-diffusion critical equation for a bare reactor. For a finite reactor of geometric buckling B, the corresponding expression for the left-hand side of equation (9.15) is called the *effective multiplication constant*, and is written as k_{eff}.

It has been assumed that the neutrons all have the same fission energy throughout this derivation. However, equation (9.15) is applicable in the general case, provided that the τ used is calculated as the average value of the Fermi age of thermal neutrons, properly weighted to allow for the fission spectrum. If the exponential is close to unity, equation (9.15) can be expanded to yield

$$k = 1 + \frac{3\pi^2}{a^2}(L^2 + \tau) \qquad (9.16)$$

where the quantity $(L^2 + \tau)^{1/2} = M$ is known as the *migration length*, and represents the distance travelled by a neutron with initial energy of 2 MeV before it is absorbed. The Fermi ages of thermal neutrons for various moderators are shown in Table 9.5.

Table 9.5 Fermi ages of thermal neutrons in various moderators

	Age (cm^2)
Water	31
Heavy water	120
Beryllium	85
Graphite	350

9.5 Thermal Reactors

Let us now consider three possible reactor assemblies: the homogeneous uranium–graphite reactor, the heterogeneous uranium–graphite reactor and the water-moderated slightly enriched uranium reactor. In the first, the fuel is distributed as evenly as possible throughout the core, while in the second it is distributed in lumps. These two examples also illustrate the way in which the various factors of importance can be extracted from the cross-section data. The third is the common American reactor type, and illustrates some of the difficulties involved in calculations on reactor mechanisms.

Homogeneous Uranium–Graphite Reactor

Consider an infinite pile for which the four-factor formula is applicable. The value of η for natural uranium and thermal neutrons is obtained from the expression $\eta = n\sigma_f(U)/\sigma_a(U)$, and n has already been given as 2·47. Table 9.1 gives $\sigma_f = 4\cdot18$ barns, and $\sigma_a = 7\cdot68$ barns as it is the sum of the fission and radiative capture cross-sections. These figures give a value of $\eta = 1\cdot344$. For a homogeneous reactor, the uranium particles are so small that it is unlikely that a collision will occur between a fast neutron and a U^{238} nucleus before the neutron leaves the particle. Once out, it will be moderated to thermal speeds before encountering another uranium particle, making fast fission of U^{238} unlikely. Consequently, it is reasonable to take $\varepsilon = 1.0$.

A chain reaction will be possible provided $k_\infty > 1$. Thus, from the four-factor formula, $pf > (1/\eta\varepsilon) = 0.744$. The only variable open to us in the homogeneous reactor (apart from enrichment of the fuel) is the ratio of uranium to moderator atoms, so we must consider its effect on the values of p and f. The fraction f, the thermal utilization factor, is the fraction of the available neutrons that is absorbed by the uranium rather than lost to the moderator and structure. As a consequence it can be simply expressed as

$$f = \frac{N_u \sigma_{au}}{N_u \sigma_{au} + N_g \sigma_{ag}} = \frac{1}{1 + \dfrac{N_g}{N_u} \dfrac{\sigma_{ag}}{\sigma_{au}}}$$

The subscripts g and u refer to graphite and uranium, respectively, and the N is the number of atoms per unit volume. The values of the reaction cross-sections are given in Tables 8.6 and 9.1 and f can be written as

$$f = \frac{1}{1 + 0.000573(N_g/N_u)}$$

As is to be expected, the thermal utilization factor decreases as the ratio of graphite to uranium increases, reflecting the greater loss of neutrons to the moderator. A graph of f against (N_g/N_u) is shown in Figure 9.9 where it can

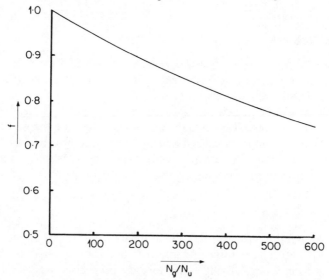

Figure 9.9. Dependence of thermal utilization factor on ratio of graphite atoms in the homogeneous uranium—graphite reator

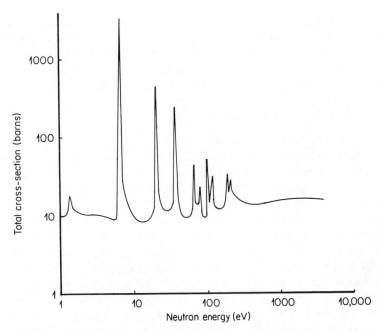

Figure 9.10. Variation with neutron energy of absorption cross-section in uranium

be seen that, even if $p = 1$, no chain reaction will be possible for values of $N_g/N_u > 500$.

The resonant escape probability, p, does not lend itself to ready calculation. Since the absorption takes place over a range of energies as the neutron slows down, p is related to the integral of the absorption cross-section over all possible energies. This absorption cross-section varies in a very complicated way as shown in Figure 9.10. The integral has been evaluated experimentally for various N_g/N_u, and some values of p, f and k_∞ are shown in Table 9.6. It is clear

Table 9.6 Dependence of multiplication factors on composition of reactor

N_g/N_u	p	f	k_∞
200	0·64	0·92	0·79
300	0·70	0·89	0·83
400	0·74	0·86	0·85
500	0·77	0·82	0·85
600	0·78	0·80	0·84

that, while k_∞ takes on its maximum value for N_g/N_u between 400 and 500, this value is only about 0·85. As a result, the homogeneous uranium–graphite chain reaction is impossible with natural uranium as the fuel.

If the fuel is enriched in U^{235}, the neutron yield per fission will be increased as will the absorption cross-section for thermal neutrons. Consequently p increases rapidly with enrichment, while f increases more slowly. For $N_g/N_u \approx 400$ the value of k_∞ exceeds unity at about twice the normal concentration of of U^{235}. Thus, a chain reaction can be sustained in a homogeneous uranium–graphite reactor provided this degree of enrichment of the fuel is achieved.

Heterogeneous Uranium–Graphite Reactor

The other approach to achieving criticality in a uranium graphite reactor using natural uranium as fuel is to arrange the fuel in the core in large lumps. This has a pronounced effect on many of the parameters of importance. For example, the resonant escape probability is increased because the U^{238} nuclei in the centre of a lump have a much smaller probability of resonant absorption than do those at the surface. The reason for this is as follows. Fast fission neutrons created in the fuel lump escape into the moderator and are slowed down gradually. When in the vicinity of another lump, those neutrons with energies near to resonant levels in U^{238} may be absorbed by the nuclei near the surface. Once inside the energy lost per collision is small making resonant absorption of neutrons entering the body of the lump unlikely. After escaping to the moderator once more, the neutrons will undergo further slowing down and will probably pass through several resonant absorption maxima before encountering another lump of fuel. The resonant escape probability will, therefore, be greater than in a homogeneous assembly of the same constitution. The larger the lump, the greater will be the proportion of fuel shielded from resonant absorption and so, other things being equal, the resonant escape probability will increase with diameter if the lumps are in the form of rods.

The heterogeneous reactor also leads to an increase in the fast–fission factor, because immediately after the fission there is a body of U^{238}-rich fuel to be traversed before moderation begins to any extent. In addition, the neutrons emitted from the fast-fission component themselves lead to an increase in the number of fast fissions. Unfortunately, in the heterogeneous reactor thermal neutrons are absorbed as they enter and penetrate the lump. This leads to a smaller flux at the centre of the lump and a decrease in the thermal utilization factor. However, this effect is more than compensated for by the increase in p and ε.

The choice of suitable lump size and separation, depends on the balance between the effects mentioned above. The actual calculation is complicated and is discussed fully in the references at the end of the chapter. However, as an example, consider a square array of parallel natural rods 2·5 cm in diameter and 11 cm apart, in a graphite block as moderator. In this case the parameters are approximately $\varepsilon = 1\cdot027$, $p = 0\cdot39$, $f = 0\cdot888$, $\eta = 1\cdot33$, $k = 1\cdot08$, $\tau = 350$ cm^2, $L^2 = 450$ cm^2, and the pile will be critical if it measures 5·5 m on the side.

Water-moderated, Slightly Enriched Uranium Reactor

This is a heterogeneous assembly, for which criticality calculations have proved to be difficult. The problem is that the ratio of moderator to fuel (N_w/N_u) is fairly low for a critical system, with the result that the lattice of fuel rods is closed-packed. The thermalization of the neutrons is, therefore, incomplete, and the reactor is not truly thermal. An important consequence is that resonant absorption in U^{235} has to be considered as well as in U^{238}. Also, hydrogen is an extremely good slowing down agent for neutrons with the result that both the Fermi—continuous slowing down—model and the two-group model (in which the neutrons are assumed to be either fast or thermal) are too inaccurate to be of any use.

It is possible to develop a modified form of the four-factor formula, however, which can be used to account for the design parameters of water-moderated reactors as long as it is appreciated that these cover a fairly limited range. The neutron cycle for this process is similar to that in the original four factor development, except for the extra x fission neutrons arising from the fission of U^{235} by neutrons which have epithermal energies (energies still slightly above the thermal range). The fast fission fraction will still be ε and mainly due to U^{238}, with the result that the absorption of one thermal neutron leads to the production of $(\eta f + x)\varepsilon$ fast-fission neutrons.

These neutrons now slow down and the process is arbitrarily divided into two parts; the first to an intermediate energy of 0·625 eV where there is a resonant absorption through radiative capture to U^{235}, and the second from 0·625 eV to thermal energy. The number of neutrons reaching 0·625 eV will be limited by the leakage of fast neutrons leaving a fraction $(1 - l_f)$, and p_8, the resonant escape probability of U^{238} down to this energy. Thus the fraction remaining is $(\eta f + x)\varepsilon(1 - l_f)p_8$. In slowing down from 0·625 eV to thermal speeds, a further fraction l_m will escape from the core by leakage, a fraction p_8' will escape resonant absorption in U^{238} and a fraction p_5 will escape absorption by U^{235}. Finally the fraction l_t of the thermal neutrons will escape from the core by leakage. The final multiplication factor is, therefore, given by

$$k = (\eta f + x)\varepsilon(1 - l_f)p_8 p_8'(1 - l_m)p_5(1 - l_t)$$

Since p_5 is the capture escape probability for U^{235}, the fraction of the neutrons reaching 0·625 eV which are absorbed is $(1 - p_5)$. Thus, if η_5 is the number of neutrons produced by fission of U^{235} per epithermal neutron absorbed, we can write the number of fission neutrons produced by this process as

$$x = \eta_5(1 - p_5)(\eta f + x)(1 - l_f)p_8\varepsilon$$

whence

$$k = \eta f \varepsilon p(1 - l_f)(1 - l_m)(1 - l_t)\beta$$

Here $p = p_8 p_8'$ represents the total resonance escape probability for U^{238}, and

$$\beta = \frac{p_5}{1 - \varepsilon(1 - l_f)p_8\eta_5(1 - p_5)}$$

If there is no leakage, this reduces to

$$k = \eta f \varepsilon p \beta$$

258

It is clear that β represents the contribution of resonance fissions in U^{235} to the neutron multiplication. It typically has a value slightly greater than unity, say 1·03.

9.6 Neutron Reflection

The expressions so far produced for reactor assemblies are for bare reactors. Normally the core of a reactor is surrounded by a blanket of moderator which acts as a reflector and redirects some thermal neutrons back into the core. This reduces leakage and decreases the minimum size of a critical assembly. There will also be an increase in the power delivered by a given mass of fuel, because, due to the reflector, the flux of neutrons at the edges of the core will be increased. This increases the average neutron flux throughout the core and hence the power capability of the reactor.

The treatment of the neutron balance in a reflected reactor is much more complicated than for a bare reactor. Many of the fast neutrons which originally escaped from the core will be returned to it, and since the absorption and diffusion properties of the reflector blanket are undoubtedly different from those of the core, calculation of neutron losses is made more difficult. To this must be added the marked effect on the spatial distribution of the thermal neutrons as illustrated in Figure 9.11.

The effect of the reflector can be allowed for by plotting a graph of the *reflector saving* (a measure of the efficiency of the reflector in decreasing the diameter of a critical assembly) against the thickness of the reflector. The reflector saving is defined as $\delta = R_0 - R_c$ where R_0 is the radius of a critical bare reactor, and R_c is the radius of the core of the critical reflected reactor. If it is assumed that the neutrons involved are all thermal, then the relationship between δ and the thickness of the reflector depends on the diffusion length in the reflector material. It is apparent that neutrons which have travelled more than some finite distance, say $2L$, are unlikely to find their way back into the core. Conse-

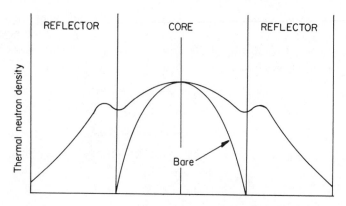

Figure 9.11 Effect of a reflector on the spatial distribution of neutrons in a reactor

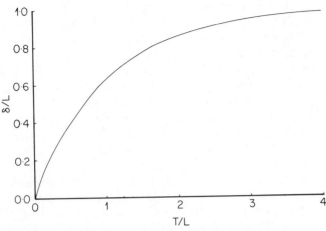

Figure 9.12. Variation of reflector savings with thickness of reflector. Both are expressed as a fraction of the diffusion length

quently it is convenient to plot δ/L versus T/L where T is the reflector thickness and L is the diffusion length. It can be seen from Figure 9.12 that there is little to be gained by increasing T beyond $2L$.

In graphite, the diffusion length of thermal neutrons is 57 cm. Therefore, if the heterogeneous uranium–graphite reactor discussed earlier is to be enclosed in a blanket 100 cm thick, the reactor saving will be about 50 cm. This has to come off both faces of the cube, so that the dimensions are reduced by 100 cm. The core size is therefore reduced to 4·5 m, and the critical mass of fuel drops from 50 tonnes to approximately $27\frac{1}{2}$ tonnes.

9.7 Fast Reactors and Breeding

There is one remaining class of reactor, the *fast reactor*, in which the neutrons are not thermalized. There is very little light material in the core, and the chain reaction is predominately sustained by neutrons having energies greater than 0·1 MeV. The main interest in these reactors lies in their ability to convert so-called *fertile* materials such as U^{238} into fissile fuel material with high efficiencies. Some fast fission parameters are shown in Table 9.7.

Table 9.7 Fission parameters for fast neutron fission

Parameter	U^{235}	U^{238}	Pu^{239}
σ_f	1·44	0·112	1·78
n	2·52	2·61	2·98
η	1·18	0·07	2·74

In the derivation of the four-factor formula, the reaction chain resulting from the resonant capture of a neutron by U^{238} was shown. This terminates in Pu^{239}, which is itself a fissile material and can be used as a fuel for a thermal fission reactor. In the natural or slightly enriched uranium thermal reactors, this conversion process represents a loss of neutrons and is, therefore, eliminated as much as possible. On the other hand, if a reactor is specifically designed to enhance these reactions it will produce more fuel while it is burning up its initial supply. In particular, if the reactor can be designed so that each fission of a fuel nucleus results in more than one conversion of a fertile nucleus to a new fissile nucleus, the possibility appears that the reactor will actually produce more fuel than it consumes.

Reactors are often classified according to the type of fissile-fertile reaction involved. *Converter* reactors are those which use one fuel and a fertile material to produce a different fuel. An example is the traditional U^{235}–U^{238} reactor which produces Pu^{239}. *Breeder* reactors are those which use a fuel which is reproduced by the neutron reactions in the fertile species. An example of this type is the plutonium-239 fuelled reactor with a blanket of U^{238} surrounding the core. Both categories of reactor can have conversion factors greater than, equal to or less than one, and are then referred to as having positive gain factor, as being self-sustaining or as having negative gain factor. The most highly favoured process for this fuel generation system is to use the fast reactors with plutonium fuel, surrounded by a blanket of uranium. There is a possibility, however, of developing a thermal reactor which can breed fuel. This is based on the other fissile isotope of uranium, U^{233}, and the absorption process in Th^{232}, to reproduce U^{233} after beta decay of the product nuclei. This process could well become important as thorium is relatively abundant, and also η, at 2·28, is higher than for U^{235}.

In a fast reactor core, few neutrons will be lost to resonant capture, and no thermal neutrons will escape by leakage. Since η neutrons are produced for each neutron absorbed in the fuel, and since one neutron is needed to maintain the chain reaction, $\eta - 1$ neutrons are available for any other processes, including conversion of fertile material. For thermal reactors, $\eta - 1$ is so close to unity that, after allowing for parasitic capture and leakage, only a very small fraction remains for conversion. In a fast reactor, on the other hand, $\eta - 1$ may be as large as 1·74 when plutonium-239 is used as fuel, so that breeding ratios of fissile nuclei produced to fissile nuclei consumed can be as high as 1·7. In fact, the actual breeding ratio may be higher than this because of the fast fission factor's contribution to the neutron balance. A typical fast breeder reactor would consist of a core containing about 25 per cent fissile material and about 75 per cent uranium-238, which is surrounded by a reflecting blanket of uranium, either natural or depleted. As a general rule the core is about 50 per cent fuel and 50 per cent coolant and structural materials.

With a thermal reactor, the calculations on criticality are simplified by the approximately Maxwellian neutron energy spectrum and by the straight-forward velocity dependence of the cross-sections. For the fast reactor, this is much more complicated. Not only is the neutron energy spectrum much wider, but it also varies according to the composition of the core. In addition, the cross-sections for fission radiative capture and other quantities required for the criticality calculations vary markedly with neutron energy. Consequently, allowance has to be made for these effects in the various calculations.

This treatment of reactors has concentrated on the condition under which a nuclear reactor can become critical, that is on *reactor statics*. The discussion has been brief and superficial. Much more detailed treatments can be found in the bibliography at the end of the chapter. Reactors themselves and the other important aspect of nuclear reactors—reactor control, or *reactor kinetics*—will be treated in Chapter 10 as will the dark side of nuclear power, radioactive waste and reactor safety.

Further Reading

Glasstone, S. and A. Sesonske, *Nuclear Reactor Engineering*, van Nostrand Reinhold Company Inc., New York, 1967.
Reactor Handbook, *2nd ed.*, Vol. IIIa, Physics, Soodak, H., (Ed.), Interscience Publishers, New York, 1962.
Salmon, A., *The Nuclear Reactor*, Longmans, London, 1964.
Segré, E., *Nuclei and Particles*, W. A. Benjamin Inc., New York, 1963.

10 Nuclear Power

It is one thing to know that nuclear fission can yield thermal or electrical power in useful amounts, but it is quite another thing to build a nuclear power station. As well as the problems of reactor criticality already discussed, there are numerous others that must be solved. The reactor must be controlled to remain just critical, heat must be transferred between the core and the boilers driving the turbines, and the fuel must be consumed efficiently. There are also safety problems. The reactor has to be shielded so that the intense radiation in its core does not endanger life in the vicinity. It must be ensured that an accident in the reactor cannot cause its core to melt and possibly penetrate either the walls or floor of the vessel, contaminating the outside area, and radioactive waste products must be safely stored. Finally there are the usual problems of large scale civil, electrical and mechanical engineering.

None of these is trivial. Each carries a penalty in terms of design and building standards, and each must be adequately catered for before a nuclear power station can be built and operated. Some of the difficulties are the standard ones which confront any large scale engineering project, while others are specific to the nuclear power industry. There is a third category of those difficulties that would exist in a conventional power station, but are 'exacerbated by the stringencies of nuclear power. For example there is always a problem in transferring heat from the furnace to the turbines in the form of steam. In nuclear power stations this is made much more difficult because the high temperatures make the material of the heat exchangers more susceptible to radiation damage, and also because there is the possibility that the reactor coolant will become contaminated by radioactive materials. This must not be allowed to leak out, and preferably, it should not be permitted to reach the turbines themselves.

There has been considerable controversy over the safety of nuclear reactors in general, and the American light water reactors in particular. Much of this criticism has been somewhat hysterical, but throughout the entire discussion, two points are repeatedly made which are of fundamental importance to the future of nuclear power. First, while there has not been a *serious* accident involving a nuclear power station, and while the accident prevention procedures of the whole nuclear power industry are painstaking and thorough, it is nonetheless true that an accident is possible. When it occurs it could have serious consequences and it is not known if the safety systems will prevent the accident from occurring. The situation under which the important safety systems will fail has not yet been discovered. In fact, a detailed report, prepared for the

United States Atomic Energy Commission (Rasmussen, 1974), shows that with the information now available, the probability of an accident to a water-cooled reactor which would involve ten deaths or more is one in 250,000 per year per plant operating. The likelihood of a core melt-down is estimated as one in 17,000 per plant per year. Second, the problem of radioactive waste disposal is a serious one. While the actual volume produced by a power station is surprisingly small, this waste material has to be stored, guarded and treated with loving care and attention for considerable periods of time—up to 500 years.

10.1 Nuclear Power Reactors

Leaving aside the fast breeder reactors for the moment, as these are not yet fully developed as commercial power producers, nuclear power reactors are classified according to their operating philosophy, and moderating material. The American industry has concentrated almost entirely on light-water reactors for commercial power production; the moderator also acts as the coolant, and the fuel is natural uranium slightly enriched in U^{235} to compensate for the extra neutron absorption of light water. At the other end of the power reactor spectrum is the British philosophy of a natural uranium, graphite-moderated heterogeneous reactor, cooled by passing carbon dioxide gas through the core. Between these two lie the heavy water moderated reactors as favoured by Canada and recently recommended for use in future British nuclear power plants. Even within one reactor type, there are different design philosophies. These are sometimes dictated by safety considerations, and sometimes by the continual search for higher and higher upper temperatures to make the conversion to electricity thermodynamically more efficient. For example, the boiling-water reactor uses a single coolant cycle. The water in the moderator boils and is heated to high temperatures in extracting heat from the reactor core. It then circulates and drives the steam turbines directly, returning to the core of the reactor when cool again. This cycle has the undesirable feature that coolant which has been in the core itself, and is therefore liable to be contaminated, is circulating through the steam turbines, possibly escaping in the event of a leak and probably contaminating the turbine blades. The pressurized water reactor, on the other hand, uses a primary coolant cycle, and a heat exchanger in which the steam for driving the turbines is generated in a secondary coolant cycle away from the core. In this way there is neither the risk of contaminated water leaking from a steam turbine, nor the risk of long term build up of radioactivity inside the turbine. Several variants of the original British graphite moderated reactors have been proposed as part of the development to higher boiler temperatures. One of these is the Advanced Gas-cooled Reactor (AGR) which has been plagued with development troubles of one sort and another. By no means have all of these been reactor difficulties, but the cost of installing AGR stations has increased and there has been pressure within the United Kingdom to discontinue them. This is reminiscent of the discussions in the

264

early 1960's when the original Magnox reactors were having similar problems and there was pressure to discontinue them also. It is perhaps noteworthy that in 1972 the cheapest electricity in the United Kingdom was produced by these very reactors.

The approximate amount of nuclear fuel required for a given power output can be readily calculated. For example, to produce 500 MW years of electricity in a plant with 33 per cent efficiency, some 0·57 tonnes/yr of fissile fuel is needed in principle. In fact, because of the loss of fissile material through radiative capture of neutrons, this figure has to be increased by about 25 per cent to 0·7 tonnes/yr. If it is assumed that fuel can be completely used in the reactor, then this corresponds to a *natural* uranium consumption of 100 tonnes/yr. In fact, nuclear fuel cannot be completely consumed in the reactor because of distortion of the fuel rods and other effects to be discussed later, and a reasonable energy extraction for one charge of uranium metal fuel elements is of the order of 3000 MW days per tonne (thermal energy). In addition, the rate at which energy can be produced is limited by the rate at which heat can be removed from the core of the reactor. This puts a limit of about 3 MW/tonne (thermal) on the power of the plant, and so produces a figure of 500 tonnes/yr of uranium fuel being required for the reactor. If a figure of £30,000 is assumed for the cost per tonne of fuel assemblies, there is an investment of fifteen million pounds in fuel—a large sum. It becomes imperative that this be used efficiently

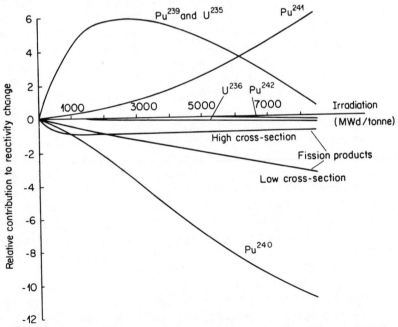

Figure 10.1 Contribution of various fission products (excluding Xe[135]) to reactivity of a natural uranium reactor as a function of total irradiation. (Salmon, 1964. Reproduced by permission of Methuen and Company Ltd.)

to extract all available energy from the fuel, and that either the irradiation time for the fuel is increased, or the fuel is recycled to remove fission fragments and manufacture new fuel elements, or preferably both. It is when viewed in this light that the breeder reactor, with its production of new fuel in a blanket around the core becomes even more attractive. However, even in the future many power stations will be driven by thermal reactors since the surplus fuel produced by the breeders must be consumed elsewhere. Not every reactor can be a breeder.

Suppose that the conversion factor C for a thermal reactor is 0·82. This is not an unreasonable factor and is approximately that of a large gas cooled graphite moderated reactor. If one U^{235} nucleus is consumed and replaced by C nuclei of Pu^{239}, then the lifetime of the reactor is extended, and although C is less than one, so that not all of the U^{238} is converted, there is still a considerable improvement in fuel economy. The conversion process leads to a chain reaction of conversions in that one initial fission produces C conversion nuclei which will undergo fission to produce C^2 nuclei and so on. Thus the total number of U^{238} atoms consumed for each U^{235} nucleus undergoing fission is $C + C^2 + C^3 + \ldots = C/(1 - C)$, and with $C = 0·82$, 4·55 new fissile nuclei will be produced per fission of an original nucleus. This results in 5·55 fissions per U^{235} nucleus, and gives a total usable fraction of the natural uranium fuel as 3·9 per cent. Unfortunately, this cannot be realized with present designs of fuel elements because of the distortion that would result. In addition to the metallurgical effects of excessive fuel burn-up, there are also limiting nuclear effects, which become apparent as a decrease in the available reactivity in the reactor. These nuclear effects are the build up of Pu^{240}, a non-fissile isotope, and the appearance of fission products within the reactor core. Some of these

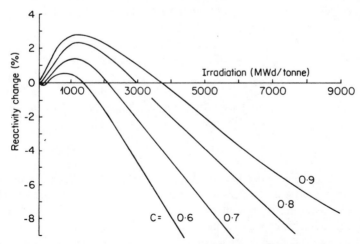

Figure 10.2. Change of reactivity of a natural uranium reactor as a function of irradiation for different values of the conversion factor C. (Salmon, 1964. Reproduced by permission of Methuen and Company Ltd.)

will have high neutron absorption cross-sections, and will therefore soak up more and more neutrons as the reactions continue. These limitations can only be overcome by inserting more fuel, or by removing the 'spent' fuel rods, which are now highly radioactive, and reprocessing them to remove the contaminants and reconstitute the uranium.

The effects of the various products of the fission process on the reactivity of the reactor are shown in Figure 10.1, where the unit of irradiation is the Megawatt Day per Tonne of fuel (MWd/T). Figure 10.2 shows the reactivity change with irradiation for a natural uranium reactor as a function of irradiation received, and for different values of initial conversion factor. This curve shows that for the natural uranium–graphite reactor discussed above, the 'poisons' have brought the reactivity back to its initial value after an irradiation of 3000 MWd/T.

10.2 Fuel Management

It is possible to extend the useful lifetime of fuel elements in the core of a reactor, and therefore to reduce fuel costs, by cycling the fuel around the core. This is possible because the rate at which fuel is exhausted is not constant but varies with the position of the fuel rod in the core of the reactor. (This is similar to the situation in which tyre wear can be reduced by cycling around the vehicle.) The variation of neutron density through the core was shown in Figure 9.12. Fuel burn-up will be faster in regions of high neutron flux. There are various schemes for cycling the fuel around the core, and these depend on reactor type and applications. Small reactors may not even cycle fuel at all, finding it cheaper and more convenient to run the system until it ceases to be critical and then to discharge and reload. On all the larger installations, however, some form of fuel cycling is essential. This aspect of reactor technology is known as *fuel management*. Some schemes in use are given below.

Centre to Outside Loading

Fresh fuel is loaded at the centre of the core, where the neutron density is highest, and is then progressively moved outwards. This produces the largest burn-up of any of the schemes, because the fresh fuel is being used where the neutron density is highest. There is the disadvantage, however, that the power density varies greatly across the core for the same reason.

Outside to Centre Loading

This is the reverse procedure in that the fresh fuel rods are introduced on the periphery of the core and are steadily moved in towards the centre. This produces a lower burn-up, but has the advantage that the power density is more evenly distributed across the core.

Bidirectional Loading

This is a system where the fuel is handled in short slugs, which are pushed through the core from one side to the other. Two complete sets of slugs are used, pushed in from opposite sides of the core. The rate at which the fuel is pushed through is determined by the rate of burn-up so that only spent fuel is discharged. The power density distribution for this type of system lies somewhere between the first two schemes.

Graded Irradiation

The core is divided into a number of regions, each of which contains fuel elements with varying degrees of irradiation. When any particular element has received its predetermined exposure, it is removed and replaced with an element of fresh fuel.

Axial Distribution

If fuel rods are simply moved around the core, there is no compensation for the variation in irradiation between the centre of the core and the top or bottom. It is possible to allow for this by dividing the fuel elements into two parts and switching these over at an appropriate time.

Two other possible schemes which do not involve movement of the fuel rods are *zonal core loading*, and *seed blanket loading*. In zonal core loading, the core is divided into radial zones which are charged with fuel of differing enrichments. The complete core remains in place for two to three years and is then replaced. In seed blanket loading, an annular ring of the core is *seeded* with fuel of very high enrichment and is surrounded on both inside and outside by blankets of natural uranium. This system produces a more even neutron flux across the centre of the core.

If natural uranium fuel elements are discharged after about 3000 MWd/T, there is still present about 60 per cent of the initial amount of U^{235}, and there is also a mixture of plutonium isotopes and fission products. If this is chemically processed to remove the fission products and possibly also the plutonium, the fuel can be used again. This type of cycling can be carried out on a continuous basis with the reactivity becoming more steady as an equilibrium distribution of the age of fuel rods is reached. After several cycles in this process, the uranium from which most of the U^{235} has been used becomes useless as a fuel and is stored to await the arrival of commercial fast fission breeder reactors. The overall burn-up before stock-piling is about 10,000 MWd/T.

10.3 Reactor Control

In practice it must be possible to start up a reactor from zero activity, let its power generation increase until it reaches the design level and then reduce the reproduction factor k so that the reactor becomes just critical at that power level. This is accomplished by incorporating *control rods*, of some material having a high absorption cross-section for thermal neutrons. By inserting or removing these rods the reproduction factor of the reactor can be reduced or increased so as to control its criticality.

This problem of reactor control is truly dynamic because the statistical behaviour of the fission process can easily produce excursions in the neutron density. If these are allowed to persist, then either the reactor will become supercritical or it will become subcritical. In addition, the build up of fission products and other poisons steadily changes the criticality conditions for the reactor. It is essential that these changes be compensated for at all times.

Control Theory

A neutron in a thermal pile spends most of its time in diffusion after moderation. Moderation may require a few microseconds while diffusion may last for several milliseconds. It is clear, therefore, that in any reasonable response time for control equipment there will be many generations of neutrons. Thus control of the pile would seem to require prohibitively fast action. Fortunately, not all the neutrons are prompt, and the delayed neutrons' contribution is important in controlling thermal neutron reactors.

If we neglect the contribution of the delayed neutrons, equations (9.9), (9.14) and (9.16) can be combined to give

$$DV^2n + \frac{D(k-1)}{M^2}\,n = \frac{\partial n}{\partial t} \qquad (10.1)$$

This expression is valid if the migration length M is short compared with the dimensions of the pile. Separating variables and using the analysis of Chapter 9 leads to

$$n = n(r)\exp{(t/T)} \qquad (10.2)$$

where

$$\frac{1}{T} = \frac{k-1-B^2M^2}{M^2/D} = \frac{k_{eff}^{-1}}{\tau_0} \qquad (10.3)$$

T is the *relaxation time* of the pile, and $\tau_0 = M^2/D$ is the mean life for the absorption of a neutron in the pile if $k = 0$. These equations show that the response of the neutron density in the pile to a change in the value of k is exponential with time. For a graphite moderated reactor, in which $\tau_0 \approx 10^{-3}$ seconds, even a value of k_{eff} as low as $1 \cdot 0005$ will lead to a doubling of the neutron density in only two seconds. This is too short for comfort.

The delayed neutrons completely change the complexion of the problem. If we assume that they form a fraction β of the total, and also that the multiplication factor k is the sum of k_p, due to the prompt neutrons only, and k_d due to the delayed neutrons only, we can write

$$k = k_p + k_d = k_p + \beta k$$

Thus

$$k_{\mathrm{p}} = k(1 - \beta)$$

If the reactor is already critical on prompt neutrons only, the delayed neutrons will have little effect on the criticality, and the doubling time will be determined by our earlier analysis. If, on the other hand, the reactor is sub-critical under the influence of the prompt neutrons, and relies on the delayed neutrons to become critical, the time constant of the pile is affected because these are produced with a different mean lifetime τ_{d}.

To allow for this, equation (10.1) must be replaced by

$$D\nabla^2 n + \frac{k_{\mathrm{p}} - 1}{\tau_0} + \frac{C}{\tau_{\mathrm{d}}} = \frac{\partial n}{\partial t}$$

where C is the density of *pregnant nuclei*, as the delayed neutron emitters are called. C is related to n by

$$\frac{\partial C}{\partial t} = \frac{k_{\mathrm{d}}}{\tau_0} n - \frac{C}{\tau_{\mathrm{d}}}$$

Solving these two coupled equations by assuming separable solutions, and exponential time terms, one can obtain a relationship between the period of the pile, T, and the so-called *excess reactivity* ρ,

$$\rho = \frac{k_{\mathrm{eff}} - 1}{k} = \frac{\tau_0}{kT} + \frac{\beta \tau_{\mathrm{d}}}{T + \tau_{\mathrm{d}}} \tag{10.4}$$

When more than one group of delayed neutrons is considered, an extra term of the same form as the last term on the right-hand side of equation (10.4) is added for each group. The excess reactivity is measured in *inhours* (inverse hours), the amount of excess reactivity that will produce a doubling period of one hour for the core.

For a given reactor fuel, the values of β, τ_{d}, etc., are all known, and so it is possible to define the reactor period in terms of the excess reactivity in inhours. As an example, consider a reactor with $\tau_0 = 10^{-3}$ sec, which is suddenly made 0·0022 more reactive. If there is only one group of delayed neutrons, which form a fraction 0·0065 of the total neutron flux, and if their lifetime is 10·25 seconds, then equation (10.4) leads to a period of 20·7 seconds. This should be compared with the period of 0·45 seconds when the delayed neutrons are absent. This is an oversimplified example, and in any real case the various delayed neutrons must all be accounted for. The effect of this is to lower the calculated period somewhat, but in this case it is still greater than 10 seconds.

When a reactor is critical on prompt neutrons alone, it is said to be *prompt critical*. This will occur when the reactivity is equal to the fraction of delayed neutrons. For U^{235} this value is 0·0065 and so the effective multiplication factor for a prompt critical thermal reactor with U^{235} as fuel is $k_{\mathrm{eff}} = 1·0065$. This is a very important condition as it is in this region that the effect of the delayed neutrons on the control of the reactor begins to decrease. As a result, special efforts must be made in reactor operation to ensure that this condition does not arise.

Fast reactors also have control problems of this type except that, because of the much shorter neutron lifetimes (10^{-7} sec), the difficulties are more acute. With this neutron lifetime, and an excess reactivity of 0·0001, the period of a fast reactor with no delayed neutrons would be $10^{-7}/10^{-4} = 0·001$ sec. This obviously renders any such reactor uncontrollable. If the delayed neutrons are included, however, the figure can be substantially increased, and for Pu^{239} it becomes 8 seconds.

Fission Product Poisoning

During the operation of a reactor, the fission products accumulate. Some of of these, notably Xe^{135} and Sm^{149}, have high neutron absorption cross-sections at thermal energies. As a result, their effect on the control of the reactor is pronounced, and they act as reactor poisons mainly by decreasing the thermal utilization factor. In fact, the fission products have little effect on the direct kinetics of the reactor because the contribution of any particular one is small, and the rate of change of fission-product concentration with time is much smaller than that of the neutron density. However, they may have an important effect on the reactivity, and this must be taken into account when designing both the reactor core and the control system.

One specific fission product is Xe^{135}, which is formed both directly and indirectly in fission chains, and which decays both by radioactive decay and by neutron absorption. The decay chain producing Xe^{135} is

$$Te^{135} \xrightarrow[1m]{\beta^-} I^{135} \xrightarrow[6·7h]{\beta^-} Xe^{135} \xrightarrow[9·2h]{\beta^-} Cs^{135} \xrightarrow[2 \times 10^6 yr]{\beta^-} Ba^{135} \text{ (stable)}$$

The thermal neutron capture cross-section of Xe^{135} is exceptionally large at $3·0 \times 10^6$ barns, and since the decay chain producing it occurs in about 6·1 per cent of fissions, it can be seen to be a very important poison.

Because the half-life of Xe^{135} is only 9·2 hours, and that of its parent I^{135} is 6·7 hours, it is clear that in any operating reactor, the xenon contamination will rise initially and reach an equilibrium value while the reactor is operating at a specified power level. After shut-down, however, the decay of the parent nucleus continues, but the most important method of removal is reduced because the neutron flux has decreased. As a consequence, the concentration of Xe^{135} will increase to a maximum some time after the reactor has been shut-down, and then decrease again as the Xe^{135} decays by beta emission to its caesium daughter. It is even possible that a reactor designed to be just critical with the equilibrium concentration of Xe^{135} appropriate to its rated power output, may be incapable of starting up again for several days after a shut-down— until the xenon poisoning has decayed sufficiently.

If the *poisoning* of a reactor is defined as the ratio of the number of thermal

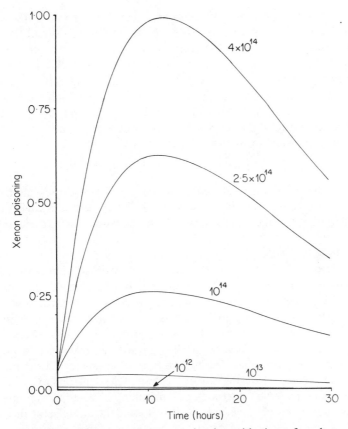

Figure 10.3. Variation of xenon poisoning with time after shut-down for different values of neutron flux

neutrons absorbed in the poison to the number absorbed in the fuel, then Figure 10.3 shows the effect of xenon poisoning after shut-down for a range of neutron fluxes. It is apparent that for low operating fluxes, the poisoning itself is low, and the rise after shut-down is negligible. However, at higher fluxes (above 10^{14} n cm^{-2} sec^{-1}), the increase is noticeable and at 4×10^{14} n cm^{-2} sec^{-1} it rises to a maximum of 1·0 after 11 hours. The obvious corollory is that a high flux reactor which is to be started up again soon after close-down must be designed with considerable built in excess reactivity, in the form of excess fuel.

The other poisons have a much smaller effect than Xe^{135}, and the largest of them, samarium-149, leads to a maximum decrease in reactivity of only 0·04 for a 2×10^{14} neutrons cm^{-2} sec^{-1} reactor. The other poisons taken together are roughly equivalent to the samarium.

Reactor Shut-down and Fission Product Power

A reactor is shut down by applying negative reactivity to it, that is by ensuring that more neutrons are absorbed than are created by fission. This results in a neutron density which is controlled by equations like (10.2) and (10.4). For a 1000 MW reactor shut down by a reactivity change of $-5\cdot5$ per cent, the neutron power will be as low as 35 MW after 10 seconds, and $2\cdot8$ MW after 100 seconds. Unfortunately, the fission products do not decay as quickly as the neutron density, and the fission product power is often an appreciable fraction of the total power of the reactor, especially after it has been running for some time. It has been found that, if a reactor has been operating at power P for time t_0, then the fission product power will be given by

$$P_d = 0\cdot07\, P\, (t^{-0\cdot2} - (t + t_0)^{-0\cdot2})$$

(Salmon, 1964). For our 1000 MW reactor, this gives 44 MW as the fission product power 10 seconds after shut-down, and even 28 hours after shut-down, the contribution to total reactor power from the fission products is still 7 MW.

Reactor Start-up and Control

The power level of a nuclear reactor is virtually proportional to the neutron flux. Consequently, control of the reactor multiplication constant ultimately controls the power level only by raising or lowering the neutron flux. Once the power has reached the desired level, k is reduced to unity. It is important to realize that the control of a reactor, especially at start-up, necessitates the measurement of two parameters. Both the power level and its rate of change with time must be monitored. The latter is the measurement of the multiplication constant since this controls the rate of increase of the neutron flux.

There are four possible methods of controlling a reactor; each involves the addition or removal of something be it fuel, moderator, reflector or a neutron absorber. Each of these methods, or a combination of them, has either been used for a reactor system, or has been proposed for one. The most common control system for thermal reactors is the use of *control rods*. These are composed of materials with high thermal or epithermal neutron absorption cross-sections, such as boron or cadmium, or, for water-cooled reactors, hafnium. It is essential that control-rod materials should be able to withstand prolonged neutron irradiation without undue change in their mechanical characteristics, and also, since they are being deliberately included so as to remove neutrons, considerable heat will be developed and they must, therefore, have good thermal conductivity. Cadmium has a very low melting point, 321°C, and is only suitable for low-temperature applications. Boron is usually used in the form of boron steel or boron carbide and aluminium. There is also the possibility of using rare earth oxides for high temperature applications. The chief disadvantage of using neutron absorbing control rods is that they constitute a waste of perfectly good neutrons. Were they not being absorbed in the control rods, they could be usefully employed generating power. This is an important aspect of power reactors with considerable excess reactivity. A variant is to use a control rod whose bottom portion is actually a fuel element so that as the control material is lowered into the core, some fuel is pushed out and vice versa. This combines the two approaches of removing fuel to reduce reactivity and removing neutrons. A third possibility is to make the control rods of some fertile material such as U^{238}, so that the absorbed neutrons perform some useful function and generate more fuel, or produce some desired isotope.

In a fast reactor, the neutron absorption cross-sections of most materials are so small that control by the use of absorbers is not feasible. Consequently, the procedure adopted here is the removal or addition of fuel, coupled with the movement of part of the reflector.

No matter which exact philosophy is used for controlling the reactor, we can still talk about *control rods*. These may be either absorber, fuel, reflector or moderator. Not all the control rods have the same function in order that the necessary range and accuracy of control can be achieved. Firstly, there are the *shim rods*, or coarse control rods, which are used to bring the reactor to approximately the desired power level. Because the rate of change of reactivity must be closely controlled during start-up, these must not be capable of moving at a high speed. Secondly, there are the *regulating rods* which are quite small, and capable of moving at high speed to counteract rapid transient changes in neutron flux. In general the reactivity equivalent of a regulating rod should be not greater than that of the delayed neutrons so that the inadvertent removal of a regulating rod cannot render the reactor prompt critical. Finally, there are the *safety rods*. These may be the shim rods but they must be capable of being dropped very quickly into the core to close the reactor down in the event of an accident. They must obviously have a high reactivity equivalent. In addition, most reactors have an extra back up feature in which boron steel shot or boric acid solution can be run into the reactor core in the event of control rods jamming.

The problems of control are most acute during start-up. This is because the reactor period can become quite short for a high-power reactor with high neutron flux and considerable excess reactivity. If this condition is allowed to persist through the prompt–critical stage, the consequences could be serious. The problem with the start up of a new reactor or one in which the fuel charge has just been replaced, is that there are so few neutrons available in the early stages of the start-up, that no measurements can be made. Large natural uranium reactors are the easiest to start-up in this respect because there are sufficient spontaneous fissions in U^{238} ($15 \, \text{kg}^{-1} \, \text{s}^{-1}$) that the resulting neutron density may be large enough to detect. In a fast reactor with no U^{238} in the core and with its physically much smaller core, it is necessary to include an artificial neutron source for start-up at least. The control rods are withdrawn very slowly, and the increase in neutron flux is monitored until it is apparent that the reactor has become critical. Thereafter the control follows a more normal pattern.

10.4 Energy Removal

One of the important, and unusual, features of nuclear reactors as sources of power, is that there is no upper limit to the power that can be generated. Theoretically, the neutron flux can be allowed to rise to any desired level and then held constant. Unfortunately, there is a limit to the rate at which the heat that is generated can be removed from the core and this determines the upper limit to the power rating of a reactor. In fact, this is such an important consideration that high flux reactor design is as much dependent on the heat-transfer characteristics as on the nuclear features. In order to increase the heat-transfer rate, both the contact area with the coolant and the coolant volume must be increased. This generally results in a reduction of the multiplication factor of the system so that any actual design must represent a compromise between the conflicting factors.

The problems of heat transfer are standard throughout the entire power industry. There are certain aspects of nuclear power generation, however, that present unusual problems. In a conventional power station, the upper temperature limit is fixed by the combustion temperature of the fuel, coal or oil. In a nuclear power station this is not so. The heat is generated by the slowing down of the fission fragments and other subatomic particles from extremely high speeds, with the associated extremely high temperatures. As a result, if the heat generated is not removed at the same rate as it is being created in the core, the temperature of the core will continue to rise until the reactor is destroyed.

There are several different heat sources in a reactor. These lead to different heat conduction problems, and each must be fully understood before the complete heat-transfer system can be produced. For example, the neutron flux is not uniform throughout a reactor core; this leads to differing fission rates and, therefore, differing heat-transfer problems in different parts of the reactor. The limit to the power that can be generated by such a reactor is determined by the problems associated with the worst part of the core, and not the average. As was shown in Table 9.2, of the 202 MeV released in fission, some 191 MeV are available as a source of heat. (The neutrinos all escape from the reactor and are, therefore, unavailable. Neutrinos constitute no health hazard as they are extremely unreactive. For example, they can pass through the Earth without interaction.) Of this 191 MeV some 175 MeV appears as the energy of fission fragments and associated beta particles. These have very short range in the fuel elements. For example, the fission fragments have a range of about 0·0125 mm and the beta particles can travel about 0·25 mm. As a result most of their energy is deposited there and very little anywhere else. The neutrons and gamma rays, on the other hand, lose their energy in the moderator or directly to the coolant. Thus there are two distinct components to the heat sources, one a diffuse heat source forming about 8·5 per cent of the total, and the remaining 91·5 per cent which is highly localized in the fuel elements themselves. In addition, the heat production will be more intense near the centre of the reactor.

There is a further complication in energy removal from a nuclear reactor. In conventional furnaces, the coolant is chosen primarily for its thermal and mechanical properties. In a nuclear reactor, however, the coolant must be capable of withstanding prolonged irradiation without deterioration, and in addition, its neutron-absorbing properties must be allowed for in the original design. This is also true for structural and other elements of a reactor. As a consequence, these are frequently of unconventional high-cost materials which usually suffer from poor thermal properties. This adds to the difficulties of cooling.

The heat-removal problem is further complicated by the high volumetric heat release rates characteristic of nuclear reactors. Nuclear reactor construction is expensive, as is the capital investment in fuel. As a result there is pressure to reduce the physical size of both the core and the containment vessel as much as possible. This means that the energy density for a given power rating is raised. Further, in a fast reactor, the core size is necessarily small, and, as can be seen in Table 10.1, the power density becomes very high indeed.

Table 10.1 Power densities in reactor cores

Power reactor type	Power density $(MW\ m^{-3})$	Specific power $(kW\ kg^{-1}\ U)$
Gas-cooled	0·53	3·0
High-temperature gas cooled	7·75	80·0
Sodium-graphite	10·25	13·0
Organic-cooled	13·75	6·3
Heavy-water moderated	18·00	32·0
Boiling water	29·00	13·0
Pressurized water	54·75	20·0
Sodium-cooled fast breeder	760·00	900·0

Coolant circuits

In power reactor systems, the heat generated is used to produce electricity. At the moment, the most efficient technique is to use a steam turbine, though there are some suggestions for direct thermionic conversion systems. If a steam turbine is used, then the heat from the reactor core must be passed efficiently to the water in the steam turbine circuit. In the boiling-water reactor this can be achieved by using the moderator as the coolant and sending it directly to drive the turbine. There are several objections to this as an operating philosophy, not least of which is the undesirability of having radioactive contaminated water circulating freely in the associated equipment outside the reactor core. An alternative approach is the pressurized-water reactor in which the moderating water serves as the primary coolant and is circulated through a heat exchanger. In this heat exchanger heat from the primary circuit is passed to a secondary water system which then drives the turbine. After the turbine there is the conventional condenser stage which removes residual heat from the secondary supply. Thus the pressurized-water reactor has three coolant circuits.

276

This is the basic configuration for most reactor systems regardless of the primary coolant. In the gas-cooled systems, the carbon dioxide primary coolant heats the water–steam secondary circuit which then drives the turbine and residual heat is absorbed by the condenser circuit. In liquid sodium fast reactors, the sodium exchanges its heat with the steam cycle and so on. In sodium cooled reactors there is formation of radioactive sodium-24, which is a gamma emitter, so that sometimes there are two sodium circuits, one through the core and radioactive, and the other heated by the primary circuit and non-radioactive. Also with sodium cooled reactors there is frequently an isolating circuit between the sodium coolant and the water–steam circuit to avoid possible accidents involving interaction between sodium and water.

Since over 90 per cent of the heat to be dissipated is generated in the fuel assemblies, these are designed to maximize the surface area exposed to the coolant. An example of a fuel element from a gas-cooled reactor is shown in Figure 10.4 where the vanes can be clearly seen.

The equation governing the flow of heat from the fuel element to the coolant is to be found in any elementary text on heat, and is

$$Q(r) = -k\nabla^2 t \qquad (10.5)$$

where $Q(r)$ is the heat generated per unit volume per unit time at the point r, k is the thermal conductivity and t is the temperature. If we assume a cylindrical fuel rod of radius a, and uniform source Q, surrounded by cladding of thermal conductivity k_c to radius b, enclosed in coolant at temperature t_2,

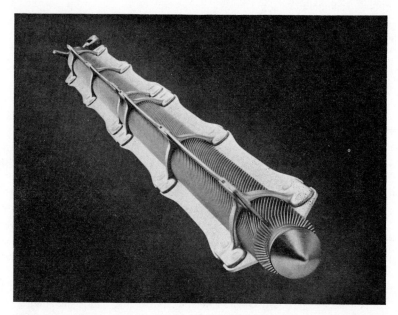

Figure 10.4. Fuel element from a gas-cooled nuclear power station. (Reproduced by permission of United Kingdom Atomic Energy Authority)

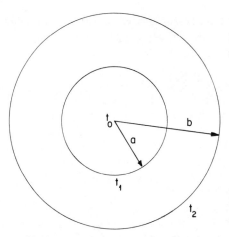

Figure 10.5. Section through fuel element

we can set up the equations in the various regions. Let us assume in addition that the temperature at the centre of the fuel element is t_0, its thermal conductivity is k and that the temperatures at the edges of the fuel and cladding are t_1 and t_2, respectively. This is shown in Figure 10.5.

Equation (10.5) can now be solved for the fuel element and the cladding to produce the solution

$$t = -\frac{Qr^2}{4k} + C_1 \ln r + C_2 \qquad 0 \leqslant r \leqslant a$$

which implies that

$$t_0 - t_1 = \frac{Qa^2}{4k}$$

since $\dfrac{dt}{dr} = 0$ at $r = 0$, and $t = t_1$ at $r = a$. Also

$$t_1 - t_2 = \frac{Qa^2}{2}\left(\frac{1}{k_c}\ln\frac{b}{a} + \frac{1}{hb}\right)$$

where h is the heat-transfer coefficient between the cladding and the coolant.

The temperature drop between the centre of the fuel rod and the coolant is, therefore, given by

$$t_0 - t_2 = \frac{Qa^2}{2}\left(\frac{1}{2k} + \frac{1}{k_c}\ln\frac{b}{a} + \frac{1}{hb}\right) \tag{10.6}$$

assuming negligible thermal resistance between the fuel and the cladding.

This expression can be used to determine the maximum specific power obtainable for a given fuel-cladding-coolant specification. For example consider a cylindrical fuel rod of diameter 25 mm in aluminium cladding 1·25 mm thick, assuming that the maximum fuel temperature cannot exceed 535°C and maximum coolant temperature does not exceed 95°C. Suitable values are, for uranium $k = 0.242$ W m^{-1} °C^{-1}, and for aluminium $k_c = 2.40$; $h = 3.41$ Wm^{-2} °C^{-1} whence equation (10.6) becomes

$$535 - 95 = \frac{Q \; 1.25^2}{2} \left(\frac{1}{0.484} + \frac{1}{2.40} \ln \frac{1.375}{1.25} + \frac{1}{3.41 \; 1.375} \right)$$

$$= 1.81 \; Q$$

Therefore, $Q = 242.9$ W cm^{-3} = 242.9 MW m^{-3}. If this is converted into the specific power, in kWkg^{-1}U, the density of uranium is 19×10^3 Kg m^{-3}, and we find $Q = 12.78$ kWkg^{-1} which is close to the value given in Table 10.1 for the Boiling-Water Reactor.

Not only is the radial distribution of heat from the fuel element of interest, but also the temperature distribution along a cooling channel within the core. This is because, obviously, the coolant heats up as it moves along the channel, so that it is important that the specific power of the fuel elements should be such that the coolant never reaches such temperatures that damage occurs. These calculations are complicated by the variation in the power density along the length of the fuel element. This effect has already been discussed in the section on fuel management, and shows itself in a greater power density in the centre of the rod than at the ends.

Certain criteria can be established which an ideal coolant must satisfy. The coolant should absorb no neutrons as this would be wasteful. It should not moderate neutrons as this could lead to a sudden increase in reactivity. The coolant should not interact chemically with any of the structural, moderator or other materials within the core, even under the effect of the intense radiation. It should be chemically stable itself, for example it should be neither explosive nor toxic. It should not dissociate under the effects of radiation. The vapour pressure of the coolant should be low so that no high pressures are built up at the operating temperature of the reactor. It should be capable of removing large quantities of heat from the reactor core for the expenditure of only small amounts of pumping power. Finally, it should be inexpensive.

No such material exists. All reactor coolants are a compromise between different aspects of these various requirements and different reactor designs. In nuclear reactors designed for power production there is a continuing search for higher thermodynamic efficiency and hence higher operating temperatures. In order to cool these high-temperature reactors, one solution is the use of sodium in liquid form. In view of the corrosiveness of sodium in the presence of impurities, and its violent reaction with water, which forms the secondary coolant, together with its formation of a radioactive isotope Na^{24} under the effects of irradiation in the core, it is clear that this material falls somewhat short of being our ideal coolant. Notwithstanding, it is a common coolant, especially for fast reactors.

A detailed treatment of energy removal would also need full consideration of the importance of turbulence in the transfer of heat to the coolant; for liquids the question of boiling is extremely important; for the fast reactors, the difference between liquid metals and normal non-metallic liquids are also of interest. These discussions are all beyond the scope of this book but are covered in detail in the further reading list.

10.5 Safety

There are three aspects of nuclear safety that must be considered. One is is the straightforward physical shielding of the surrounding area from the harmful effects of radiation during normal operation of the reactor, or after some conceivable reactor accident or malfunction. Another is the disposal of radioactive waste. The third is the less specific problem of ensuring that, if a reactor malfunction should occur, then appropriate preventative action is taken immediately, and automatically if possible, to ensure that the malfunction is repaired, or if this is impossible that the reactor is closed down. These are very serious considerations with all nuclear installations, and their importance cannot be understated. The accident record of the nuclear power industry is exemplary, better than other industries. Nevertheless, the seriousness of the possible implications of a reactor accident must always impose severe restrictions on the operating conditions and safety precautions of the industry. This is perhaps underlined by the accident early in 1974, at Flixborough, in England, in which a large modern chemical plant exploded, killing over 30 people and causing extensive damage in the surrounding area. Although this was not a nuclear plant, the chemical industry also has stringent safety regulations and it was believed that precautions were adequate, but the event still occurred.

Reactor Shielding

All high-energy radiations are harmful to living tissue. This is true apparently without exception, and is the reason why all personnel involved in the radiation industry, from radiographers to high energy nuclear physicists, must take precautions to avoid exposure, and to monitor that exposure which they do receive. If the living tissue happens to be malignant then it may be damaged or killed, producing an overall benefit to the body, but, in general, people must be protected from exposure to such radiations.

The nuclear reactor is a rich source of high-energy radiations of all kinds, neutrons, protons, electrons, gamma rays and X-rays. These must all be kept within a containment vessel which is called a biological shield. This is usually a thick skin of concrete, some 2 m thick, in which the dangerous radiations are absorbed. Unfortunately, the absorption of this radiation energy causes heat to be generated in the shield material, and this must be kept within limits which are tolerable to the material of the shield. For example, input fluxes greater than about 30 mW cm^{-2} will set up temperature gradients in Portland and barytes concrete such that the strength of the concrete will be significantly reduced. Because of this, a further shield of a material which can withstand the higher temperature gradients must be placed in front of the concrete to protect it. This is called a *thermal shield*. This material must have a stringent set of characteristics. It must be capable of withstanding high temperatures, it must have a high thermal conductivity, a high atomic number, for scattering electromagnetic radiation, and also a high neutron capture cross-section. Once again, it should also be stable, easy to fabricate and cheap. Two obvious possibilities are steel and lead, but others have been suggested including steel or graphite impregnated with boron, or lead alloyed with cadmium. Another possibility is boral, which is a mixture of aluminium and boron carbide rolled between aluminium sheets and which has good neutron attenuation properties.

It must also be appreciated that the reactor coolant comes out of the reactor containment vessel to supply steam to the turbines. This is another possible source of radiation that must be protected. Most coolants become radioactive to some extent and the pipework carrying the primary coolant must be shielded to protect personnel. The coolant also picks up radioactive contaminants from the various materials with which it interacts directly within the core.

The most penetrating radiations are high energy gamma rays, and these set the limits of shielding that are required for adequate protection. It is almost impossible to eliminate absolutely all radiation leakage, but it is fairly straightforward to reduce it to very low, tolerable levels.

Reactor Safety

The whole topic of reactor safety is difficult, complicated, and, to a certain extent, contentious. There are some, mostly within the nuclear industry, who say that an accident of a serious nature in a nuclear power station is extremely unlikely because of the extensive safety precautions that are taken. On the other hand, there are those, some within the industry, but most outside it, who claim that the ultimate risks are so great, and the safety precautions so untested, that the nuclear power programmes of all nations should be stopped. This criticism has been particularly levelled against the light water moderated reactors favoured by United States companies.

It is difficult to compare the hazards of nuclear reactors with those of other industrial activities because the potential damage is so much greater. This has led to the development of a new safety philosophy in which attempts are made to foresee possible accidents and to prevent their occurrence. This foresight is built upon calculations, hypotheses and experimental research. Herein lies the main criticism of the whole concept of nuclear safety protection. Since an accident is defined as something unforeseen, attempts to foresee it must automatically fail. At the moment the primary responsibility for safety lies with the designer, the constructor and the operator as each of their phases becomes relevant. Thus, the designer must take all possible precautions to ensure that the fundamental design is 'safe'. The constructor must ensure that the plant as fabricated satisfies the appropriate safety regulations, and the operator must ensure that appropriate safety precautions are taken while the plant is operational.

In all discussions of reactor safety, a clear distinction must be drawn between *normal* and *abnormal* operation. Thus, normal operation is the carrying out of all operations that can be seen as necessary and can be expected in operating a plant. Included in this would be starting up, closing down, reloading, repairing instruments and valves, etc. It can be expected that during normal operation a valve seat may leak but if the valve case in a primary circuit cracked this would not be regarded as part of normal operation and would be classed as abnormal. Similarly, it can be expected that, during normal operation, the cladding of a fuel element will corrode or crack. The simultaneous failure of many fuel elements would, however, be inconsistent with adequate design and would be classed as abnormal. It would be an error to always examine nuclear safety in terms of the maximum credible accident that can be foreseen. This is by definition an unlikely event. Much more important are the various smaller accidents that can occur. Nuclear runaway, in which the core heats up until the fuel melts or vaporizes, is unlikely. However, local overheating is much more probable, and, in fact, is quite common. This can occur through several different mechanisms. There may be deposits on the heat-transfer surfaces of the fuel elements through corrosion for example; these will reduce the heat removal rate and allow the local core temperature to rise. Coolant channels may be partly or completely blocked by dirt or other foreign matter, or by deformation of the fuel elements. Flux peaking, which can happen through the replacement of a control rod by water, for example, but can arise through other mechanisms which are not fully understood, can lead to local *hot spots*. Flux oscillations within the core can be serious; EBR-I showed oscillations due to an interaction between the temperature coefficient of the reactivity, and a bowing of the fuel elements—the reactor was operated successfully in spite of this effect. Oscillations have been observed in various reactor systems and pose one topic for which designers and operators should be alert. By far the most serious accident for the light-water reactors is loss of coolant. Even if the reactor were instantly shut down, there is still sufficient decay heat in a large power reactor to melt the core. Emergency cooling systems have been designed to flood the core in the event of coolant failure, but these can only be properly tested in the event and the event has not yet occurred. Further, if they fail, or if metal–water reactions are induced, the high radioactivity levels would drive workers from the immediate vicinity and no further measures other than those already planned for could be taken.

Another serious aspect is the possibility of chemical reactions. Comment has already been made on the possibility of a sodium–water reaction in liquid metal cooled reactors. There are other possibilities. For example, in the Windscale accident (Windscale, United Kingdom, 1957) a fire occurred involving the graphite moderator and the uranium fuel. A uranium–graphite fire is difficult to extinguish. Water was used successfully at Windscale, but there are hazards involved with its use. It is also possible that under extreme conditions aluminium will burn in air or react with water as will uranium metal or zirconium. Organic liquids are used for cooling some reactors and, like sodium, vaporized hydrocarbons can explode or burn vigourously with air. Fire control is well understood, however, and this is merely a matter of taking appropriate preventative measures.

The accidental release of radioactive material in severely hazardous amounts from the core of a reactor is feasible. Consequently it is customary to provide suitable containment in the form of a biological shield—an outer barrier meant to contain radiation and which can tolerate small pressure increases. It is difficult, if not impossible, to provide complete protection against leakage of this type and so some is tolerated. For the more pressing problem of ensuring that the core and coolant remain inside the reactor assembly, stringent regulations have been laid down by the various controlling bodies. For example, the American pressurized water reactor vessels must satisfy the ASME Pressure Vessel Code, Section VIII (1956), or the API Code Section I.

Confinement of radiation represents the attempt to contain radioactive emanations and to remove them from air, water or other efflux from a reactor. It is possible to use filters, scrubbers or absorbers for this purpose. Once again in the Windscale incident, the filters contained all but a small amount of iodine, and the release of strontium-90 was negligible.

Other factors which affect reactor safety and cause concern both to the anti-nuclear power lobby and also within the nuclear power industry, are the effects of population distribution, the importance of ensuring that any possible radioactive leakage does not find its way into the water supply through ground water, and the importance of ensuring that nuclear reactors are placed on geologically safe sites so that, for example, they are not split open by an earthquake.

Radioactive Waste

All nuclear power stations will produce radioactive waste as a necessary by-product. The very nature of the fission process ensures that radioactive nuclei will be created in large numbers. Some of these, for example xenon-135, have very short half-lives and will, therefore, decay away quickly causing no long term biological hazard. Others are less obliging. In handling radioactive wastes there are two different problems that must be faced. Initially the fuel elements have to be removed from the core of the reactor by remote handling after irradiation, taken to some suitable site and reprocessed to remove usable fuel and some of the useful radioactive isotopes that have been created. This necessitates the transportation of highly radioactive material on public highways, railways and even internationally. The containers which are used for carrying the fuel elements are, therefore, designed to contain the radiation— achieved by making the walls thick—and also to withstand all feasible accidents and to be fireproof. Once the containers arrive at the reprocessing centre, the fuel elements are remotely handled and stored under water to allow some of the short-lived isotopes to decay. They are then fed into a machine which strips off the outer can and slices the fuel rods into short lengths. These are dissolved in solvents and care is taken to avoid criticality problems. The solution is then chemically processed to extract the useful material and to leave the radioactive waste. The waste matter is then stored or discharged depending on its character. Low-level short lifetime liquid wastes can be discharged safely into the sea, but high-level long lifetime wastes are held in stainless steel high integrity storage vessels and it is envisaged that they may have to be stored in this way for several hundred years in some cases.

The inventory of radioactive isotopes from fission reactors is extensive, but the principal problems lie with Cs^{137}, Sr^{90}, Tc^{99} and Pu^{238}. The half-lives of the caesium and strontium isotopes are about 40 years, implying long storage periods before the activity has reduced to tolerable levels.

Liquid wastes have until now been stored in tanks either underground or on the surface. This has proved satisfactory, but there are disadvantages. The demand for storage space will increase, and there is a risk of leakage. These two disadvantages make the search for other forms of storage more attractive. One of these is to concentrate the wastes, preferably into a solid, and to store the relatively small volumes of concentrate in selected locations.

One of the problems with high-activity storage is that the beta and gamma radiation is absorbed in the stored material and causes heating. This means that water-cooling circuits must be supplied for the liquid wastes. In the case of solid wastes, the form of solid used must be capable of withstanding the highest temperature reached in this way. Possibilities suggested have been evaporation of the water, addition of the solutions to concrete, fusion into ceramics or glass, evaporation and calcination in a kiln or on a fluidized bed. Another possibility is to bury the wastes in carefully chosen areas where they cannot contaminate the ground water supply. This has proven difficult, and several schemes of this type have been either discontinued or refused development permission. A further possibility is to dump the waste in the ocean. Too little is known of the exact behaviour of ocean currents to make this a safe process, and so only low-level wastes are disposed of in this way. It has also been suggested that the waste materials could be sealed in containers and buried in salt deposits deep below the surface. This has met with serious criticism on the grounds of possible future dissolving of the salt with a resultant hazard.

There are two other sources of radioactive waste. There is the solid waste that arises from the use of air and water filters and their contamination, and there are the gaseous wastes such as nitrogen, iodine, krypton and xenon. The gases are released when the fuel element is dissolved and can be removed by fairly conventional means. The iodine can be removed by reaction with silver to form silver iodide; the nitrogen, in the form of oxides, will combine with water to form nitric acid, and, if necessary, the inert gases can be adsorbed onto carbon or silica gel at low temperature. Solid wastes are usually buried.

Without a doubt, the radioactive waste disposal problem is the most serious obstacle to the development of nuclear power programmes. All of the other objections relating to the safety of reactor installations and so forth are primarily engineering ones and any one failure, while damaging on a local scale, will have little global significance. Each particular accident will supply information which will make future installations safer. Not so with waste, however. As nuclear power stations proliferate, the waste disposal problem can only increase.

These are the various aspects that influence the choice of nuclear power reactor. Other factors are also important; price, construction difficulties, availability of enriched fuel or heavy water, and acceptable standards of safety for the construction of the pressure vessel. These lead to different philosophies of construction in different parts of the World. This is best exemplified by the emphasis on the light-water reactor for power production in the United States, and the gas-cooled reactor in the United Kingdom. Both countries have maintained research and development programmes in all other types of reactor but nevertheless clear philosophies appeared in the preference of one type over the other. What are the differences, and why should such clear differences occur? We can best answer this by looking at the characteristics of the principal types of reactor.

10.6 Light-water Reactors

This is the American choice and has come in for considerable criticism, on grounds of safety. It exists in two principal variants, the Boiling-water Reactor (BWR), and the Pressurized-water Reactor (PWR). Both have been described as 'gigantic nuclear kettles'.

Pressurized-water Reactor

In the PWR power station (Figure 10.6) there is a primary coolant circuit which contains water at high temperature and pressure, typically 270°C and 2000 psi. This circuit is followed by a secondary, steam generating, circuit which then supplies the turbine. Thus there is full isolation of the radioactive coolant from the external components. The vapour pressure of water determines the upper limit to the efficiency of these reactors, and the practical limit seems to be 2000–2500 psi, giving them a maximum efficiency of some 28–30 per cent.

Difficulties arise through the corrosive nature of water at high temperatures. These are common problems in steam plants and are overcome by the use of special materials. Nonetheless, the problems are more acute in nuclear plant, and in the cladding materials the appearance of corrosion products may lead to further radioactive species through neutron and gamma-ray bombardment. This would increase maintenance problems and perhaps also shielding difficulties. The neutron capture cross-sections of the corrosion-resistant materials restrict the choice of suitable materials for use.

The advantages of the PWR are that the design is simple, and experience has shown them to be safe and dependable. They are easily controlled and are

REACTOR   HEAT EXCHANGER   TURBINE   GENERATOR

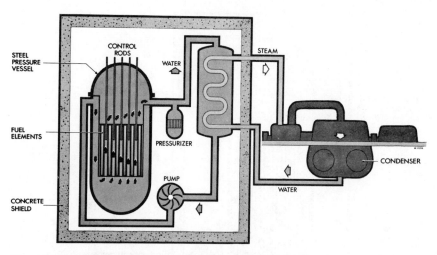

Figure 10.6. Schematic diagram of a Pressurized-Water Reactor. (Reproduced by permission of United Kingdom Atomic Energy Authority)

stable because overheating expands the water and decreases the multiplication factor, so reducing the power density and cooling the core again. In addition, they are 'demand following' in that an increase in turbine demand decreases the temperature of the returning coolant, so increasing the reactivity and power density of the reactor.

Boiling-water Reactor

The BWR (Figure 10.7) represents the other philosophy in water cooling. If the coolant is allowed to boil in the core of the reactor, producing steam which is fed directly to the turbine, not only is the separate heat exchanger of the PWR eliminated, but also the heat removal efficiency is increased as the heat is being removed through the latent heat of evaporation of water and not through the sensible heat of a temperature increase. Consequently the coolant circulation requirements are reduced. It was felt for many years that the boiling-water reactor would not be suitable for power-station operation because fluctuations in the rate of steam production would cause wide local fluctuations in reactivity and would lead to difficulties in control. However, steam is a poorer moderator than water because of its lower density, and, consequently, any increase in power density will be accompanied by an increase in the steam fraction of the coolant and a subsequent reduction in the reactivity. That is the reactor is self-regulating. This was tested in BORAX-I in 1953–1954, and was found to be so. In order to determine the magnitude of fluctuation that would indeed be self-regulating, BORAX-I was subjected to larger and larger

286

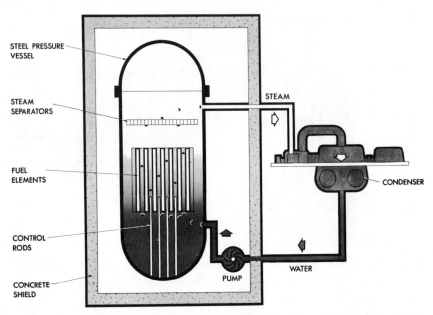

STEEL PRESSURE
VESSEL

STEAM
SEPARATORS

STEAM

FUEL
ELEMENTS

CONDENSER

CONTROL
RODS

CONCRETE
SHIELD

PUMP

WATER

Figure 10.7. Schematic diagram of a Boiling-Water Reactor. (Reproduced by permission of United Kingdom Atomic Energy Authority)

sudden increases in reactivity and ultimately the power excursion arising from a rapid excess of 4 per cent destroyed the reactor.

Boiling-water reactors are now common in the power producing industry. Their designs are similar to the pressurized-water models and the fuel is enriched to 1·44 per cent-U^{235} or more. Because there is no need for the secondary heat exchanger they may be operated at slightly lower temperatures and Zircaloy or stainless steel may be used for the fuel element cladding.

These two reactor types now dominate the world market. They are probably the cheapest to install in terms of capital cost. They are compact and use water as both moderator and coolant; the properties of water in power systems are well understood. They suffer from the limitation that the steam produced is saturated, and so requires wet-steam turbines. One approach to improving the efficiency has been to use an oil-fired superheater as at Indian Falls, New York, to superheat the steam and allow dry steam turbines to be used. The oil supplies about 40 per cent of the power station's thermal energy input. The possibilities for nuclear superheating have been, and are being, investigated in an experimental BORAX-V system and two prototype power stations, Pathfinder Station, near Sioux Falls, South Dakota, and BONUS in Puerto Rico. Significant reductions in power costs are to be expected from large scale power stations of this type, but there are considerable difficulties to be overcome before this becomes a reality. For example, the superheating requires a different technology because, while the superheating region is an integral part of a boiling-water reactor, the coolant in this region is a gas rather than a liquid. Severe changes in conditions are, therefore, found between the two regions of the reactor core.

10.7 Heavy-water Reactors

Heavy-water reactors are the first obvious alternative to the light-water types insofar as they are still liquid cooled and can be either pressurized-coolant or boiling-coolant types—at least in principle. They have come into public notice with the successful operation of the large CANDU (Canadian Deuterium Uranium) power plant at Pickering, Ontario, and the decision in the United Kingdom to use the SGHWR (Steam Generating Heavy Water Reactor) plant operating at Winfrith Heath in Dorset as the basis for the next phase of the United Kingdom nuclear power stations. These two reactors typify the two philosophies in heavy water reactor design. CANDU uses heavy water both as moderator and coolant, while SGHWR uses heavy water as the moderator and light water as the coolant. The main advantage, from an engineering viewpoint, of the heavy-water reactor is that the moderating lengths for the neutrons are larger than in the light-water case. This means that the pile is not so closely packed and that high-pressure pipe can be used for the coolant circuit instead of a pressure vessel. The engineering problems are, therefore, reduced.

CANDU

The CANDU reactor has the moderator in a large tank at atmospheric pressure, and the heavy-water coolant together with the fuel elements are contained in high-pressure piping. The pressure tubes are horizontal, and a unique feature is the provision of a special system for bidirectional charging and discharging of the fuel elements while the reactor is under power. This necessitates coupling and uncoupling the fuelling machine to the channel containing heavy water at high pressure, 100 bar, and a temperature of nearly 300°C. This caused much trouble during the development stages. The fuel is natural uranium oxide pellets clad in zirconium. The CANDU reactor has been extremely successful and has given the lie to many of the critics of the earlier prototype.

SGHWR

This reactor (Figure 10.8) has the same atmospheric pressure heavy water calandria as CANDU, and similar pressure tubes containing the fuel elements and the coolant. The coolant, however, is light water, and the tubes are vertical. Fuel element interchange must be accomplished off load. The use of light water in the coolant channel necessitates the use of slightly enriched fuel which once

REACTOR   **TURBINE**   **GENERATOR**

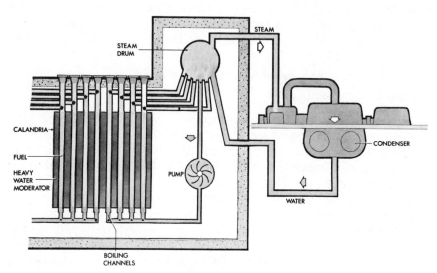

Figure 10.8. Schematic diagram of a Steam Generating Heavy Water Reactor. (Reproduced by permission of United Kingdom Atomic Energy Authority)

again is uranium oxide pellets clad in zirconium alloy. About 11 per cent of the water in the coolant circuit is converted into steam and the steam and water are separated in steam drums, the steam passing to the turbine, and the water back to the core directly. The boiling channel outlet temperature is 281°C, and the steam conditions are 278°C and 61·4 bars. SGHWR is therefore something of a cross between CANDU and the BWR.

The heavy-water reactors are somewhat more expensive to build than their light-water competitors because some 20 per cent of the capital cost must be invested in the supply of heavy water. In addition, the SGHWR requires enriched fuel which adds further to the cost. It is probable, however, that they are inherently safer than the light-water plants, because the pressure problems are easier to solve with pipes than with pressure vessels. CANDU is now proven on the scale of large power plants, but SGHWR has yet to prove itself in this sphere, as Winfrith Heath operates at only 100 MW gross electrical power.

The only reactor known to us which operates as a boiling heavy water reactor is the Halden Boiling Heavy Water Reactor in Norway, operating at 575 psi and at a heavy steam temperature of 250°C. The thermal power of the reactor is 10 MW, and in order to maintain the criticality of the reactor at high temperatures, slightly enriched fuel must be used.

10.8 Gas-cooled Reactors

Gas-cooled reactors were suggested as far back as 1943, in the United States, but were discarded in favour of water-cooled types because of fears regarding the leakage of the chosen coolant, helium. A plant was built in the United Kingdom, at Windscale, using air as the coolant and graphite as the moderator to supply plutonium to Britain's nuclear weapons programme. Because of the successful operation of this reactor, it was a natural extension to adopt it for the Calder Hall plutonium and power generating plant. This time carbon dioxide was used as the coolant, and once again graphite was the moderator. Calder Hall became the prototype for the *Magnox* Reactors which now form the basis of the United Kingdom nuclear power supply. The name arises through the use of a magnesium alloy called Magnox as the cladding material for the fuel elements. The Magnox stations have a fairly low efficiency as power stations go, and in the effort to increase the efficiency of power production by raising the steam temperature various advances in gas-cooled reactors have been either developed or suggested (Reactors UK, 1972). The first example of these is the Advanced Gas-cooled Reactor (AGR), which has been built in the United Kingdom and has been beset with difficulties. Five are being built and all are behind schedule. They have proved expensive in construction, and this type has been dropped from British plans for future reactor plant. The ultimate development in the gas-cooled line is the High-Temperature Reactor (HTR), which uses helium as the coolant and no metal in the core other than the fuel. Graphite is used both as structural material and canning. The outlet temperature of the coolant may be as high as 750°C.

MAGNOX

The Magnox Power Stations (Figure 10.9) commissioned in the United Kingdom, range from Calder Hall, commissioned in 1956 and generating 50 MW per reactor at an efficiency of 19 per cent, to Wylfa, commissioned in 1970, and producing 590 MW per reactor at an efficiency of 31·4 per cent. Calder Hall contained some 110 tonnes of uranium fuel in a core 9·45 m in diameter and 6·4 m high, with 1113 tonnes of moderator and reflector. Wylfa contained 595 tonnes of fuel and 3740 tonnes of graphite in a core 17·4 m in diameter and 9·14 m high. Calder Hall produced the coolant at 7·9 bars and 345°C, and the Wylfa plant was designed to operate at 27·5 bars and 402·3°C, but has now been down-graded slightly to eliminate some of the corrosion problems that were found in mild steel components. It would be possible to overcome this corrosion problem by the use of other materials in further models, but this would add to the cost of construction. The Magnox stations have performed extremely well after some initial teething troubles, and the appearance of these corrosion difficulties after five to ten years is to be expected as part of the cost of development. The steam temperature at the turbine has risen from 320°C (15·5 bars pressure) at Calder Hall, to a design 396°C (49·2 bars) at Wylfa. The fuel, being natural uranium, avoids the difficulties and expense of acquiring enriched fuel, and has been found to be economically satisfactory, so much so that in 1971 nuclear power stations produced the cheapest electricity in the

REACTOR   HEAT EXCHANGER   TURBINE   GENERATOR

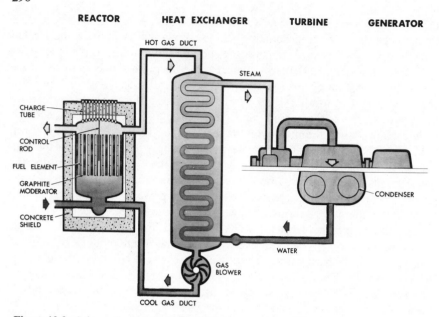

Figure 10.9. Schematic diagram of a Magnox Reactor. (Reproduced by permission of United Kingdom Atomic Energy Authority)

United Kingdom (Marsham and Pease, 1973). Gas-cooled reactors have been less popular outside the United Kingdom because they are more expensive to construct than their light-water competitors. The use of uranium metal as the fuel element put an upper limit on the temperature of the core and on the fuel burn-up, limiting it to about 3000 MWd/tonne.

A problem with all gas-cooled reactors is the leakage of coolant. In Calder Hall this was as large as two tonnes of carbon dioxide per day, but has now been reduced to less than 0·5 tonnes per day. This leakage, coupled with the high cost of helium, makes it economically sensible to try to attain higher temperatures using carbon dioxide rather than helium as the coolant despite its increased chemical activity with graphite at higher temperatures. This led to the development of the AGR.

AGR

The Advanced Gas Cooled Reactor (Figure 10.10) has functioned satisfactorily in its prototype form at Windscale in the United Kingdom. It was commissioned in 1962, and known as WAGR. The reactor power is 105 MW (thermal) and 33 MW (electrical) representing an efficiency of 31·4 per cent. The fuel is 3·1 per cent enriched uranium oxide pellets clad in stainless steel and the outlet conditions of the carbon dioxide coolant are 500–575°C and 19·65 bars. The steam conditions are 454°C and 45·85 bars. The moderator is graphite. Unfortunately the satisfactory performance of the prototype AGR has not

REACTOR   **TURBINE**   **GENERATOR**

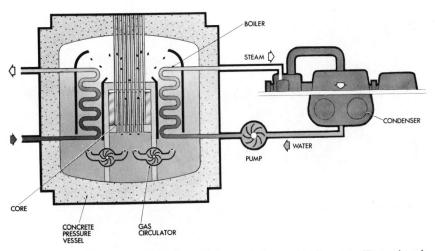

Figure 10.10. Schematic diagram of an Advanced Gas-cooled Reactor. (Reproduced by permission of United Kingdom Atomic Energy Authority)

yet been repeated in the full scale commercial models. These are all designed at about 600 MW or slightly more, with steam conditions comparable with those in a fossil fired power station, but none has yet come on stream. They have been plagued by constructional and other difficulties, but it is hoped that the first of them will be on power during the winter of 1975–1976, some three years late.

HTR

The HTR (Figure 10.11) was envisaged as an extremely high temperature steam producer, with the temperature of the coolant exhaust reaching 750°C. This would provide steam at 540°C and 100 bars at the turbine, giving a net thermal efficiency of 35 per cent. The High Temperature Gas-cooled Reactor at Peach Bottom, Pennsylvania, was designed to operate at these levels and to produce electricity, while the 12 European nations' reactor, DRAGON, at Winfrith in the United Kingdom, was designed to test the reactor, operating at a thermal power of 20 MW with no electricity production. The two reactor systems are very similar in many ways, and both are designed to examine both the feasibility of high-temperature operation using helium as the coolant and graphite as the moderator, and also the possibilities for thermal neutron converter reactors operating on the thorium–uranium-233 basis to produce a supply of fissile fuel from non-fissile thorium. In the American reactor the fuel was 13·8 per cent enriched in U^{235}, while in Dragon, some of the fuel elements are 36–95 per cent enriched fuel, and some are experimental elements containing

292

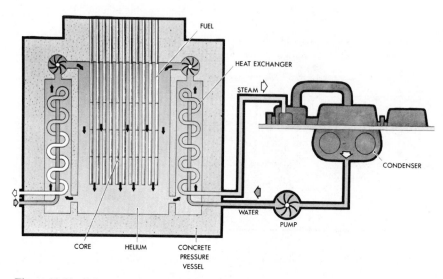

Figure 10.11. Schematic diagram of a High-Temperature Reactor. (Reproduced by permission of United Kingdom Atomic Energy Authority)

a variety of materials. Dragon has facilities for handling either pellet, pebble-bed type fuel or more conventional pin fuel elements. The fuel is in the form of coated particles of carbides or oxide, with a complex coating of carbon and silicon carbide covering the particle. Thus the coating is ceramic and operating temperatures of 1250°C are possible as a normal feature. Excursions to 2000°C for short periods appear to cause no damage. Fuel burn-up is in the region of 100,000 MWd/tonne at the operating temperature of 1250°C. The core size of DRAGON is approximately 1·07 m diameter and 1·6 m long.

An important consequence of the feasibility of the HTR is the possibility of using closed cycle gas turbines directly, eliminating the need for a steam generating stage. For conventional steam systems the boiler design can be more compact, and the efficiency will be improved. A further long term possibility is the supply of process heat for the chemical and steel industries.

Gas-cooled plant is expensive compared with light-water reactors and is physically larger because of the longer moderating length of neutrons in graphite than in water. Only the Magnox reactors use natural uranium fuel so that even this cannot be said to be an advantage. It is true, however, that the possibilities for extension to extremely high temperatures give the type a long-term advantage especially if the possibilities for process heat are realized.

10.9 Fast Reactors

These reactors (Figure 10.12) are of interest because of their breeding charac-

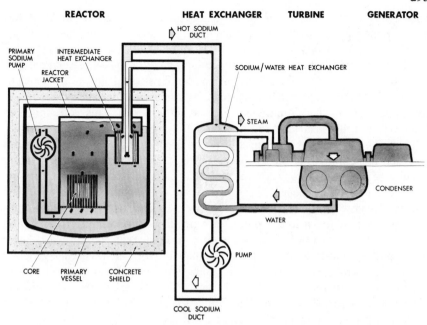

REACTOR   HEAT EXCHANGER   TURBINE   GENERATOR

Figure 10.12 Schematic diagram of the Prototype Fast Reactor. (Reproduced by permission of United Kingdom Atomic Energy Authority)

teristics. No effort is made to slow the neutrons but, because collisions between them and structural or fuel materials are unavoidable, some slowing is inevitable and average energies before capture are of the order of several hundred kilovolts. Because high temperatures are possible in fast reactors, sodium appears to be the most suitable coolant at the moment.

The core of a fast reactor is small and the power density high (see Table 10.1). Thus it is essential that the coolant should have good thermal properties. The specific power of the core is high since this favours the doubling time for breeding, and also improves fuel economics. Because of this, the limitation imposed by fission products is smaller, and there is more flexibility in the choice of constructional materials.

It is important that the temperature dependence of the reactivity changes due to the coolant should be negative because that of the fuel is positive. This is to maintain stability, but should not be overdone as otherwise the reactor will need too much excess reactivity for starting up from cold. As was discussed in Chapter 9, fast reactors can be controlled satisfactorily using delayed neutrons in the same way as thermal reactors. However, with plutonium-239 as fuel the delayed neutron fraction is small so that careful attention must be given to the design of the control system.

The development of the fast breeder reactor for commercial electricity production can be exemplified by the British experience (Reactors UK, 1972). The Dounreay Fast Reactor (DFR) was commissioned in 1959 after earlier zero-energy experiments with an experimental reactor ZEPHYR. DFR has now supplied power to the electrical system for over 12 years, and while its power is low by power station standards at 14 MW (electrical), it provided the experimental test bed for development and analysis of commercial designs. In 1974, the Prototype Fast Reactor was commissioned and is designed to produce 250 MW (electrical) and to act as the test bed for the first truly commercial scale version, the Commercial Fast Reactor (CFR). This is designed at 1300 MW output (electrical) and is to be in full production in the early 1980's.

The DFR experiments were successful, and the PFR, now operating, has been designed to demonstrate the key engineering, technical, economic and safety features of the much larger commercial plants. There is a small reactor core, and there are no pipe penetrations into the tank below the sodium level. The reactor is below ground and surrounded by a leak jacket. The coolant is unpressurized, and the primary pumps are protected by radiation shields and may be withdrawn from the top of the tank for maintenance and repair. There is a secondary, non-radioactive, sodium coolant circuit which is used to generate steam. Temperatures are sufficiently high that normal dry-steam turbines, appropriate to fossil fuel plant conditions with reheat, can be used and the efficiency is about 41 per cent. The outlet temperature of the liquid sodium coolant is 560–600°C, and the steam conditions at the turbine stop valve are 513–538°C and 159·5 bars. The fuel is mixed oxides of plutonium and uranium clad in stainless steel and made up into clusters of thin pins.

Experience from DFR, the experimental small scale plant, suggests that it will be possible to achieve very high fuel burn-up factors so that low fuel charges will be achieved. The hoped for 100,000 MWd/tonne seems to be assured, and many experiments have exceeded this burn-up (Marsham, 1973). DFR isolated some of the difficulties that had to be overcome, including the swelling of steel under fast neutron bombardment and the necessity of ensuring that there was no entrainment of gas in the coolant. It is now up to PFR to demonstrate the reliability of large commercial plant.

These are the principal reactor types of interest to the nuclear power industry. There are other designs including the organic-cooled types which use molten organic materials as the coolant, and there are fluid fuel types in which the fuel is either a solution or in the form of a molten metal. There are advantages and disadvantages to both categories, but they do not seem to be attracting much enthusiasm as potential power producers.

References

ASME Pressure Vessel Code, Sections I and VII, ASME, New York, 1956.
Marsham, T. N., *Atom*, No. 201, p. 150, UKAEA, 1973.
Marsham, T. N. and R. S. Pease, *Atom*, No. 196, UKAEA, 1973.
Rasmussen, N., USAEC Draft Report, 1974.
Reactors UK, Fifth Edition, UKAEA, 1972.
Salmon A., *The Nuclear Reactor*, p. 94, Longmans, London, 1964.

Further Reading

Yevick,J. G. and A. Amorosi, (Eds.), *Fast Reactor Technology—Plant Design*, M.I.T. Press, Cambridge, Mass., 1966.

Glasstone, S. and A. Sesonske, *Nuclear Reactor Engineering*, Van Nostrand Reinhold Company Inc., New York, 1967.

Moore, J., D. Higs, N. Bradley and I. T. Rowlands, *Status of the Steam Generating Heavy Water Reactor*, *Atom*, No. 195, p. 7, 1973.

Reactor Handbook, 2nd ed., Vol. I—Materials, C. R. Tipton, (Ed.), Interscience Publishers Inc., New York, 1960.

Reactor Handbook, 2nd ed., Vol. II—Fuel Reprocessing, S. M. Stoller and R. B. Richards, (Eds.), Interscience Publishers Inc., New York, 1961.

Reactor Handbook, 2nd ed., Vol. IIIb—Shielding, E. P. Blizard and L. S. Abbott, (Eds.), Interscience Publishers Inc., New York, 1962.

Reactor Handbook, 2nd ed., Vol. IV—Engineering, S. McLain and J. H. Martens, (Eds.), Interscience Publishers Inc., New York, 1964.

Salmon, A., *The Nuclear Reactor*, Longmans, London, 1964.

11 Nuclear Fusion

After an initial period of great enthusiasm, followed by disenchantment as the experimental difficulties became apparent, nuclear fusion is now being viewed once again with interest and indeed, hope, as the ultimate solution to man's energy supply problems. Fusion relies on the increasing stability of light atoms as the atomic number increases, and in particular, on the extremely high binding energy of the helium nucleus. As a consequence, if two lighter nuclei fuse together to form helium, energy will be released, This has been achieved in the hydrogen bomb, but not in controlled laboratory experiments.

Strictly speaking, fusion has been achieved in the laboratory. Using particle accelerators, protons, deuterons and tritium nuclei have been accelerated to the necessary high speeds to penetrate the potential barrier protecting the nucleus and to undergo fusion. It is from these experiments that we know the conditions necessary for fusion reactors to work, and also the cross-sections for the different possible reactions. However, despite its extremely great importance in gaining insight into the processes involved, the use of accelerators is not applicable to power production.

From measurements of the parameters for the various possible reactions, it has become clear that fusion is most likely to be achieved between deuterium (H^2) and tritium (H^3). This reaction is likely to proceed under the least difficult experimental conditions, and also has one of the largest cross-sections. The most highly prized reaction is that involving deuterium alone, but this is only likely to proceed under much more difficult conditions. Before looking at either of these reactions in any detail let us examine the physical conditions under which fusion may be possible.

11.1 Physical Conditions for Thermonuclear Reactions

The basic force that must be overcome in fusion reactions is the Coulomb repulsion of the two positively charged nuclei. Classically the energy needed by a particle to surmount this barrier is given by $Z_1 Z_2 e^2 / 4\pi\varepsilon_0 R_0$, where Z_1 and Z_2 are the atomic numbers of the two nuclei, e is the charge of the electron and R_0 is the distance between nuclear centres at which the attractive forces become dominant. That is R_0 is the minimum separation at which Coulomb's law still holds. For light nuclei R_0 can be taken as 5×10^{-15} m and so the energy to surmount this barrier becomes approximately $10^{-14} Z_1 Z_2$ J. A simple calculation leading to an estimate of the temperatures involved in nuclear fusion

296

reactions can be formed by setting this energy equal to $\frac{3}{2}kT$. This leads to $T \approx 5 \times 10^8$ K for the fusion of hydrogen isotopes. In fact, this argument is too simplistic, even though it does produce about the correct temperature for thermonuclear fusion to take place. There are several reasons why it is not satisfactory. Experiments with accelerated nuclei show that barrier penetration is an important factor in fusion, and the there is no clear threshold energy above which fusion can occur. Instead, nuclear reactions can take place at energies considerably below those corresponding to the top of this barrier, and the cross section for the fusion of two nuclei can be reasonably represented as a function of their total energy referred to the centre of mass by

$$\sigma (W) \approx \frac{\text{constant}}{W} \exp \left(- \frac{2^{3/2} \pi^2 M^{1/2} Z_1 Z_2 e^2}{4 \pi \varepsilon_0 h \, W^{1/2}} \right) \qquad (11.1)$$

where M is the reduced mass of the interacting nuclei (Gamow, 1938; Atkinson and Houtermans, 1928). This expression is finite even for relatively small energies W, but increases rapidly as W increases.

The experimental systems seen as giving the best chance of production of power from fusion sources are the so-called thermonuclear systems in which the fusion fuels are heated together to the extremely high temperatures indicated by our simple calculation and obtain the necessary energy from thermal agitation. In such a system, the energies of the interacting nuclei form a Maxwellian distribution, and this has an effect both on the characteristics of the reacting material, known as a *plasma* since it will be a gas made up of an equal distribution of positive ions and electrons, and on the actual temperature that must be reached to enable the reaction to proceed. In this Maxwellian distribution, the number of nuclei per unit volume, $dn = N(W)dW$, whose energies lie between W and $W + dW$, will be

$$dn = N(W)dW = \text{constant} \times \frac{W^{1/2}}{T^{3/2}} \exp (- W/kT)dW \qquad (11.2)$$

Here T is the *kinetic temperature* and relates to the energy distribution among the particles rather than the *temperature*, which would include radiation contribution as well.

The probability $R(W)$ that a fusion reaction will take place in the energy range W to $W + dW$ will be given by

$$R(W) = \frac{dn}{dW} \sigma (W) \approx \frac{C}{W^{1/2} T^{3/2}} \exp \left(- \frac{2^{3/2} \pi^2 M^{1/2} Z_1 Z_2 e^2}{4 \pi \varepsilon_0 h W^{1/2}} - \frac{W}{kT} \right) \qquad (11.3)$$

where C is a constant. At a given temperature, the Maxwellian distribution ensures that a small fraction of the nuclei in the plasma will have very high energies. In addition, $\sigma(W)$ increases steadily as the energy of the interacting particles increases. The result is that the probability of fusion takes on a finite value even though the plasma temperature may be significantly below any reasonable feasible temperatu.e, and that $R(W)$ shows a maximum at some energy W_m significantly greater than that appropriate to the peak in the Max-

298

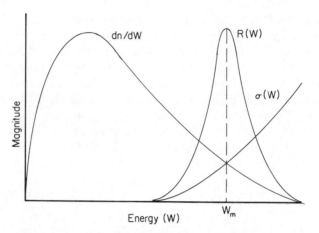

Figure 11.1. Maxwellian distribution, fusion reaction cross-section, σ, and probability R of a fusion reaction's occurring

wellian distribution. This is illustrated in Figure 11.1. The total reaction rate is determined by the common area under the dn/dW and $\sigma(W)$ curves, so that it is apparent that any observed thermonuclear reactions will be due to a small fraction of the nuclear collisions that take place. As the plasma temperature is increased and the Maxwellian peak moves to high energies, the peak representing $R(W)$ increases in both height and width, showing the advantages of increasing the temperature.

To a good approximation, the variation in $R(W)$ with W is determined by the exponential term and it is straightforward, therefore, to estimate W_m by determining the value of W for which this is a maximum. It is found to be

$$W_m = \left(\frac{(2M)^{1/2}\pi^2 Z_1 Z_2 e^2 kT}{4\pi\varepsilon_0 h} \right)^{2/3}$$

This is usually expressed in keV and referred to the kinetic temperature. This is possible because energy is related to temperature through Boltzmann's constant ($1 \text{ keV} \equiv 1\cdot16 \times 10^7 \text{ K}$). If we consider a reaction involving deuterium nuclei only, then W_m can be expressed as $W_m = 6\cdot3\ T^{2/3}$ with the temperature in keV. At a temperature of 1 keV, therefore, most of the fusion reactions will be due to nuclei having a kinetic temperature of $6\cdot3$ keV.

It is clear that the reaction rate in a fusion reactor depends on the number densities of the two reacting species, and on the reaction cross-section. Since, as we have seen, this cross-section varies with energy, there is an implicit dependence on the speed of the nuclei. There is also an explicit dependence because if one species is assumed to be moving at speed v through a stationary lattice, then the reaction rate depends on the number of collisions per second, which in turn depends on the speed of the particle. For two species, with number densities n_1 and n_2, and a reaction cross-section σ at relative speed v, the reaction rate will be $R_{12} = n_1 n_2 \sigma v$ reactions per unit volume per second.

Since there is a range of speeds available from the Maxwellian distribution, and since σ is velocity dependent, the reaction rate that is of interest in practice is averaged over all possible speeds. It is, therefore, given by $R_{12} = n_1 n_2 \overline{\sigma v}$. If there is only one species, then $n_1 = n_2 = n$, but each interaction is counted twice, so that $n_1 n_2$ is replaced by $\frac{1}{2} n^2$. The average $\overline{\sigma v}$ can be determined in specific cases by integrating over the Maxwellian distribution, but the dependence of σ on v must be known.

Possible Nuclear Fusion Reactions

In the search for fusion reactions emphasis is placed on the very light nuclei which have the lowest Coulomb barriers. The most favourable reactions appear to be those involving deuterium, and it is unfortunate that the possibilities involving hydrogen itself appear to have such low cross-sections that they are unrealistic candidates for power production systems. There are four possible reactions. Two of them involve only deuterium itself, another involves deuterium and tritium, while the fourth involves deuterium and an isotope of helium He^3. The four reactions are

$$_1D^2 + _1D^2 \longrightarrow _2He^3 + _0n^1 + 3\cdot27 \text{ MeV (the neutron branch)}$$

$$_1D^2 + _1D^2 \longrightarrow _1T^3 + _1H^1 + 4\cdot03 \text{ MeV (the proton branch)}$$

$$_1D^2 + _1T^3 \longrightarrow _2He^4 + _0n^1 + 17\cdot6 \text{ MeV}$$

$$_1D^2 + _2He^3 \longrightarrow _2He^4 + _1H^1 + 18\cdot3 \text{ MeV}$$

Of these, the fastest is the deuterium–tritium reaction, which is therefore preferred for the development of fusion reactors. It is interesting to note that, if the conditions favouring the deuterium–deuterium reaction can be attained, the proton branch will produce tritium to feed a further deuterium–tritium reaction, while the neutron branch will supply helium for the slower deuterium–helium fusion. This reaction is unfavourable until the temperature is as high as 100 keV, when its rate approaches that of the deuterium–deuterium reactions. Unfortunately, tritium is not a common material, being radioactive with a half-life of 12·3 years. It must, therefore, be manufactured and this is achieved by neutron bombardment of Li^6, in the reaction

$$_3Li^6 + _0n^1 \longrightarrow _2He^4 + _1T^3 + 4\cdot6 \text{ MeV}$$

The period of usefulness of the deuterium–tritium fuel cycle therefore depends on the World's reserves of lithium.

If sufficiently high temperatures can be reached to allow all four reactions to proceed, then the consumption of six deuterons would lead to a total energy release of 43·2 MeV. If the temperature were less than 50 keV so that the D–He^3 reaction was unavailable, this is reduced to 24·9 MeV for the consumption of five deuterons. Initially, however, only the D–T reaction will be possible releasing 17·6 MeV.

300

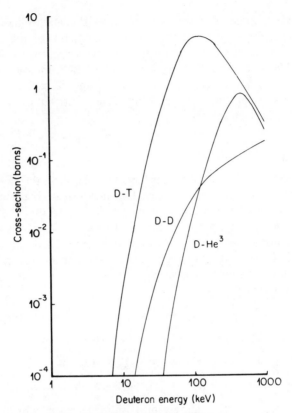

Figure 11.2. Variation with energy of fusion reaction
cross-section

Reaction Cross-sections

The reaction cross-sections for the various fusion reactions have been measured in accelerator experiments (Arnold and coworkers, 1954; USAEC Report, 1957). These are shown in Figure 11.2, and the cross-sections for each of the individual D–D reactions can be taken as half of the value on the graph, at least up to 120 keV. The peaks in the cross-sections are due to resonance phenomena common in nuclear reactions. It is also important to know the average $\bar{\sigma}v$, and these are listed in Table 11.1 (Tuck, 1954, 1955; Thompson, 1956, 1957; Boulègue and coworkers, 1958; Wandel and coworkers, 1959). They are also shown in Figure 11.3. The effect of the Maxwellian distribution can be seen from these two figures. Consider the D–T reaction at 10 keV. From Figure 11.2, the cross-section is $1 \cdot 7 \times 10^{-3}$ barns $= 1 \cdot 7 \times 10^{-31}$ m^2. The reduced mass of the D–T system is approximately three-fifths of the mass of deuterium, that is 2×10^{-27} kg. The relative speed v is therefore given by $\frac{1}{2}Mv^2 = 10$ keV and is approximately $1 \cdot 3 \times 10^6$ m s^{-1}. Thus $\bar{\sigma}v = 2 \cdot 2 \times 10^{-25}$ m^3 s^{-1}. Figure 11.3 shows it to be 10^{-22} m^3 s^{-1}, an improvement by a factor

Table 11.1 Variation of $\overline{\sigma v}$ with temperature

Temperature (keV)	D–T cm³ sec⁻¹ × 10⁻¹⁹	D–D cm³ sec⁻¹ × 10⁻¹⁹	D–He³ cm³ sec⁻¹ × 10⁻¹⁹
1	0·07	0·002	6×10^{-7}
2	3·0	0·05	2×10^{-4}
5	140·0	1·5	0·1
10	1100·0	8·6	2·4
20	4300·0	36·0	32·0
60	8700·0	160·0	700·0
100	8100·0	300·0	1700·0

greater than 400. Consequently, the effect of the Maxwellian distribution is to increase the reaction rates considerably when compared with the expected values on the basis of the cross-section data alone.

In any system of interacting nuclei, it is important to know the *reaction mean free path*. In the case of fusion this is the mean free path between fusions.

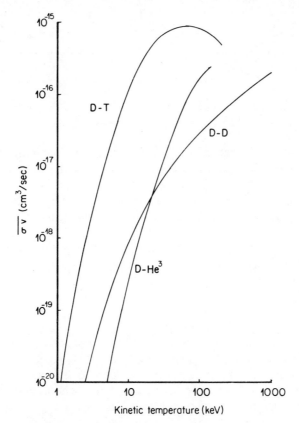

Figure 11.3. Variation of $\overline{\sigma v}$ with kinetic temperature for fusion reactions

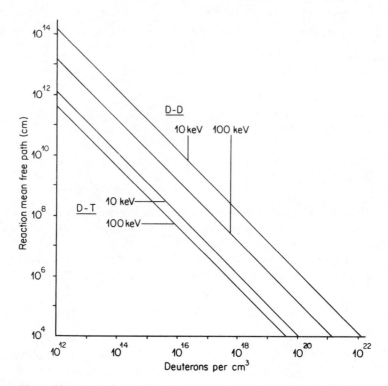

Figure 11.4. Variation of reaction mean free paths with density for D–T and D–D fusion at two kinetic temperatures

If the system contains n reacting nuclei per unit volume, and if the cross-section is σ, then the mean free path λ is equal to $1/n\sigma$. For the Maxwellian energy distribution, σ must be replaced by $\bar{\sigma}$ averaged over all velocities at a given kinetic temperature. Figure 11.4 shows the values of λ for the D–T and D–D reactions as a function of number density for two kinetic temperatures, 10 and 100 keV. As an example, for a 10 keV D–T reaction the mean free path of the deuteron is 10^9 cm assuming that the density is 10^{15} nuclei per cm^3, that is the deuteron will travel 10,000 km before reacting.

Thermonuclear Power Density

These then are the conditions necessary for fusion to occur. The temperature must be high enough, and the plasma must be contained to allow the reaction to proceed. It is important to know, however, what power density can be expected from a fusion reaction under specified conditions. If Q joules are released per reacting pair of nuclei, and if there are n_1 and n_2 of the two species in unit volume, then the rate of energy release is

$$R_{12} = n_1 n_2 \overline{\sigma v}\, Q \text{ watts per unit volume}$$

If we consider the D–T reaction, and put $n_D = n_T = n$, $Q = 3\cdot5$ MeV $= 1\cdot6 \times 10^{-13}$ J and consider the temperature as 10 keV when $\overline{\sigma v} = 10^{-16}$ cm^3 s^{-1}, we find the power density to be

$$R_{DT}(10 \text{ keV}) = 6\cdot2 \times 10^{-29} n^2 \text{ W cm}^{-3}$$

Fission reactors are operated at power densities of about 100 W cm^{-3} so that similar perfomance levels would necessitate operating such a fusion reactor at number densities of about 10^{15}–10^{16} deuterons per cm^3. It is important to note that an increase in the density of the reacting nuclei leads to a much greater increase in the power density since there is a square-law dependence. Thus, if it were possible to construct a fusion reactor to operate with densities similar to those of standard atmospheric temperature and pressure, 10^{19} nuclei cm^{-3}, the power density would become greater than 10^{10} W cm^{-3}, and the thermal power produced by a modern 1000 MW plant operating at 40 per cent efficiency, would be produced from a volume of only $0\cdot25$ cm^3, which is obviously impractical.

If the number density is reduced, it is clear that the power density will drop very rapidly, becoming uninteresting in power production terms. On balance it would seem that the particle density for a power reactor should be about 10^{15} nuclei cm^{-3}.

11.2 Energy Losses in Fusion

Competing with the fusion energy release, and subtracting from it, are certain energy losses. Some of these are avoidable, or at least reducible, and can be adjusted by careful design of the reactor. Falling into this category are the various instabilities in the plasma, which lead to loss of power either by reducing the temperature of the plasma, or by allowing it to disperse completely. There are also various losses that occur outside the reaction volume and which once again can be allowed for or reduced. These include ohmic losses in the magnetic field coils and associated circuitry, and the normal inefficiency of conversion of heat into electricity. It is important that the plasma instability problems are overcome because otherwise the hopes held forth for fusion will prove to be illusory.

There are certain losses that are inherent in the system, however, and about which little can be done. The dominant cause of these losses is *bremsstrahlung* which arises because the plasma consists of equal numbers of positive and negative charges in a state of great agitation. As the highly energetic charges move about the plasma, they are continually in collision with each other. In these collisions, energy is lost and appears as radiation emitted by the decelerating charges. Also, if the ions in the plasma are not completely stripped of their electrons, atomic transitions of the normal type can take place and will again lead to the absorption and emission of radiation. There is a temptation to assume that, since we are concerned with extremely high 'temperatures', black-body radiation will be important. In fact, this is not so. The use of the term *temperature* implies a thermodynamic equilibrium that does not exist because of the very small number densities involved in possible fusion reactor designs. The consequence of these low densities is that the system is not opaque to the bremsstrahlung and so cannot be a black body. As a result black-body thermodynamics is inapplicable. The term *kinetic temperature* was used deliberately to distinguish it from thermodynamic temperature.

It can be shown (Glasstone, 1960) that the rate of emission of energy from a plasma through bremsstrahlung is given by

$$P_{br} = 5\cdot35 \times 10^{-31} n_e \sum (n_i Z^2) T_e^{1/2} \ W cm^{-3} \tag{11.4}$$

where T_e is in keV, n_e is the number density of electrons, n_i is the number density of ions of atomic charge Z and the summation is taken over all ionic species. If the fusion reaction is to become self-sustaining, then the power produced from the fusion process must exceed the losses through bremsstrahlung. Equation (11.4) simplifies in the case of the D–D and D–T reactions to

$$P_{DD(br)} = 5\cdot35 \times 10^{-31} n_D^2 T_e^{1/2} \ W \ cm^{-3}$$

$$P_{DT(br)} = 2\cdot14 \times 10^{-30} n_D n_T T_e^{1/2} \ W \ cm^{-3}$$

assuming a 50 per cent mixture in the D–T case. The temperature at which the reaction becomes self-sustaining is known as the *ideal ignition temperature*, and can be seen from Figure 11.5 to be 4 and 36 keV, respectively, for the D–T and D–D reactions assuming 10^{15} nuclei per cm³. These temperatures correspond to $4\cdot65 \times 10^7$ K and $4\cdot18 \times 10^8$ K, respectively. It is clear that the D–D reaction is an order of magnitude more difficult to attain than the D–T reaction.

At the high temperatures involved in fusion reactions, the bremsstrahlung radiation from electron–electron collisions cannot be neglected, and in addition the presence of impurities will have an important contaminating effect, because they will have higher Z values. The bremsstrahlung varies with the square of Z and so all impurities must be kept to a minimum. In addition to bremsstrahlung radiation, which is mostly in the ultraviolet to soft X-ray region of the spectrum, and therefore does not represent a total loss to the system since it will be captured by the containment vessel and converted into heat, there is also *cyclotron radiation* loss to be considered. This arises through the motion of the charges in the strong magnetic fields proposed for their containment. In these fields, the charges will describe curved orbits of one sort or another, and will therefore emit radiation by virtue of their accelerations. For the ions, this component is insignificant, but for the electrons it may exceed the bremsstrahlung. The cyclotron radiation will be mostly in the infra-red and microwave regions, and so will be at least partly absorbed back into the plasma—it will, therefore, not represent such a great energy loss. A further problem is that the power loss through cyclotron radiation rises as the square of the kinetic temperature. For temperatures less than about 5 keV, cyclotron radiation losses are less than those due to bremsstrahlung. For temperatures greater than this they

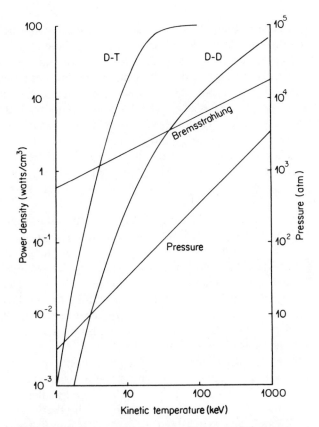

Figure 11.5. Power density from fusion and losses to bremsstrahlung as a function of kinetic temperature for 10^{15} nuclei per cm³. The ideal ignition temperature is given by the point where the power density exceeds the losses. Also included is the pressure in the reactor

rise rapidly. The effect on the D–T ignition temperature is to raise it to about 7 keV, but it makes the D–D ignition forbiddingly high. One approach that is receiving attention is the possibility of making the reactor sufficiently large that the cyclotron radiation is effectively reabsorbed. This leads to exceedingly large reaction volumes for D–D fusion reactors.

11.3 The Production of Power from Thermonuclear Reactions

Two criteria have now been established for the successful operation of a deuterium–tritium fusion reactor. The temperature must be high enough for the power extracted to exceed the power input and the losses, and the number density of the reacting nuclei must be sufficient to ensure an adequate overall reaction rate. Possible reactors could be designed for either continuous or pulsed operation. Continuous reactors would require that the temperature could be held at the necessary high value and that the particles could be contained inside the reaction volume, while power is extracted from it, in an entirely continuous and stable fashion. Since it has not yet proved possible to stabilize the plasma sufficiently even to extract power on an intermittent basis, this must be taken as an unlikely development in the short term. The pulsed mode appears to hold out the greatest hope for success. In this system, a quantity of fuel would be injected into the reaction volume, compressed and heated rapidly until the ignition temperature is reached, and the plasma so produced held stable for long enough for the fusion reaction to proceed. If sufficient fuel can be heated in this way, and if the plasma can be held for a sufficient length of time, then the amount of power extracted through the fusion reactions can exceed the power input.

There are certain limitations in addition to the temperature that are placed on the operation of such a system. If the plasma is sustained at the temperature T for a time t, and if it contains $2n$ particles per cm^3 (n fuel atoms injected), then, assuming a Maxwellian temperature distribution, the average energy per particle will be $\frac{3}{2}kT$ and the energy needed to achieve the temperature T will be approximately $3nkT$ J cm^{-3}. The thermonuclear power produced and the bremsstrahlung power loss in the time t can be represented by P_{th} and P_{br} respectively, and it can be assumed that all of the bremsstrahlung will be absorbed in the heat collecting system of the reactor. As a result, the energy liberated, assuming that the plasma is cooled to the initial (presumably room) temperature, will be $(P_{th} + P_{br})t + 3nkT$ J cm^{-3}. If it is assumed that 33 per cent efficiency can be achieved in the conversion of this energy, then the condition for the operation of a successful fusion reactor will be that one-third of the above energy is greater than $P_{br}t + 3nkT$ J cm^{-3}. That is, after rearrangement, that

$$\frac{P_{th}/3n^2kT}{(P_{br}/3n^2kT) + 1/nt} > 2$$

We know that P_{th} and P_{br} both vary as n^2 so that only T and the product nt determine the value of this inequality. T must be greater than the ideal ignition temperature, and nt can be determined by either numerical or graphical solution of the minimum value of the inequality using the expressions already developed for P_{th} and P_{br}. It has been found that for the D–D reaction nt must be greater than about 10^{16} atoms s cm^{-3} and for the D–T reaction $nt > 10^{14}$ atoms s cm^{-3}. This is known as the *Lawson criterion* (Lawson, 1957) and represents the turn around point at which a fusion reactor begins to produce more useful power than it absorbs. Obviously the exact value of nt at which this occurs depends on the efficiency with which the power can be extracted, but this will probably be reasonably close to 30 per cent, typical of present day power plant.

The neutrons released in the fusion reactions are also an important consideration in feasibility calculations for two reasons. First, they can be used to convert Li^6 into tritium if a blanket of lithium is formed around the fusion reactor.

Second, they could mimic breeder reactors by converting uranium-238 into plutonium-239 or thorium-232 into uranium-233. Since one neutron is produced for the consumption of one tritium nucleus in the fusion, and since the absorption of neutrons by the lithium blanket cannot be assumed to be totally efficient in producing new tritium, it is apparent that the fusion reactor cannot be totally self-sufficient in its tritium supply. It is possible that this deficiency could be supplemented to a certain extent by including materials such as beryllium-9, bismuth-209 or lead-207 in the blanket and using neutron multiplying (n, 2n) reactions. Even with these, it is probable that all D–T fusion reactors will need to have tritium supplied to them from an external source. They will therefore be dependent on the existence of fission reactors as neutron sources for the conversion of Li^6.

11.4 Approaches to Controlled Fusion

We now know the problems that must be solved experimentally for a pulsed nuclear fusion reactor to operate and produce power from deuterium and tritium as the fuel. The temperature must be above the ignition temperature (about 7 keV (8×10^7 K) to allow for bremsstrahlung losses), the density of nuclei should be about 10^{15} per cm^3 to allow reasonable power densities and the Lawson criterion must be satisfied; that is, the product of the number density and the containment time must be greater than about 10^{14} atoms s cm^{-3}. This corresponds to a containment time of one-tenth of a second.

The fundamental problem of containment is that the plasma must not be allowed to come into contact with the walls of the containment vessel. If this were allowed to happen then not only would there be intense local heating, which would produce conditions beyond the capabilities of ordinary materials to withstand, but also there would be a loss of energy from the hot particles in the plasma. These cooled particles would return to the plasma reacting region, and would serve to lower the temperature. In addition, the absorption of some of the plasma particles by the surface of the containment vessel would lead to sputtering and evaporation. This would result in the appearance of large impurity concentrations in the plasma, and would lead to a considerable increase in the bremsstrahlung losses.

Magnetic Containment

The solution to the problem clearly cannot lie in the nature of the walls and other means of confinement must be sought. The most promising appears to be the use of magnetic field of suitable geometries. Electric fields can be suggested as a possibility, but they must be ruled out almost immediately for two reasons. Firstly, a plasma consists of a mixture of positively and negatively charged particles. If a system of electrodes can be established which will contain one set, then particles with the opposite charge cannot be contained because the electrodes will have an opposite effect on them. Second, Earnshaw's theorem

(Post, 1956) shows that there is no position of stable equilibrium for a charged particle in an electrostatic field, no matter how complex the structure. As a consequence, it is impossible for a plasma to be contained by electrostatic methods.

The only apparent alternative is the magnetic field. If a magnetic field can be formed which is closed on itself in all directions, then charged particles will always be deflected when they try to move radially outwards and will never be able to escape if the field is strong enough. It is fairly straightforward to estimate the necessary field strengths. If a plasma consists of n_i ions with velocity $\mathbf{v}_i$, n_e electrons with velocity $\mathbf{v}_e$ and if the ions are assumed to be hydrogen isotopes, then the pressure gradients arising through the forces exerted by a magnetic field $\mathbf{B}$ are

$$\nabla p_i = n_i e \mathbf{v}_i \times \mathbf{B}$$

$$\nabla p_e = - n_e e \mathbf{v}_e \times \mathbf{B}$$

Thus the overall pressure gradient is

$$\nabla p = ne(\mathbf{v}_i - \mathbf{v}_e) \times \mathbf{B}$$

if the number densities of ions and electrons are assumed to be equal. The quantity $ne(\mathbf{v}_i - \mathbf{v}_e)$ is the net rate of movement of charge and is the current density $\mathbf{j}$ Thus

$$\nabla p = \mathbf{j} \times \mathbf{B}$$

Now, from Maxwell's equations, $\nabla \times \mathbf{B} = \mu_0 \mathbf{j}$ where $\mu_0 = 4\pi \times 10^{-7}$ H m^{-1} so that

$$\mu_0^{-1}(\nabla \times \mathbf{B}) \times \mathbf{B} = \nabla p$$

If we assume that the confinement is to be purely radial in a straight magnetic field, the left-hand-side of this equation simplifies to

$$-\frac{1}{2\mu_0} \nabla B^2,$$

so that

$$\nabla\left(p + \frac{B^2}{2\mu_0}\right) = 0$$

making

$$p + \frac{B^2}{2\mu_0} = \text{constant}$$

The quantity $B^2/2\mu_0$ represents a *magnetic pressure*, and is the energy density of the magnetic field. If the plasma is contained by an external field $\mathbf{B}_0$, then the pressure at the outside of the plasma must be equal to zero, and so $p + B^2/2\mu_0 = B_0^2/2\mu_0$. It should be noted that B represents the field at any point within the plasma. If the plasma is fully diamagnetic so that all magnetic field is excluded, then B will be zero, and the pressure drop at the surface will be a maximum. In this case $p_{max} = B_0^2/2\mu_0$ and an estimate of the fields involved in containment can be made. The kinetic pressure is nkT where n is the particle density of the plasma. Assuming 2×10^{15} particles per cm³, at a temperature of 10 keV, the pressure will be $1 \cdot 6 \times 10^6$ W m⁻² leading to a containment magnetic field of 2 Wb m⁻². This is well within the reach of present day technology. Various possible geometries for the magnetic fields have been suggested. The simplest is the *Pinch effect* which is exactly related to the mutual attraction between two current carrying wires. The fuel is contained in a tube and a heavy electric current is passed through it. The effect of this current is to produce many charges moving parallel to each other with high speeds. As a result, they experience the same mutual attraction as current carrying wires and the gas condenses towards the axis. With the condensation, the pressure and temperature increase, as does the degree of ionization, and with it the current and the strength of the self-constriction. Unfortunately there are too many instabilities for this system to be successful and the duration of the pinch is quite short. All the other magnetic containment systems rely on the application of an external magnetic field and various possibilities have been *stellarator, astron, magnetic mirror* and the latest and apparently the system with the greatest hope of success, *Tokomak*.

The magnetic mirror systems (Figure 11.6) have a straight reaction volume with axial magnetic fields and rely on the creation of much stronger magnetic fields at the ends than at the middle. These stronger fields are supposed to constrain the charged particles of the plasma to return to the reaction volume.

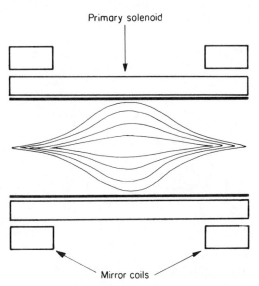

Figure 11.6. Magnetic mirror plasma confinement system

They do succeed in this, but, unfortunately there is a 'hole' along the axis of the field through which the plasma can leak. Adiabatic compression of the plasma to reach the ignition temperature was suggested through the longitudinal movement of the mirrors, so as to constrict the reaction volume, and the simultaneous increase of the current through the central solenoid to obtain radial compression.

The Astron system consists of another axial field, into which are injected highly energetic electrons. It is suggested that these electrons will form themselves into a circulating layer and will produce closed field lines through interaction with the axial field. This closed field system will then provide plasma containment.

The alternative approach is to use an external solenoid to provide the confining magnetic field, and to bend the reaction volume round on itself to form a torus. There have been several different toroidal confinement systems, with various winding geometries aimed at eliminating or controlling the instabilities that can arise. It is in the field of understanding these various instabilities that most progress has been made since 1962 when the ZETA toroidal pinch experiments of the UKAEA seemed to seriously threaten the whole of thermonuclear development. The most successful confinement system to date, and the one that shows most promise of extension to reactor conditions, is the Russian developed Tokomak. It is a toroidal pinch discharge stabilized by a very strong applied magnetic field. A typical system is sketched in Figure 11.7. The toroidal compo-

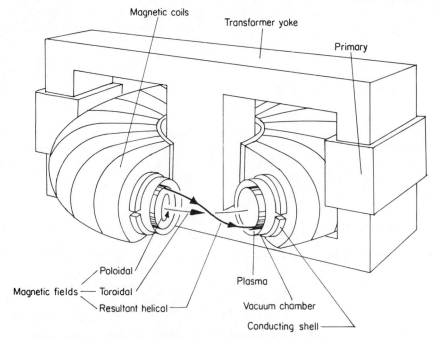

Figure 11.7. Schematic drawing of a Tokomak toroidal confinement system

Figure 11.8. The United Kingdom Atomic Energy Authority CLEO–TOKOMAK device. (Reproduced by permission of United Kingdom Atomic Energy Authority)

nent of the magnetic field in the torus is produced by the circular array of windings, and a poloidal component arises through the pinched rotating current in the plasma. The resultant magnetic field is helical as shown in the figure. This system is still prone to instabilities as the plasma expands radially. To overcome this, the designers of Tokomak enclosed the reaction volume in a conducting toroid whose effect is to produce forces opposing the growth of the instability as the poloidal magnetic lines of force around the plasma expand and generate eddy currents in the conductor.

Several Tokomak systems have been built or are being built around the World, and Figure 11.8 shows the UKAEA installation at Culham. The performance of the Russian T-3 machine as measured in a joint British–U.S.S.R. study is indicative of the success being achieved (Marcham and Pease, 1972; Anashin and coworkers, 1971). Plasmas have been sustained for one-tenth of a second, electron temperatures of 30×10^6 K have been reached and ion temperatures of 6×10^6 K. Abundant thermonuclear reactions in deuterium have been identified, containment times have reached 10–20 ms, and nt has approached 10^{12}. The design parameters of the various Tokomak installations are intended to be complementary so that, for example, some of them are specifically intended to try to reach high temperatures. Two such devices are Alcator at Massachusetts Institute of Technology, and the Texas Tokomak at the University of Texas. This latter has recently produced temperatures of 200×10^6 K (17 keV) though for extremely short containment times. Nevertheless, this is the first machine to report the attainment of temperatures above the ignition temperature for deuterium–tritium fusion.

In the stellarator system, confinement is again attained through the use of a toroidal structure. The difference between it and Tokomak is that Stellarator has no toroidal current through the plasma, but rather helical multipole windings are placed on the torus. The merit of this is that all of the containment forces are supplied by external fields making the variation of experimental conditions more easy. It may, therefore, make a better test-bed for reactor development. It is, however, much more expensive to construct than Tokomak.

Inertially Confined Plasmas

Another possible approach to controlled fusion is to use the experience of the hydrogen bomb and try to develop smaller scale explosive heated inertially confined plasmas that will produce power. This is possible because the number densities of solid materials, at about 5×10^{22} atoms cm^{-3}, require a containment time of only 0.2×10^{-8} seconds or greater for the Lawson criterion to be satisfied. Since the velocity of disassembly should be about the speed of sound at 10^8 K, that is about 10^8 cm s^{-1}, is apparent that pellets of about 1 cm diameter would be necessary, and about 10^8 J of heating energy would be required. This must be supplied in less than the containment time, so that heating currents greater than 10^{16} W are essential. This is a formidable task.

Calculations made at Livermore Radiation Laboratory, United States of America, (Nuckolls and coworkers, 1972) have indicated the possibility of significant advances being made in the compression of solid hydrogen using spherically symmetric irradiation with high-power lasers. A major problem is the avoidance of a strong shock wave which would reduce the maximum compression that could be achieved, and instead, a carefully shaped pulse of incident energy can be used to ensure an implosion compression that is close to adiabatic. It is believed that densities greater than those at the centre of the Sun may be achieved. This must be an area of intensive experimental effort in the next few years, and not the least of the effort will be devoted to the development of intense laser beam equipment in which the pulse time is extremely short, between 10^{-12} and 10^{-9} seconds.

Despite the fact that there does not seem to have been sure experimental confirmation of the Livermore predictions as yet, it must be emphasized that high-power lasers have been used for plasma heating and, indeed, there have been reports (Nature, 1974) of the detection of thermonuclear neutrons from laser-compressed deuterated targets. In these experiments, a density increase of 80 times was claimed when two opposing Nd-laser beams were focused onto the solid-state target. While the number of neutrons is small at about 10^{-8} of the break-even yield and the laser energy is only about 100 J, the results are encouraging and experiments involving laser energies from 1 to 10 kJ are being planned.

11.5 Thermonuclear Engineering

With the renewed enthusiasm for fusion as a power source, brought on by the success of the Tokomak devices and the predictions regarding laser induced fusion, attention has turned to the problems of large-scale engineering that must be solved if a commercial fusion reactor is to be built. Many of the problems are qualitatively similar to those encountered in either conventional power stations or in breeder reactor plant. These include heat transfer, thermal stress, radiation damage from the neutron flux and shielding, both thermal and biological. Some of these are less acute than in conventional nuclear plant. For example, the thermal loadings will be lower and, because of the extremely small amounts of fuel in a reactor at any one time, the risks from an accident are minimal. A typical proposed toroidal reactor, see Figure 11.9, with a doughnut some 5·5 m in radius and a section some 1·25 m radius, would contain less than one gram of fuel at a density of 10^{15} atoms per cm^3. In addition, the constraints set by the breeding of tritium are less than those in the breeding of plutonium. The fuel is very inexpensive and fuel processing is slight. However, there is the major problem of magnetic containment, and this involves a large capital cost. The reaction torus will be 2·5 m high and 13·5 m in outer diameter. This will be surrounded by a breeding blanket of lithium for the production of new tritium fuel; the blanket will be penetrated and surrounded by the heat-absorbing medium which will conduct heat to the heat exchangers and hence to the turbines for producing electricity. This blanket must be surrounded by a heat shield to prevent heating in the magnet coils which then must surround the whole reactor. There must also be a biological shield to protect personnel from any radiation leakage. As a result of all of this a fusion reactor will be a fairly large machine.

The extrapolation from present reactor experiments to power producing designs is rather tenuous and large-scale apparatus is already under construction in Russia, the United States and Germany. This will bridge the gap by providing both operating conditions close to those to be expected in a real

314

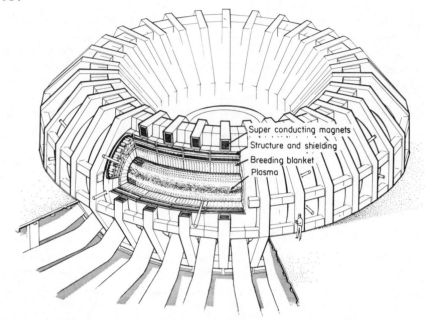

Figure 11.9. Proposed toroidal fusion power reactor. The scale is indicated by the human figure at the bottom right of the diagram. (Reproduced by permission of United Kingdom Atomic Energy Authority)

reactor and experience in the design and fabrication difficulties of such large thermonuclear devices.

11.6 Radioactive Hazards of Fusion Reactors

The inventory of radioactive materials arising through the use of fusion reactors for power production is naturally much less than that from the use of fission. While the tritium fuel itself is radioactive, this is consumed and there are no radioactive products of the reaction itself. However, the fusion reactor does produce a flux of excited neutrons. When these strike the material surrounding the plasma they may well induce nuclear reactions in it, and may generate radioactive waste. It is a task for thermonuclear engineers to reduce this radioactive inventory to a minimum by the choice of structural materials that do not produce long-lived radioactive isotopes under neutron bombardment. Two materials are widely considered as suitable, niobium and vanadium. Niobium produces only two long-lived radioactive isotopes Nb^{93m} and Nb^{94}, and vanadium produces none, though impurities in the vanadium would lead to a certain radioactive contamination. This would be several orders of magnitude less than the niobium activity, which in turn is several orders of magnitude less dangerous than fission reactors. Nonetheless, there is a certain hazard which poses the same containment problems on a much smaller scale.

It is tempting to think of vanadium as the solution of this problem in the blanket structure but unfortunately its thermal characteristics are somewhat less attractive than its neutron characteristics. It would mean operating at lower temperatures than would niobium, and would, therefore, yield a lower thermodynamic efficiency. However, its radioactive inventory is so attractive that the other objections to its use may prove to be secondary considerations.

11.7 Summary

No attempt has been made in this chapter to provide an authoritative and complete account of the physics and technology of thermonuclear fusion. Instead, an attempt has been made to provide a summary of the elementary considerations essential to an understanding of the underlying principles together with a brief account of the progress being made towards its achievement. An excellent picture of this progress can be seen in the United States Congressional Hearings documents listed in the Further Reading list at the end of this chapter. Much of the necessary theory for understanding the nature of plasmas and their instabilities is to be found in textbooks on Plasma Physics and some of these are included in the list.

References

Anashin, J. M., E. P. Gorbunov, D. P. Ivanov, S. E. Lysenko, N. J. Peacock, D. C. Robinson, V. V. Sannikov and V. S. Strelkov, *Soviet Physics JETP*, **33**, 1127 (1971).
Arnold, W. R., J. A. Phillips, G. A. Sawyer, E. J. Stovall and J. L. Tuck, *Phys. Rev.*, **93**, 483 (1954).
Atkinson, R. d'E. and F. Hootermans, *Z. Physik*, **54**, 656 (1928).
Boulègue, G., P. Chanson, R. Combe, M. Feix and P. Strasma, *Proc. Second U.N. Conf. on Peaceful Uses of Atomic Energy*, **32**, 409 (1958). The D–D reaction values in Table 1 are a factor of ten too large.
Gamow, G., *Phys. Rev.*, **53**, 598 (1938).
Glasstone, S. and R. H. Lovberg, *Controlled Thermonuclear Reactions*, p. 25ff, D. van Nostrand Co. Inc., Princeton, New Jersey, 1960.
Lawson, J. D., *Proc. Phys. Soc. (London)*, **B70**, 6 (1957).
Marsham, T. N. and R. S. Pease, *Atom*, No. 196, 1972.
Nature, **251** Sept. 13, 99 (1974).
Nuckolls, J., L. Wood, A. Thiessen and G. Zimmerman, *Nature*, **239**, No. 5368, 139 (1972).
Post, R. F., *Ann. Rev. Nuclear Sci.*, **9**, 367 (1956).
Thompson, W. B., *UKAEA Report AERE T/M-138*, 1956; *Proc. Phys. Soc. (London)*, **B70**, 1, 1957.
Tuck, J. L., *USAEC Report LAMS-1640*, 1954.
Tuck, J. L., *USAEC Report LA-1190*, 1955.
USAEC Report LA-2014, N. Jarmie and J. D. Seagrave, (Eds.), 1957.
Wandel, C. F., T. H. Jensen and O. K. Hansen, *Nuclear Instr.*, **4**, 249 (1959).

Further Reading

Coppi, B. and J. Rem, *Scientific American*, **227**, 65 (1972).
Delcroix, J. L., *Plasma Physics*, (2 Vols.), John Wiley and Sons Inc., New York, 1968.

316

Emmett, J. L., J. Nuckolls and L. Wood, *Scientific American*, **230**, 24 (1974).
Glasstone, S. and R. H. Lovberg, *Controlled Thermonuclear Reactions*, D. van Nostrand Co. Inc., Princeton, New Jersey, 1960.
International Atomic Energy Agency, Vienna, *Plasma Physics and Controlled Nuclear Fusion Research*, Unipub Inc., New York, 1966, 2 Vols. 1968, 2 Vols. 1971, 3 Vols. 1972, Nuclear Fusion Supplement.
Lubin, M. J. and A. P. Fraas, *Scientific American*, **224**, 21 (1971).
Tanenbaum, B. S., *Plasma Physics*, McGraw-Hill Book Co., New York, 1967.
U.S. Congressional Hearings, Subcommittee on Research, Development and Radiation of the Joint Committee on Atomic Energy, Congress of the United States. Current Status of the Thermonuclear Research Program in the United States. Nov. 10th and 11th 1971, Parts 1 and 2. U.S. Government Printing Office, Washington, 1971.

12 Solar Power

The previous chapters of this book have been concerned with the conventional sources of power: coal, oil, peat and nuclear energy. These have been the dominant power sources for industrial society, and will continue to be so for the foreseeable future. There are other possibilities, however, including wind power and hydroelectric power. These will be dealt with in the next chapter, but since most of the possible alternatives to conventional fuel sources rely ultimately on the incoming radiation from the Sun, we will first examine the possibilities for direct utilization of the Sun's radiant energy.

In Chapter 2, the input of solar energy to the Earth was discussed in some detail, and it was shown that the total received was 1.8×10^{14} kW, some 20,000 times the present average artificial power consumption. This suggests that the direct conversion of the Sun's radiation could supply all of our energy needs for a very long time. To a good approximation, the radiation spectrum arriving at the Earth from the Sun is characteristic of a black body with a temperature of 6000 K. This means that, if the collector and storage medium used to trap this radiation are theoretically perfect, with no heat losses and with materials capable of withstanding the temperatures, it is in principle possible to attain this temperature. Thus, assuming that such efficient collection of the energy supplied is possible, solar energy converters share one problem with nuclear fission reactors—the power must be extracted at such a rate that the temperature in the collector does not rise above that which can be tolerated by its structural materials. In fact, this is usually not a limitation, because the energy flux from the Sun is low at about 600 W m^{-2} at the Earth's surface, on average. Thus collection areas are very large, and heat losses great. In the French solar furnace, at Odeillo, in the Pyrenees, however, temperatures as high as 3800 K have been attained using a large array of focusing mirrors to concentrate the Sun's energy onto a fairly small furnace. Usually the incoming energy is lost through reflection, conduction away from the collecting medium, reradiation and convection from the surface, and steps have to be taken to counteract these losses.

12.1 Equilibrium Temperature of Irradiated Matter

Let us consider a more restricted collector consisting of a thin flat plate lying on a perfect insulator, and completely thermally isolated from its surroundings. Such a plate will lose energy only by radiation. If the solar flux onto this

317

plate is P, and if the plate will absorb a fraction α of this radiation (α is called the *absorptivity* of the material), then it will heat up until it reaches an equilibrium temperature determined by the Stefan–Boltzman law (equation (2.3)), such that the re-emitted radiation just balances the absorbed radiation. Thus,

$$\alpha P = \varepsilon \sigma T^4$$

where ε is the appropriate value of the *emissivity* of the plate. From this it is clear that the equilibrium temperature of the plate is given by

$$T^4 = \frac{\alpha}{\varepsilon} \frac{P}{\sigma} \tag{12.1}$$

and the highest equilibrium temperatures will be reached with collectors having high α/ε ratios. That is, for plates which absorb as much as possible and re-emit as little as possible of the incoming radiation. The ratio α/ε can be as high as two or three, but is usually about unity. Such an absorber is called a *neutral absorber*. It should be emphasized that the absorptivity relates to the solar spectrum, that is to black-body radiation at temperature 6000 K, while the emissivity relates to the temperature of the absorber. A black body will have identical values of α and ε at any given temperature.

An estimate of the magnitudes involved can be gained by considering such an idealized neutral absorber subject to an incoming energy flux of 625 watts m^{-2}, which is less than the average flux in the tropics. Using $\alpha/\varepsilon = 1$, we find that the equilibrium temperature is 322·5 K = 49·5°C.

This calculation makes no allowance for the interaction between the absorber and the surroundings. If we assume still that there is perfect insulation behind the plate, and that the only interaction is with the atmosphere through convection, we can make a reasonable allowance for this. Air in contact with the warm surface is heated, rises and is replaced by cool air; thus there is a steady stream of cooling air across any heated surface. On a calm day, and for small temperature differences between the plate and the surrounding air, this will remove about 4 watts m^{-2} per K difference between the plate and the air. If the air is moving, then the rate of heat removal is much greater, for example becoming about 30 watts $m^{-2} K^{-1}$ at a wind speed of about 10 m s^{-1}.

A further effect is the presence of long wavelength radiation in the atmosphere. This arises through the greenhouse effect discussed in Chapter 2, and is diffuse in character. It is still present even under complete cloud cover, and represents an energy input of between 100 watts m^{-2} under clear sky conditions, and 300 watts m^{-2} under damp cloudy conditions. Thus this effect counters the loss of heat due to convection, and serves to help make our earlier estimate of the equilibrium temperature of the plate reasonably correct.

It is possible to improve the efficiency of our flat plate collector by two means. The absorbers can be selectively chosen so that their α/ε ratio is high, and convection can be cut down by placing a transparent cover over the plate. It is possible to produce oxide coatings, usually of nickel and copper, which will give a polished metal plate an absorptivity of about 0·9 and an emissivity for long wavelength radiation of about 0·1. The ratio is, therefore, about nine and our equilibrium temperature would rise by about a factor of $\sqrt{3}$ to 558 K (285°C). When allowance is made for convection, and the long wavelength component, however, this figure is reduced to approximately 400 K (127°C). If a cover plate is used, this reduces convection losses. Most glasses will transmit about 90 per cent of solar radiation, but only about 10 per cent of long-wavelength radiation. Thus there is a greenhouse effect raising the temperature of the air within the enclosure, and reducing the convection losses from the plate. In addition, the temperature of the cover plate itself is raised through the absorption of the long-wavelength component. Let us consider as an example a cover plate which is totally transparent to incoming solar radiation, totally absorptive to long-wavelength radiation, and which has an emissivity of one. This is protecting an absorber of absorptivity α and emissivity ε. The temperatures are T for the absorber, T_c for the cover plate and T_a for the surrounding air. The convection losses between the absorber and the cover can be taken as h watts m^{-2} K^{-1} and those between the cover and the air as h_c. Then, at equilibrium, heat is being lost as fast as it comes in and so the power loss from the absorber is

$$P_1 = \alpha P \tag{12.2}$$

where P is the incoming solar power. The exchange of heat between the absorber and the cover must be P_1, and is made up of two components, the convective term, proportional to the temperature (within limitations), and the radiation term governed by the Stefan–Boltzman law, so that it is

$$P_1 = h(T - T_c) + \varepsilon\sigma(T^4 - T_c^4) \tag{12.3}$$

The exchange between the cover and the air is similar, and includes a term for the absorption of P_a, the incoming long-wavelength radiation from the atmosphere. Thus,

$$P_1 + P_a = h_c(T_c - T_a) + \sigma T_c^4 \tag{12.4}$$

These three equations will define the equilibrium temperatures of the two plates for any given conditions. For example, suppose that $\alpha = 0·9$, $\varepsilon = 0·9$, $P = 600$ watts m^{-2}, $T_a = 300$ K, $h = h_c = 4$ watts m^{-2} K^{-1} and the plate is one square metre in area. Then the equilibrium temperature for the absorber will be 367 K (94°C). If we use a selective absorber with $\varepsilon = 0·1$, this temperature goes up to 427 K (154°C). These represent a significant improvement over the exposed collector. This figure needs to be revised somewhat to allow for the non-ideal behaviour of the cover plate and its absorption of some of the incoming radiation. Because of this absorption significantly better performances are not achieved merely by increasing the number of covering plates. Rarely is any further improvement gained after the second.

The temperature in the collector can be raised if the solar flux onto it can be increased. This can be achieved by focusing the light collected over a large area onto a small one. Such focusing is possible only with the direct solar radiation, and not with the diffuse, long-wavelength component. The factor by which the collector flux is increased is obviously the ratio of the focusing mirror area to the absorber area, and is known as the *concentration ratio*. For a system of mirrors and a flat plate absorber such that the concentration ratio is two, equations (12.2), (12.3) and (12.4) must be solved again with $P = 1200$ watts m^{-2} and all the other conditions held to their previous values. This would produce temperatures of 428 K (155° C), and 539 K (266°C) for the neutral and selective absorbers, respectively. This type of concentration can be achieved very simply by having flat mirrors inclined at an angle to the absorber and the Sun's rays in such a way as to reflect more light onto the absorber. It must be remembered that in order to achieve a concentration ratio of two, rather more than twice the collection area must be normal to the Sun's rays because mirrors are not 100 per cent efficient, and will always absorb some of the radiation.

The ultimate in achieving high concentration ratios is the parabolic mirror. Because the Sun is so far away, the radiation received by the mirror is almost parallel. Consequently, a parabolic mirror can focus it onto a very small focal volume. If this volume is made the collector, very high concentration ratios can be achieved. Unfortunately there is a limit. The Sun's rays are not exactly parallel. Because of the finite size of the Sun's disc, there is a divergence of $\gamma = 0.0093$ radians (32 minutes of arc) which is the angle subtended by the Sun's disc at a point on the Earth. This divergence sets the limit on the smallest size of focus that can be achieved with a given mirror. It is straightforward to show that for the axial component of the incoming radiation, assuming that it is a converging beam of half-angle $\gamma/2$, the size of the image at the focus of the mirror will be $2f \tan(\gamma/2)$ where f is the focal length. This is equal to $f\gamma$ for such small angles of deviation, and a mirror of focal length one metre will produce an image of the Sun approximately one centimetre in diameter at the focus. For light gathered from the non-axial elements of the mirror, however, the picture is more complicated as shown in Figure 12.1. The image formed around the focus is larger because the rays have a longer distance to travel, and also, it is inclined at an angle to the axis of the parabola. The consequence is that for an actual mirror, the size of the focal region is somewhat larger than $f\gamma$ and also has a finite volume. However, as a fair approximation we will take $f\gamma$ as determining the width of the focal region.

There are two types of parabolic reflectors: *axially symmetric* and *cylindrical*. In the axially symmetric type, the reflector is a paraboloid of rotation and forms a dish. If the focal length is f, and the diameter of the dish is d, then the concentration ratio will be determined as

$$\text{Concentration ratio} = \frac{\text{Cross-sectional area of dish}}{\text{Cross-sectional area of focal region}}$$

This is given by $(\pi d^2/4)/(\pi f^2 \gamma^2/4) = (d/f\gamma)^2$. Thus a dish of diameter one metre, and focal length one metre, will have a concentration ratio of about 10^4.

The equilibrium temperature of a small collector placed at the focus of this

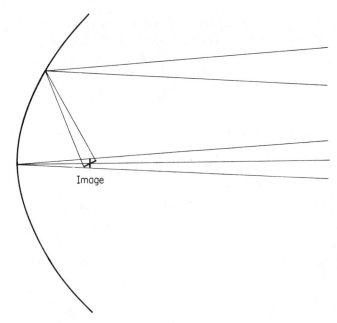

Figure 12.1. Formation of the image at focus of a parabolic
collector

reflector can be calculated from equations (12.2), (12.3) and (12.4) taking care
to correct h for the new size of the focal region. If this is done, however, it
becomes apparent that the losses due to convection are small and that radiation
losses are the main limiting factor in determining T. Equation (12.1), therefore,
is the appropriate equation to use, with P being replaced by $P \times$ concentration
ratio. In our example, the equilibrium temperature, therefore, becomes 3745 K
(3472°C) if $\alpha/\varepsilon = 1$.

For the cylindrical parabolic collector, the concentration ratio is much
smaller. In this case, the width of the initial collector area is d, and its length
is l, giving a cross-sectional area for the mirror of dl. The area of the focal
region is now approximately $f\gamma l$ since it must be a line. Once again, the concentration ratio will be the ratio of these areas, and is therefore $d/f\gamma$. This is the
square root of the value for the dish reflector.

If we consider a one metre focal length mirror with maximum width one
metre, and length one metre, the concentration ratio will be approximately 100,
and the equilibrium temperature 1009 K (736°C), which is considerably above
the operating temperatures of most power stations.

Two features are apparent from this rudimentary analysis. First, the concentration ratios for cylindrical reflectors are only the square root of those for
dish-type reflectors. However, the equilibrium temperature varies with the
fourth root of the concentration ratio so that a reduction of a factor of 100 in
this ratio leads to a reduction in the equilibrium temperature by a factor of
just over three. Second, the cylindrical reflector lends itself to a type of cons-

truction readily suited to power production, It can be made very long, increasing the collection area and the amount of coolant that can be heated. There is no advantage in constructing excessively large reflector assemblies if the same total mirror width d can be achieved with a number of smaller assemblies since the concentration ratio varies linearly with d. This is useful, as an array of smaller assemblies can be constructed more cheaply and in such a way as to offer less wind resistance than one large single reflector.

Further, since the concentration ratio is so insensitive, the quality of the mirrors need not cause excessive concern.

12.2 Heating by Solar Energy

It is clear from the above discussion that the equilibrium temperatures of bodies exposed to the Sun's light can be quite high. However, there are other important features that must be considered, for example what is the time needed for reaching equilibrium, and at what rate can energy be removed from the collector?

Time to Reach Equilibrium

This depends on the thermal capacity L of the system. Thermal capacity is defined as the specific heat of the material multiplied by the mass of material present. It therefore represents the amount of heat needed to raise the temperature of the material by one degree (K). In broad terms, therefore, the time for an absorber to reach its equilibrium temperature will be given by $Pt = L\Delta T$ where P is the incoming energy flux, L is the thermal capacity of the system and ΔT is the temperature difference between ambient temperature and the equilibrium temperature. Thus if the incoming flux is 600 W m^{-2}, and the thermal capacity of the absorber is 1·5 Wh m^{-2} K^{-1}, which represents a copper sheet approximately 2 mm thick, the equilibrium temperature of 320 K (as determined earlier) will be reached from an ambient temperature of 300 K in approximately 3 minutes. Unfortunately this approximation is too rudimentary. We must include the losses in these calculations since these vary with the temperature of the absorber, and both the radiative, T^4, term and the convective, T, term will introduce an effect whereby the rate of increase of the temperature will decrease as the temperature increases. The equation can be readily written for the exposed flat plate absorber by modification of equations (12.2), (12.3) and (12.4). The solar influx of heat as absorbed by the plate is still $P_1 = \alpha P$ as in equation (12.2), but the heat loss when the temperature of the plate is T is given by

$$h(T - T_a) + \varepsilon\sigma T^4$$

Consequently, the heat is absorbed by the plate at a rate

$$H = \alpha P + P_a - h(T - T_a) - \varepsilon\sigma T^4$$

where P_a is the long-wavelength influx.

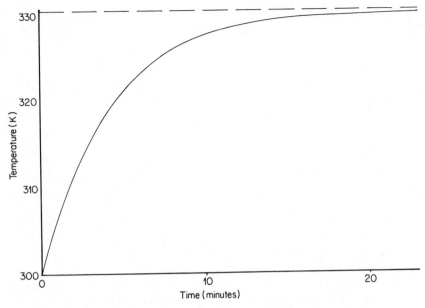

Figure 12.2. Variation in the temperature of a flat plate collector with time

However, the rate at which heat is absorbed by the plate is given by $L\dfrac{\mathrm{d}T}{\mathrm{d}t}$, where L is the thermal capacity of the plate, and so

$$L\frac{\mathrm{d}T}{\mathrm{d}t} = \alpha P + P_a - h(T - T_a) - \varepsilon\sigma T^4$$

This can be rewritten to give the time necessary to reach some temperature T from the ambient temperature T_a as

$$t_T = \int_{T_a}^{T} \frac{C\,\mathrm{d}T}{(-T^4 - AT + B)} \tag{12.5}$$

where

$$A = h/\varepsilon\sigma;\ B = (\alpha P + P_a + hT_a)/\varepsilon\sigma;\ C = L/\varepsilon\sigma$$

A graph of t against T is shown in Figure 12.2 for our example of a neutral absorber with $\alpha = \varepsilon = 0\cdot9$, $L = 1\cdot5$ Wh m^{-2} K^{-1}, $h = 4$ W m^{-2} K^{-1}, $P_a = 200$ W m^{-2} and $P = 600$ W m^{-2}. It is possible to solve the coupled equations (12.2), (12.3) and (12.4) suitably modified to describe the corresponding problem for the shielded flat plate absorber, but the problem is appropriately more difficult. Nonetheless, it is clear that equilibrium temperatures are reached quite quickly, and, in particular, we can assume that the absorber temperature will follow modest changes in solar intensity without undue delay.

Extraction of Heat from Absorber

If the absorber system is used to supply heat to some other body, then

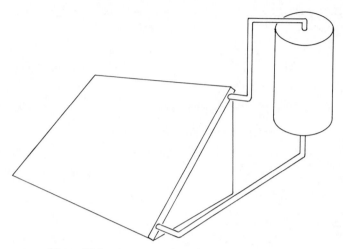

Figure 12.3. Flat plate heater with storage tank

effectively the heat losses of the absorber are increased. This lowers the equilibrium temperature, and reduces the thermodynamic efficiency of the subsequent processes. In addition, the actual relationship between the absorber and the heat-transfer medium, or coolant, will be complex because the coolant is not at the same temperature over the whole of the absorber surface.

The simplest heat-transfer system is the flat plate absorber constructed in such a way that the coolant covers the whole of the back of the plate, The coolant would circulate by convection, and would be transferred to a larger storage tank. Such a system is outlined in Figure 12.3. It is clear that under convective flow, the temperature of the water entering the heater at the bottom will be lower than that leaving at the top. Nonetheless, an estimate of the performance of such a heater can be gained by assuming that the absorber surface is at a uniform temperature. If we assume the same conditions as led to equations (12.2), (12.3) and (12.4), and if we assume that the coolant is at temperature T and removes energy at a rate P_e, then we can rewrite equation (12.2) as

$$\alpha P = P_1 + P_e \qquad (12.6)$$

reflecting the effect that part of the heat absorbed by the plate is lost through convection to the cover, and part is removed by the coolant. This equation can be rewritten as

$$\alpha P' = P_1$$

where $P' = P - P_e/\alpha$. The remaining equations are unchanged, and so the equilibrium conditions of the system can be estimated by solving equations (12.6), (12.3) and (12.4) for different rates of removal of heat from the absorber or for different equilibrium temperatures. A graph showing the equilibrium temperature for different rates of power production is shown in Figure 12.4 for different values of the incident solar flux and for a neutral absorber. The

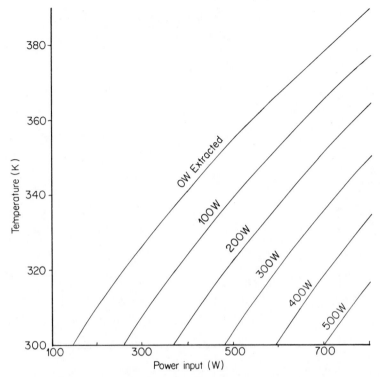

Figure 12.4. Equilibrium temperature of a flat plate collector with $\alpha = \varepsilon = 0.9$, as a function of insolation for different rates of energy removal

corresponding results for the selective absorber used earlier with $\varepsilon = 0.1$, are shown in Figure 12.5. These curves emphasize that as the temperature of heat extraction increases, the amount of heat that can be removed (P_e) decreases for a given insolation and collector geometry. Thus there is a balance that must be struck between desired operating temperature and desired power output.

All solar heaters are faced with the problem that their energy source moves during the day and vanishes completely at night. As a result they either have to be designed in such a way that they track the Sun's path so as to maximize their output, or they must be placed in an optimum position and with an optimum orientation to take the best advantage of their stationary position. It is a relatively straightforward matter to extend the material in Chapter 2 to obtain the angle θ between the Sun's rays and the normal to the flat plate collector. Using the definitions of Chapter 2, Figure 2.8, we can define the collector as having a zenith η, and an azimuth (relative to local noon) ρ. Equation (2.1) has already given the zenith angle of the Sun, z, and the solar azimuth A is given by

$$\sin A = \cos \delta \sin h / \sin z \qquad \text{(Robinson, 1966)}$$

Since the hour angle is $h = 15t°$ where t is the time in hours since local noon,

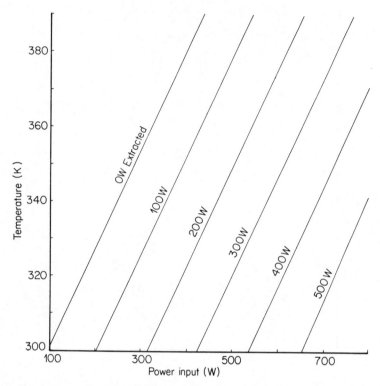

Figure 12.5. Equilibrium temperature of a flat plate collector with $\alpha = 0.9$ and $\varepsilon = 0.1$ as a function of insolation for different rates of energy removal

and the angle of declination can be approximated by $\delta = 23.5° \sin 2\pi d/365$ where d is the number of days since mid-summer, the Sun's zenith and azimuth can be readily calculated for any time in the year. Also, by determining direction cosines for the Sun and the normal to the collector, it is simply shown that

$$\cos \theta = \cos \eta \cos z + \sin \eta \sin z \cos (\rho - A)$$

If the collector is inclined to the solar direction at an angle θ, the power intercepted by it is $I \cos \theta$, where I is the incident solar intensity. It is now possible to examine the performance of a given stationary collector as a function of time of day and for different times of the year, using the insolation data of Chapter 2.

This is presented in Figure 12.6. If the calculation is extended to include the conversion efficiencies of the collector, as in Figure 12.4, we find that a one square metre collector in the United Kingdom could be expected to deliver about $1\frac{1}{4}$ kWh of heat per day at 40°C, even in winter under clear air conditions. However, solar heaters in British conditions must be expected to supply much less heat than their design figure, because of cloud cover.

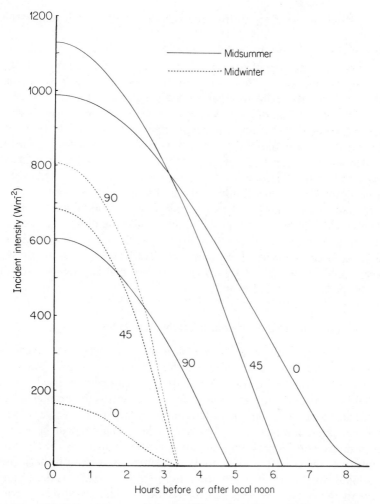

Figure 12.6. Insolation at latitude 55° on flat plate collectors at three zenith angles as a function of time since local noon. Data are presented for both midsummer and midwinter. Allowance is made for atmospheric effects by using the equivalent air–mass model of Chapter 2 with $\mu = 0{\cdot}1$

Radiative Cooling

Another aspect of solar heaters appears when the calculations that produced Figure 12.4 are extended to very low incident radiation levels, and low long wavelength contributions. These conditions will prevail under clear sky conditions at night for example, and result in the equilibrium temperature of the collector falling below the ambient temperature. As a result, the solar heater will remove heat from the heating system and will function as a cooler. This could be an advantage in a hot climate, but in a cold climate the heater would have to be valved off at night to prevent loss of heat. In the hot climate, the valving off would be done during the day to prevent heat gain within the enclosure.

Both of these possibilities reflect one of the important aspects of any solar heating system—storage of the collected energy. This is a problem not only with solar heating systems, but also with many other energy sources. Because of this it will be dealt with in greater detail in a later chapter. However, it is worth mentioning some of the possibilities briefly here. Water is a useful heat storage medium. It has a good thermal capacity, and a fairly poor thermal conductivity. It is a liquid and so can be used as the heat-transfer medium. It wets most surfaces, and so gives good thermal contact with the collector. Its boiling point is fairly high for low temperature storage systems and is adequate for heating purposes. One disadvantage is that its freezing point is also fairly high, so that there is danger of its freezing in the outside ducts at night or in extremely cold weather. Other possibilities are the use of water or some other suitable fluid (even air) as the heat-transport medium, and to store the heat in some other material. Possibilities are materials such as paraffin wax, or Glauber's salt ($Na_2SO_4 \cdot 10H_2O$), which have low melting points and store the energy in the latent heat of fusion.

12.3 Conversion of Solar Energy into Mechanical Work

So far we have considered the conversion of solar energy into usable heat. In this form it can be used directly, with high efficiency, to heat buildings or water, or to dry crops, or to cook. Alternatively it can be converted into some other usable form of energy and the basic conversion is to mechanical work. As the efficiency of a heat engine is determined by the temperature range over which it operates, most solar energy driven heat engines will be extremely inefficient unless considerable concentration is used to increase the equilibrium temperature of the collector. If we make the simple assumption that equation (12.3) holds throughout the entire operating region of solar furnaces, and that it represents the convective losses correctly, then we can investigate the dependence of the equilibrium temperature on concentration ratio simply by raising the insolation P. In fact, the approximation is not too bad at the higher temperatures because the dominant loss is then from radiation, and consequently the actual approximation used to represent the convective loss is unimportant. It is only in the narrow region in which the convection term is approximately equal to the radiation term, and in which the approximation for convection is invalid, that the curve is inexact. The results are shown in Figure 12.7, and it is clear that for an insolation of 400 W m^{-2}, the temperature does not reach the 645 K of a nuclear power station until the concentration ratio is about 20. Even at this temperature the maximum possible conversion efficiency from heat to mechanical energy is only 35 per cent.

In principle it is possible to use any of the external heat source thermodynamic cycles for solar engine operation. Thus the Stirling and Rankine cycles are both suitable. The practical difficulties lie in designing suitable regenerators

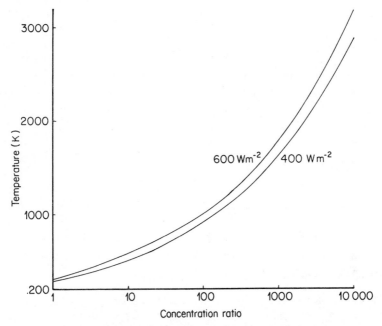

Figure 12.7. Equilibrium temperature of a collector as a function of concentration ratio for two values of insolation

and in producing devices which will allow the expansion and compression stages to take place at almost constant temperature. In fact, if we expect engine efficiency to be approximately half of Carnot efficiency, we will find that the overall efficiency of a flat plate collector driven heat engine will be only some 5 per cent at best, producing a power output of 30 W m^{-2} for our 600 W m^{-2} input used earlier. If concentrators are used to raise the temperature of the collector, the efficiency of the engine can be increased dramatically to approximately 30 per cent with a concentration ratio of 1000. A schematic heat engine based on the Rankine cycle is shown in Figure 12.8, and uses solar energy to drive a shaft through a turbine.

Refrigeration is also possible using solar energy as a heat source. The simplest such device is the absorption refrigerator in which the refrigerant is dissolved in another fluid (for example, ammonia dissolved in water) at a low temperature. This is pumped to a high pressure and the fluids separated by heating. Heat is then removed from the refrigerant in a condenser, the refrigerant is expanded through a throttle valve back to low temperature and pressure and redissolved in the carrier fluid. This is a less efficient refrigeration process than the more usual compression cycle (see Chapter 16), but has the advantage of simplicity. It requires a source of heat for the separation stage and a small power supply for the pump. Because of this it lends itself readily to solar source operation. It is possible to operate entirely from thermal energy using, for example, the Electrolux system (Roberts and Miller, 1958).

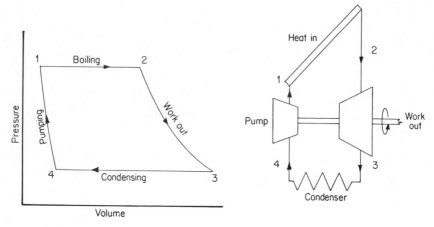

Figure 12.8. Schematic solar heat engine based on the Rankine cycle

12.4 Conversion of Solar Energy into Electricity

The conversion of solar energy into electricity is probably the ultimate aim of most energy conversion engineers. Electricity represents the best in 'convenience fuels', requiring the minimum of effort from the customer, and solar energy represents the ultimate in convenience energy sources requiring the minimum effort from the power generation utility for its conversion. No fuel has to be carried around and there is no waste material that has to be disposed of. Unfortunately, the conversion of solar energy to electricity on a large scale is not a simple task.

In conventional power production, the energy of the fuel is converted into electrical energy in a three-stage process: the fuel is burnt to produce heat, the heat is then converted into mechanical energy and thence to electricity. Electricity generators are extremely efficient and to a good approximation we can say that the conversion from mechanical into electrical energy is achieved without loss. For the conversion of solar energy into electricity, therefore, the limiting process is the collection and conversion into mechanical energy as discussed in the last section. There it was mentioned that the ultimate efficiency of such a system, with a concentration ratio of 1000, is about 30 per cent. The area of collector necessary for power generation on a commercial scale is formidable. At 600 W m^{-2} insolation, and an efficiency of 30 per cent, a plant of 500 MW output would require a collector area of just less than one square mile. In addition, since this is at a concentration ratio of 1000, the entire collection area must be capable of tracking the Sun, probably by using heliostats. This is a formidable technical problem. A further limitation to power production in this way is the intermittent nature of the energy supply. Large scale generating plant is not easily run up and down on a daily basis. Consequently, a system must be developed whereby excess heat is trapped during the sunny period, and stored for use after this period. As a result, the collector area is greatly increased, and storage tanks have to be introduced to the power plant site.

These additions greatly increase the cost and complexity of solar-electricity production by conventional means, and almost completely ensure that, apart possibly from a few prestige projects, other methods of solar energy–electricity conversion will have to be found. These will mostly apply to small-scale projects such as domestic or small industrial installations.

There are two possibilities: direct conversion of the heat collected from the Sun—*thermoelectricity;* and direct conversion of the Sun's light as it is collected —*photoelectricity.*

Thermoelectricity

The best known thermoelectric generator is the *thermocouple* which relies for its operation on the *Seebeck effect*. If two pieces of different metals are joined at one end, and if this joint is kept at a different temperature from the other ends then a potential difference V is established between the two unattached ends. The magnitude of this potential difference is given by

$$V = S\Delta T$$

where ΔT is the temperature difference across the metal wires, and S is known as the Seebeck coefficient. The polarity of the developed voltage depends on the particular metals used, and also on whether the junction is hotter or colder than the free ends of wire. For metals this effect is typically very small, about $10\mu VK^{-1}$, and so the Seebeck effect has not been regarded as a suitable source for large scale electricity production. It has been reserved almost exclusively for temperature monitoring instruments though some simple devices do exist. With the discovery that S can be very large in semiconductors, however, new possibilities arise. Typical values are greater than $150\mu VK^{-1}$, and in some cases the figure can be as high as $1000\mu VK^{-1}$. The potential offered by cascading semiconductor Seebeck junctions in series to increase the voltage developed across the chain, and in parallel to increase the current-producing capabilities becomes attractive.

Unfortunately not only the Seebeck coefficient is important in considering the efficacy of a particular material for thermoelectric generation applications. Two other particular effects have a dominant influence. Firstly, there is the problem of thermal conduction from the hot junction to the cold junction through the semiconductor material itself. This will be reduced by decreasing the thermal conductivity of the material. Secondly, there is the problem of Joule heating of the semiconductor material as the electric current is passed through it. The overall classification of a particular semiconductor as a possible thermoelectric generator material is dependent on the relative values of all three characteristics. The Thomson effect in which the Joule heating is enhanced when current is flowing through a conductor in which there is a temperature gradient is also a contributor to the performance, but we will neglect it for simplicity. If the semiconductor material has a resistance R and is carrying a current I and the temperatures of the ends are T_1 and T_2, where $T_1 > T_2$, we can write down a simplified relation for the efficiency of a thermoelectric device. Since a potential difference appears across the junction when it is heated or cooled, it is reasonable to expect that a junction will in fact heat or cool if a potential difference is established across it. This was found experimentally to be true and is known as the Peltier effect. It can be shown that the energy transferred at a junction is dependent both on the temperature of the junction and on the current being passed. It is given by

$$Q = STI$$

where S is the Seebeck coefficient. Also the rate of heat transfer along the conductor is given by $Q = K(T_1 - T_2)$ where K is the thermal conductance. A practical assembly will be composed of sets of alternate junctions, one set being at the high temperature and the other at a low temperature.

We are now in a position to write down the equations for the power balance at the hot and cold junctions. At the hot junction, energy will be lost through the Peltier effect, and thermal conduction. It will be gained through some fraction f of the Joule heat passing in from the conductor. We can write, therefore, that

$$Q_1 = ST_1I + K(T_1 - T_2) - I^2Rf$$

At the cold junction the power balance will be

$$Q_2 = ST_2I + K(T_1 - T_2) + I^2R(I - f)$$

The useful power available from the thermoelectric device is therefore

$$P_2 = Q_1 - Q_2 = S(T_1 - T_2)I - I^2R$$

and the efficiency of energy conversion is

$$(P_2/Q_1) = \frac{S(T_1 - T_2)I - I^2R}{ST_1I + K(T_1 - T_2) - I^2Rf}$$

The value of this depends on the temperature difference between the hot and cold junctions, the thermal conductivity and electrical resistivity of the material and the current passing through it. For convenience f may be taken as one-half. The current can be adjusted by changing the value of the load resistance imposed on the junction. If this is denoted by R_L, then the efficiency will be a maximum for

$$R_L = R(1 + ZT_m)^{1/2}$$

where

$$Z = S^2/(KR)$$

$$T_m = \tfrac{1}{2}(T_1 + T_2) = \text{mean temperature}$$

and will be given by

$$E = (P_2/Q_1) = \frac{n-1}{n+r}(1-r)$$

where $n = R_L/R$, and $r = T_2/T_1$.
Since $R_L/R = (1 + ZT_m)^{1/2}$, the efficiency depends only on the quantity Z and the temperatures of the junctions. We can write E as

$$E = 1 - r - \left(\frac{1-r^2}{n+r}\right)$$

when it is clear that for a fixed value of r, E increases with n, ultimately becoming $1 - r = (T_1 - T_2)/T_1$, the thermodynamic efficiency, when n is infinitely large. Now n depends on Z which involves the thermal conductance and electrical resistance of the material. If we change to the more general parameters, electrical resistivity ρ and thermal conductivity k, and let each arm of the thermoelectric connector have a length l, we can define Z as a figure of merit, given by

$$Z = S^2/4k\rho$$

Z can be regarded as a measurable quantity for any pair of thermoelectric materials. The higher the value of Z, the better a pair of materials will be as a thermoelectric source. The highest values attainable so far seem to be about 0.003 K^{-1}, and this only at certain temperatures; a much more typical value for a cascade of semiconductor elements is 0.0005 K^{-1}.

Suppose such a device is used with a collector of equilibrium temperature 900 K and a sink temperature of 300 K. This is equivalent to a concentration ratio of 50 with a 500 W m^{-2} source. Then the Carnot efficiency is $\frac{2}{3}$, $r = \frac{1}{3}$, $T_m = 600$ K and $n = 1.14$, resulting in a final efficiency of 5.9 per cent—or about 30 W m^{-2} if $Z = 0.0005$ K^{-1}. This is about the same efficiency as a solar driven mechanical engine. It is worth noting that, if the figure of merit can be increased to about 0.005 K^{-1}, the efficiency of this collector would become $\frac{2}{7}$ or almost 30 per cent. Whether such high figures can be attained without restrictions in the operable temperature range is still debatable. If we accept limitation of of the upper temperature to about 400 K, however, and use a material with $Z = 0.002$, then $n = 1.3$, and the efficiency becomes about 3 per cent. Since this is some 50 per cent better than the efficiency of conversion in plants, and since these temperatures can be reached without complicated concentrators, this is an attractive proposition. However, it must be borne in mind that the same restrictions on final equilibrium temperature, and power removed from the collector, apply as in the case of solar heaters, so that the final usable efficiency will be lower than this, probably less than one per cent.

The other realizable approach to thermoelectric conversion of solar energy is to use *thermionic emission*, in which the Sun's radiation heats the cathode of a diode valve-like assembly. Electrons liberated from the cathode would drift to the anode and return via some external load circuit. It is immediately apparent that such a system must involve energy losses, both through radiation and also through some electrons being released from the anode and drifting back to the cathode. This will occur if heat lost by the cathode is allowed to raise the temperature of the anode unduly, and it constitutes a net loss to the useful current passed by the device. The ultimate limit in the efficiency, therefore, is represented by the rejection of heat from the hot cathode to the cold anode. Unfortunately other factors limit the efficiency before this limit is reached.

The number of electrons that will be emitted from the surface of a material at temperature T is given by the Richardson equation

$$j = AT^2 \exp\left(-\phi/kT\right) \tag{12.7}$$

where j is the current density, A is a constant, k is the Boltzmann constant and ϕ is a paramater of the material known as the *work function*. The performance of some typical electrode materials is shown in Table 12.1, where it is

Table 12.1 Emissive properties of possible electrode materials at different temperatures

Material	ϕ	Thermionic current density (A m^{-2})				
	(eV)	500 K	1000 K	1500 K	2000 K	2500 K
Tungsten	4·5			0·1	25	$6·5 \times 10^3$
Caesium on tungsten	2·1		100	5×10^6	4×10^7	
Caesium on silver oxide	1·0	25	10^7			

clear that high work functions imply low current densities, and that the current density increases rapidly with temperature. High temperatures have to be reached before the current becomes usable at about 10 A cm^{-2} (10^5 A m^{-2}). In addition, there is a limit imposed by the Joule heating of the electrode itself when high current densities are available.

If the anode is maintained at a temperature T_a which is assumed to be low enough that there is no back emission of electrons, that is that its Fermi level is much lower than that of the cathode, then we can write down the balance equation corresponding to equation (12.3) for the flat plate collector. This is

$$P \times CR = j_c(\phi_c + 2kT_c) + \varepsilon\sigma T_c^4$$

For a current density j_c, energy must be supplied to the cathode at a rate $j_c\phi_c$ and because of the statistical distribution of electron energies at temperature T_c, a fraction will have energy greater than the necessary ϕ_c. This fraction is usually estimated to be equivalent to the addition of a further $2kT_c$. Equation (12.7) can now be used to substitute for j_c resulting in an equation relating the equilibrium temperature of the cathode and the incoming insolation, as

$$P \times CR = (\phi_c + 2kT_c)AT_c^4 \exp\left(-\phi_c/kT_c\right) + \varepsilon\sigma T_c^4$$

The potential difference across the generator is given by the difference in the work functions at the cathode and the anode, and is therefore $\phi_c - \phi_a$ which is typically about one volt. This creates the problem that extremely high currents must be handled in order to generate large amounts of power.

The very existence of these large currents produces yet another problem, *space-charge limitation*, in which the cloud of negatively charged electrons in the space between the anode and the cathode of the generator actually prevents further electrons from being emitted by the cathode and so limits the current that can be drawn. Certain steps can be taken to minimize this effect, for example the space between the anode and cathode can be made very small, of the order of 1 μm, so that there are very few electrons in the space between anode and cathode. Another possibility is to deliberately introduce positive ions into the space to soak up some of the space charge.

Large currents are a nuisance. The losses in transmission of low voltage large current supplies are severe, and in addition there are magnetic fields produced which themselves cause problems. To minimize these effects, thermionic generators can be cascaded in series so that each produces a fairly small current and the voltage across the whole assembly is large.

If a thermionic generator is used to drive a mechanical engine, then in principle the overall efficiency of the combination will be greater than that of a direct solar powered engine. This is because the upper temperature of an engine is limited by the materials, and we may assume that it is not greater than 1300 K. In a thermionic generator, on the other hand, the upper temperature limit may be much higher with a corresponding increase in efficiency.

The work function of cathode materials for thermionic generators must be less than about 3 eV so that the electron emission is high at reasonable temperatures. One of the most commonly used materials is the caesium coated or impregnated tungsten cathode with a work function slightly greater than that of bulk caesium at about 2 eV. It is instructive to examine the variation of T_c with input power for this value of ϕ_c. This is shown in Figure 12.9, and it is clear that because the emission of energy increases rapidly with cathode temperature, the cathode temperature is itself fairly independent of insolation. A feature of thermionic generators, therefore, is that extensive care in limiting the upper temperature of the cathode is unnecessary. What is necessary, however, is to vary the load according to the current generated so as to produce a *matched load* such that the potential across it is always $\phi_c - \phi_a$. This leads to complexity in the load circuitry, though charging of a lead–acid storage battery provides a good load for this purpose. The lead–acid accumulator will accept a wide range of charging currents while maintaining the potential across its electrodes almost constant. As such it qualifies as a matched load.

In common with heating panels of various sorts, thermoelectric generators require concentration of the solar radiation to achieve high equilibrium temperatures in order to increase the efficiency of the device—or to make it work at all. The question must be asked, whether or not it is possible to use solar

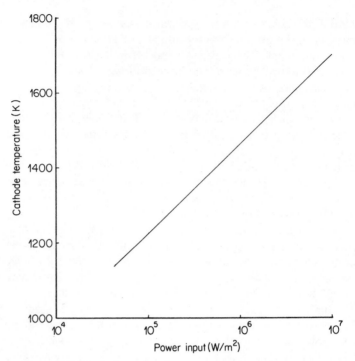

Figure 12.9. Variation of cathode temperature with input power for $\phi_c = 2\text{eV}$

radiation directly, without first producing heat, to generate electricity in useful amounts.

Photoelectricity

The photoelectric effect was discovered in 1887 by Hertz, who found that an electric charge was developed on a metal plate exposed to light. In addition, the charge was not induced until the frequency of the light reached some value, critical for each metal. The effect was explained in 1905 by Einstein in terms of the quantum theory earlier proposed by Planck. Light was envisaged as travelling in wave packets or photons each with energy appropriate to the frequency of the radiation. Thus high-frequency light carries more energy than low-frequency light. On striking a material and being absorbed, the light gives up its energy to the electrons of the material. If it has enough energy, an electron may be removed from its parent atom and escape from the target material, but, if the light photon does not have enough energy to liberate an electron, no effect is observed. The currents that could be developed from such devices, and the efficiencies of conversion, used to be extremely low, but with the development of semiconductor materials, greater hopes are now held out for photoelectric solar energy conversion systems.

The important feature of the solar energy spectrum is no longer its equilibrium temperature, but becomes the distribution of the intensities of various wavelengths throughout the spectrum. This spectrum was shown in Figure 2.13, and can be laid out in tabular form as in Table 12.2. Also since we are no longer

Table 12.2 Distribution of solar intensity by wavelength

Wavelength interval (μm)	Percentage of solar energy in interval
< 0·3	0
0·3–0·5	17
0·5–0·7	28
0·7–0·9	20
0·9–1·1	13
> 1·1	22

interested in the temperature, it must be apparent that the thermodynamical temperature restriction on efficiency of conversion does not apply.

When atoms become packed close together to form a solid, their electrons begin to interact. As a consequence, the distinct energy levels which existed in the free atom, become broadened and blurred and develop into *energy bands* which characterize the solid material. The two bands that are of importance in determining the electrical and optical properties of solids are the *valence band*, the highest energy band for bound electrons, and the *conduction band*, in which electrons are free to move around the solid, migrating from atom to atom. Solids can be characterized according to the size of the energy gap or *band gap* between the two energy bands. Insulators will have a very high band gap so that it is extremely difficult for electrons to jump from the valence band into the conduction band. Metals are characterized by having the conduction and valence bands overlapping, or by having partially filled valence bands, so that there is a constant supply of electrons, free to move under the influence of any applied electric field. There is another category of solid, the semiconductor, in which the band gap is large enough that electrons are prevented from jumping between the valence and conduction bands in large numbers, and yet not so large that the transfer is impossible. Such materials can be either *intrinsic semiconductors*—having the desired characteristic themselves—or *extrinsic semiconductors*, in which the effective band gap is reduced by deliberately introducing impurities which have energy levels lying in the band gap of the parent material, and which will either easily give up an electron to the conduction band of the semiconductor (*n*-type) or absorb one from the valence band (*p*-type). This latter process leaves a 'hole' in the valence band which then behaves similarly to an electron in the conduction band and is free to migrate.

Let us now examine the behaviour of one particular photoelectric material, silicon. Silicon is an extremely common element. It is the basic ingredient of sand in the form of silica, SiO_2. Silicon has a band gap of about 1·1 eV. This corresponds to a wavelength of about 1·1 μm. Consequently solar radiation of wavelength greater than about 1·1 μm (some 22 per cent of incoming solar energy) will fail to excite electrons in silicon from the valence band into the conduction band and will, therefore, fail to cause the material to conduct electricity. Radiation with energy close to 1·1 eV will have a high probability of inducing an electron to transfer, and as the energy of the incoming photon increases, the electron will be excited to higher and higher energies within the conduction band. It will be free to migrate, but will have a high kinetic energy and will cause heating of the solid silicon. Thus, a photon of wavelength 0·6 μm has energy 2·07 eV, and will give the electron about 1 eV more energy than it needs to get into the conduction band. This represents a loss in two respects. First, 2 eV of energy is being used up to do the useful work which only needs 1·1 eV, and second, the silicon is being heated and must be artificially cooled, or else its photoelectric characteristics will change in a detrimental way. If the heat is allowed to accumulate, the silicon may eventually melt.

It we take the fraction of the energy converted in the various wavelength bands from Table 12.2, we can write down the maximum possible energy conversion in silicon at each wavelength in Table 12.3, and show that the limiting

Table 12.3 Energy conversion in silicon

Wavelength interval (μm)	Efficiency of conversion	Proportion recovered (%)
< 0·3	—	—
0·3–0 5	0·36	6
0·5–0·7	0·55	15
0·7–0·9	0·73	15
0·9–1·1	0·91	12
> 1·1	0	0
Total		48

conversion efficiency is 48 per cent which is interestingly high. Unfortunately this figure appears to be unattainable in practical devices and the limit appears to be around 25 per cent, with silicon solar cells having efficiencies typically around 10 per cent.

Any real device will consist of the photosensitive element and an external load circuit, and it is of interest to consider the form of the characteristic I–V curve for different insolations (see Figure 12.10). There are two limiting conditions: the *open-circuit voltage*, E_g, achieved when the photoelectric element is illuminated and no load is placed across it, and the *short-circuit current*, I_s, achieved when the terminals of the device are connected together. The open-circuit voltage is small for low illumination because there is a small number of electrons excited, and consequently a small amount of charge transfer and a small potential difference. The upper limit on this is set by the voltage which can be established across the semiconductor–metal contact, and this is limited by the

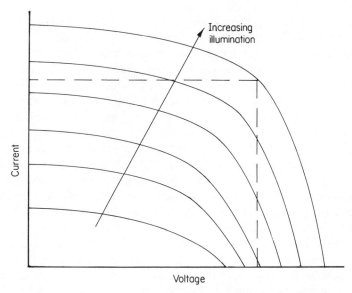

Figure 12.10. Characteristic curves of a photoelectric generator

electrical balance in the materials. The short-circuit current is directly proportional to the illumination. Between these two extremes a compromise must be reached. There is a finite load resistance, which can draw current only as limited by Ohm's law. The potential difference across the load has a definite upper limit, and so there is an upper limit to the current in the load.

The power produced by a photoelectric device is the product IV taken from the characteristic curve in Figure 12.10. For any given value of the illumination there will be a maximum value of this power given by the area of the largest rectangle that can be drawn within the particular characteristic curve. It is clear that the current will be not much less than the short-circuit value, and its actual value depends on the efficiency with which the electrons can be collected before they relax back into the valence band, or 'recombine' with the hole. The evidence is that this efficiency might be about 70 per cent. It is reasonable to expect the potential difference across the load to be something less than one-half of E_g, the maximum possible, so that the power available externally would be about 0.45×0.7 or 32 per cent of the ideal value. That is about 32 per cent of 48 per cent of the incoming solar flux.

Once again it is important to match the solar cell to its load, and once again, we have only a low-voltage device; for silicon generators the voltage will be about 0·6 V. If higher potentials are needed then solar cells must be cascaded in series in the same way as thermionic generators.

The excess heat that is deposited in the solar cell material is an embarrassment because it contributes to the losses in the electron–hole transport in a catastrophic way. For example, an increase from 20 to 100°C in the temperature of the cell would typically reduce the power output to one-third. It is essential, therefore, to conduct radiation in the useless part of the spectrum away from the cell and to maximize the heat dissipation of the cell itself.

Silicon has traditionally been the principal material of solar cells. Its use is limited by the difficulties in producing pure silicon, and in growing suitable single crystals; the cost is several pounds per square centimetre of exposed surface. There are other contenders, one of which is cadmium sulphide, now widely used in photographic exposure metres. When used in the same way as silicon, cadmium sulphide shows no real advantages as a photoelectric material; when deposited as a thin film on copper, however, the change is dramatic because the band gap is reduced, so extending the range of the solar spectrum that can be used effectively. Much current work is being devoted to these thin-film devices and interest is also being shown in thin-film silicon cells.

Photoelectric generators have a high conversion efficiency for solar radiation. Their main limitation at the moment is their extremely high cost. Installation charges for photoelectric power generators using existing materials would be two orders of magnitude greater per installed watt, than for conventional power generating plant. It is for the future to show whether or not such devices can be made economic on an industrial scale.

12.5 Prospects for Solar Power

What then are the prospects for solar power? In Northern Europe, with extensive cloud cover throughout the heating season, and with great need of reliable heating systems, the prospects are not good. Certainly solar heaters could be used for supplying hot, or at least warm, water and for supplementing domestic heating systems on suitable days. Nonetheless, complete heating supplies, not dependent on solar radiation, will be essential. Solar heating can only be regarded as a fuel saving device, and possibly an expensive one at that.

In more equatorial regions, and in most of the desert regions of the Earth, the position is completely different. With insolation of 600 W m^{-2} or better as commonplace, the heating of domestic water supplies by simple and cheap flat plate heaters, and the introduction of a measure of air-conditioning by radiative cooling at night, become realistic possibilities. Also solar stills can help to provide a supply of pure water in such regions if a supply of impure water (e.g. salt water) is already available.

On the larger scale, industrial energy supplies require large areas of solar collectors, and, in order to achieve high thermodynamic efficiencies, a measure of concentration using focusing devices. In addition, much experimental work is necessary on the development of selective absorbers. There is also a need for heat-storage facilities, and the net result is that very large areas would have to be devoted to such solar power plant. It is a matter of considering economics, ecological requirements and social need, and balancing these against the economics and other factors associated with the use of our expendable fuels in deciding whether to implement large scale solar power generation. Certainly the technology is not beyond our capabilities, though it may be beyond our imaginations.

Finally, even on a very low level domestic scale, solar heating can be usefully adopted in northern climates by sensible design of housing. South-facing windows and thick curtains will allow heat to be collected during the day, with an equilibrium temperature that is uncomfortably high, and kept within the house for part of the evening, by the simple expedient of opening and closing the curtains at appropriate times.

References

Robinson, N., *Solar Radiation*, p. 34, Elsevier Publishing Corporation, Amsterdam 1966.
Roberts, J. K. and A. R. Miller, *Heat and Thermodynamics*, p. 357, Blackie and Son Ltd., London, 1958.

Further Reading

Brinkworth, B. J., *Solar Energy for Man*, Compton Press, Salisbury, 1972.
Robinson, N., *Solar Radiation*, Elsevier Publishing Corporation, Amsterdam, 1966.
Solar Energy (journal), published for Solar Energy Society, Arizona State University, Tempe, Arizona.
Zarem, A. M. and D. D. Erway, *Introduction to Utilization of Solar Energy*, McGraw-Hill Book Co., New York, 1963.

13 Other Natural Power Sources

In the previous chapter we discussed the harnessing of solar radiation for heating and for electricity production, but solar energy does not simply manifest itself on Earth as direct and diffuse radiation. The uneven heating effect of the Sun over the surface of the Earth causes circulation in the atmosphere and in the oceans. The energy required to produce atmospheric circulation is converted into the power of the wind which in turn excites water surfaces into wave motion. Ocean currents result from the uneven heating of the seas. Moreover, there is a transfer of matter between the seas and the atmosphere. Solar energy is taken up by surface water which then evaporates and rises high into the atmosphere. There it is transported over large distances by the winds to be eventually returned to the Earth's surface as precipitation. Some of the precipitated water falls on high ground and then flows back towards the sea losing potential energy in the process. Thus some solar energy is converted into the flow energy of rivers and streams.

There are other natural sources of power which are not directly attributable to the Sun. The Moon's influence on our planet can be seen in the periodic rise and fall of the sea level, the tides. Strictly speaking, the effect involves three bodies, the Earth, the Moon and the Sun. However, lunar tides are of much more importance as a potential power source and we can view the effect of the Sun as a perturbation on these tides. The Earth itself has an internal energy source. Since its formation some four and a half thousand million years ago our planet has cooled considerably but the temperature of its core is still exceedingly high (about 5000°C). Thus there is a source of thermal energy just beneath our feet, geothermal power.

It is accepted that the fossil fuels (which were themselves formed by solar energy) are an exhaustible resource. Similarly, nuclear fission depends on what we believe to be fairly limited supplies of suitable elements. So, apart from the prospect of effectively inexhaustible power from nuclear fusion, the natural power sources represent our only truly infinite energy resource. Thus astrophysicists predict that the Sun will continue in its present equilibrium state for 5×10^9 years in which case wind, water and wave power are certainly infinite by any human time scale. Similarly, the orbital parameters of the Earth–Moon system are only slowly varying on an astronomical time scale which means that tidal power can be bracketed with these infinite power sources. Finally, the Earth is cooling at such a slow rate that even if we tap significant

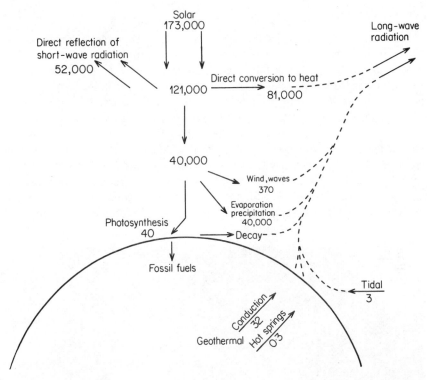

Figure 13.1. Approximate breakdown of the Earth's energy budget. (The units are 10^6 MW)

quantities of its stored thermal energy, geothermal power will be available indefinitely.

All of these natural power sources (barring geothermal power) share the common disadvantage of being extremely variable. They may fluctuate erratically, as does the wind, or regularly, as do the tides. This must be borne in mind during the ensuing discussion on the possible methods for tapping these resources. One aspect, that of energy storage, will be neglected in this chapter but this is only because it warrants a full treatment in the following chapter.

An approximate breakdown of the magnitudes of the various natural energy flows involved in the Earth's overall budget is presented in Figure 13.1. Tidal and geothermal sources provide only a fraction of one per cent of the energy input to the surface of the planet, the Sun providing the rest. Apart from direct conversion into sensible heat the largest concentration of solar power is in the evaporated water circulating in the atmosphere, two orders of magnitude greater than the power used to generate the winds and the ocean currents. We shall now discuss the natural sources of power, other than direct solar power, in order of their potential capacities: water power, wind power and the associated sources of waves and ocean currents, tidal power and geothermal power.

344

13.1 Water Power

It is clear from Figure 13.1 that the main use to which solar energy is put at the Earth's surface is the evaporation of water from seas and lakes. In fact 20 per cent of the total incoming solar radiation is used to lift water vapour into the atmosphere. Much of the consequent precipitation falls over the oceans and over low ground and is therefore lost as a power source but water power is still the largest concentration of solar power produced by natural processes.

Man's use of water power dates from at least Roman times. The horizontal water wheel first appeared in the first century B.C. and was used for grinding cereals. By the fourth century the vertical wheel had appeared and by the 16th century the water wheel had become the prime mover behind the industrialization of Europe. During the 18th and 19th centuries water power was extensively used in the United States in saw mills, textile mills and so on. Figure 13.2

Figure 13.2. A 19th century water wheel at Wellbrook, Northern Ireland. (Reproduced by permission of the National Trust. Photograph by Sean Watters)

is a photograph of an advanced 19th century water wheel at Wellbrook, County Tyrone, Northern Ireland, which is still operational today. It was used to drive a beetling mill and operate a battery of hammers which 'beetled' the fibres of linen to produce the material's characteristic sheen and polished appearance.

In contrast, the generation of hydroelectric power is a comparatively recent development. The first major hydroelectric installation was originally conceived in 1870 but was only commenced in 1886 at Niagara Falls, North America. The plant was finally operational on April 1st, 1895, with an output of 5000 hp (approximately 3750 kW). One year later power was sent from Niagara to Buffalo, 22 miles away, another technological first (Oliver, 1956). On a much smaller scale, one of the first hydroelectric installations in Europe was in County Antrim, Northern Ireland, and was used from 1883 to power the Giant's Causeway tram (Figure 13.3). Today, Ireland has an installed hydroelectric capacity of 220 MW, Great Britain has 1286 MW and the United States has more than 50,000 MW.

The technology of hydroelectric power generation is quite standard by now. Basically, flowing water is dammed and then diverted through a mechanical device to convert the water's kinetic energy into rotational energy which can then be converted into electrical energy in a generator. The choice of the particular mechanical device to be used depends on the height through which the water falls. For a large head, the Pelton wheel is the most suitable; for a medium head, say 100 metres, the Francis turbine is the best choice; for a small head, the Kaplan turbine is to be preferred.

The Pelton wheel consists of a disc, with a number of cups fixed round its edge, mounted on a shaft (Figure 13.4). The cups are known as buckets and have a ridge in the centre which splits the incident jet of water. The surface of the bucket is so shaped that the water runs smoothly round it and out the lower edge without splashing. If the wheel were fixed, the water jet would have its velocity reversed because of the shape of the buckets. Its potential for doing work would, therefore, be unchanged. Alternatively, if the wheel were to rotate so that its rim moves at the same velocity as the incident jet, no work would be done on the wheel. The greatest efficiency will be achieved when the water falls from the buckets having lost all of its incident speed. This will be so when the bucket velocity is half the jet velocity. Thus for a given head of water the optimum rim speed for the wheel is fixed and the speed of rotation must be varied by changing the size of the wheel. If the volume of water is large, more than one Pelton wheel may be mounted on the same shaft with more than one jet playing on each wheel. The power output from Pelton wheels may be as low as a fraction of a kilowatt or as high as thousands of kilowatts. For example, the Big Creek installation in California, with a head of 750 m, has an output of 55,000 kW.

When the head is low or when very large quantities of water have to be handled, the Pelton wheel is replaced by a turbine, similar to the steam turbines described in Chapter 7. The Francis turbine (Figure 13.5a) was the first inward-flow turbine to be designed (1849). In this design the water flows along a radius

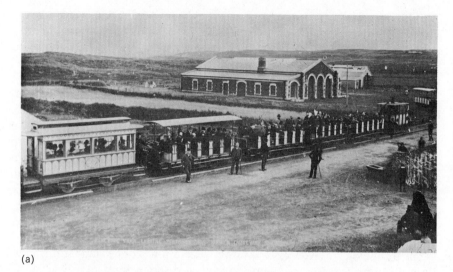

(a)

(b)

Figure 13.3. (a) The Giant's Causeway electric railway, Portrush, Northern Ireland. (b) The hydroelectric power station at Bushmills which powered it. The power station was opened in 1883. (Reproduced by permission of the Trustees of the Ulster Museum)

Figure 13.4. The runner of a Pelton Wheel installed at the Poatina power station, Australia. It is designed for a maximum output of 82,000 hp at a speed of 600 rpm and a net head of 807 m. (Reproduced by permission of Boving and Company Ltd.)

of the machine and emerges along the shaft with the blades taking up almost all of the water's energy in turning it through a right angle. Kaplan turbines (Figure 13.5b), on the other hand, have movable blades which can be adjusted to maintain a high efficiency over a wide range of loads. As mentioned above they are suited to very low heads, the design size of the turbine increasing as the head decreases. Kaplan turbines with runners of up to 8 m diameter have been developed. Typical large outputs for these turbines are 55,000 kW for the Kaplan and 130,000 kW for the Francis.

An estimate can be made of the water power available from a given source if the flow rate F and the head H are known. Thus

$$P = \rho FgH$$

Taking ρ as 1000 kg m^{-3} and g as 9·8 m s^{-2}, the equation reduces to

$$P = 9 \cdot 8 FH \text{ kW}$$

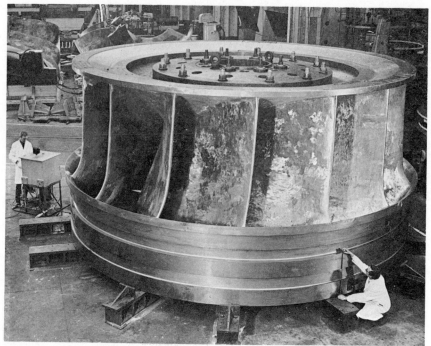

(a)

(b)

(c)

Figure 13.5. (a) The runner of the Francis turbine for El Chocon power station in Argentina. It is manufactured in solid stainless steel in three segments which are secured by shrunk-in keys. The total weight of the unit is 185 tonnes and it can produce 278,000 hp at a speed of 88·3 rpm and a net head of 58·4 m.
(b) A Kaplan runner being installed at the Waipapa power station, New Zealand. It can produce 24,000 hp at a speed of 125 rpm and a net head of 16·5 m.
(c) The assembly of the spiral casing, guide vanes and covers for unit No. 1 at the Foyers pumped storage scheme (see Chapter 14) on Loch Ness, Scotland. The unit is a reversible Francis turbine capable of generating 150 MW from a head of 162·8 m at a speed of 273 rpm. When pumping it delivers water at 77 m³ s⁻¹ against a head of 183 m. (All reproduced by permission of Boving and Company Ltd.)

with F in m³ s⁻¹ and H in metres. It is informative to calculate the potential power that might be extracted from a typical small 'back garden' stream. For a flow rate of 1 m³ s⁻¹ and with a head of one metre the potential power is approximately 10 kW. Now suppose that the annual rainfall is one metre (about 40 inches). Assuming it all runs off, then an area of one square kilometre would lead to a mean flow rate of 10^6 m³ yr⁻¹ or 1/30 m³ s⁻¹. Thus a collection area of A km² will he necessary to produce 10 kW of power averaged through the year, where

$$P = 10 = 9 \cdot 8HA/30$$

A collection area of 30 km² is therefore necessary in order to produce 10 kW of power in our small stream with a head of one metre.

From a knowledge of stream flow records throughout a given country, an estimate can be made of that country's potential hydropower capacity. This is obviously a painstaking process and subject to error and the results, as listed in Table 13.1, can be taken as order of magnitude estimates only. It is interesting

Table 13.1 World water power capacity

	Potential (1000 MW)	Per cent of World total	Developed (1000 MW)
North America	313	11	59
South America	577	20	5
West Europe	158	6	47
Africa	780	27	2
Middle East	21	1	—
South East Asia	455	16	2
Far East	42	1	19
Australasia	45	2	2
U.S.S.R., China, etc.	466	16	16
	2857	100	152

(Reference: M. King Hubbert, 1969, in *Resources and Man*, Publication 1703, Committee on Resources, National Academy of Sciences—National Research Council, W. H. Freeman and Co., San Francisco, 1969. Reproduced by permission of the National Academy of Sciences.)

to express a nation's potential water power capacity in terms of its population. On this basis Norway heads the hydropower league with a per capita potential capacity of 13 kW, followed by Africa with 4, the U.S.S.R. with 2, Canada with 1·3, the United States with 0·8 and the United Kingdom with 0·2.

In interpreting these figures and in comparing water power with other resources it must be borne in mind that the efficiency of electricity generation in hydroelectric plant is extremely high, typically 80–90 per cent. This is to be compared with the best thermal plants with efficiencies of 40 per cent. The basic resource is essentially free and can be used for other purposes after it has passed through the hydroelectric power station. However, hydroelectric installations are capital intensive and this must be balanced against the low running costs (Chapter 18).

Although hydroelectric power production is essentially clean and non-polluting, it can have a number of detrimental effects on the environment. The first of these is purely visual. Major hydroelectric schemes, by their very nature, are usually located in mountainous, and, therefore, scenic, regions. The second effect is that the dams associated with such schemes block the flow of sediment in the stream. Reduced turbidity in the water behind the dam, increased evaporation losses due to the formation of a constant open area of surface water and changes in the water chemistry may all combine to enhance the deposition of sediment. Apart from the interference with the natural processes, this may also curtail the useful life of the dam as the reservoir silts up, in which case hydropower cannot be regarded as an infinite resource. At present there appears to be no solution to this problem. This blocking process also affects the flow of nutrients in the stream. The classic example is the Aswan dam in Egypt which hoards the water of the Nile in Lake Nasser. The rich nutrients which were deposited along the banks of the Nile during the summer floods are now left behind in Lake Nasser and the farmers of the Nile are forced to use artificial fertilizers for the first time.

13.2 Wind Power

The power of the wind has been harnessed by man for many centuries. Windmills which were used to grind wheat are still to be seen in many countries and similar devices were common in America where they were used to pump water from deep wells. Of course, the crowning glory of wind technology was the clipper ship, used to speed tea and other goods from India to Britain. In recent years there has been renewed interest in the application of wind power to electricity generation and it is this that we seek to evaluate in this section.

It is interesting to note in passing the extent to which wind power and water power have been used directly. Table 13.2 presents data for the United States

Table 13.2 Direct United States wind and water power consumption

	(10⁹ kWh)		(10⁹ kWh)
1850	67	1940	25
1860	92	1945	23
1870	85	1950	16·5
1880	88	1955	14
1890	83	1960	11·2
1900	79	1962	10·1
1910	84	1964	9
1920	106	1966	8
1930	46	1968	6·7
1935	35	1970	5·6

(Reference: J. C. Fisher, 1974, *Energy Crises in Perspective*, John Wiley and Sons, Inc., New York. Reproduced by permission of John Wiley and Sons Inc.)

between 1850 and 1970. The maximum annual consumption, approximately 100×10^9 kWh, was reached around 1920, since when direct consumption has declined to a mere 5×10^9 kWh.

It is a simple matter to calculate the potential magnitude of wind power. For a uniform air stream moving with velocity v (in m s^{-1}), the mass of air crossing unit area per second is ρv, where ρ, the density of air, is 1·29 kg m^{-3}. This means that there is an energy flow of $\frac{1}{2}\rho v.v^2$ per second so that the power P which could potentially be extracted from the wind using a device of unit cross-sectional area is

$$P = \tfrac{1}{2}\rho v^3$$

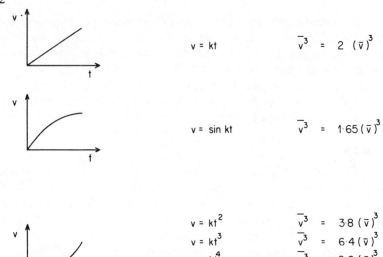

Figure 13.6. Relationship between the mean wind speed, $\bar{v}^3$, and the mean cube wind speed, $\bar{v^3}$, for some simple temporal variations of wind speed

Thus wind power varies as the cube of the wind velocity.

Very little accurate information is available as to the magnitude of P in different localities. Detailed meteorological records are often available for average wind speeds and for the frequencies with which certain speeds occur (see later), but it is important to emphasize that the average power $\bar{P}$ is not simply related to the average wind speed. Thus the average speed $\bar{v}$ when cubed does not equal the average value of the cube of the wind speed, $\bar{v^3}$. In fact the available wind power can be seriously underestimated by using the figure for $\bar{v}$ in the above equation for P. We can illustrate this by a few simple examples.

Let us suppose that the wind speed increases from zero to a maximum value according to some simple function of time $f(t)$. In the first case (Figure 13.6), v increases linearly with time. The average wind speed is given by

$$\bar{v} = \frac{1}{\tau} \int_0^\tau kt \, dt$$

$$= \tfrac{1}{2} k\tau$$

On the other hand,

$$\bar{v^3} = \frac{1}{\tau} \int_0^\tau k^3 t^3 \, dt$$

$$= \tfrac{1}{4} k^3 \tau^3$$

Figure 13.7. An anemometer.

Therefore,

$$\overline{v^3} = 2 \times (\bar{v})^3$$

For the other functional forms chosen, $\overline{v^3}$ is underestimated by a factor of between 1·65 and 10 if it is simply equated to $(\bar{v})^3$. Obviously these models are somewhat divorced from reality and in truth the wind speed will tend to rise and fall gradually but with sharp fluctuations—gusts—superimposed on this pattern. The examples quoted should, however, be enough to convince the reader that there is no substitute for direct measurements of v^3 and hence $\overline{v^3}$.

Wind speeds are usually monitored by cup anemometers (Figure 13.7). The rotation of the cup assembly drives a small generator and produces an electrical signal which is easily recorded. An alternative method is to use a Pitot tube or

Prandtl tube. This is a small tube inserted in the air flow such that the entrance to the tube faces the wind, Thus there will be high pressure static air at the entrance while the air flowing rapidly past will be at low pressure. A head of liquid, monitoring this pressure differential, can be calibrated directly in terms of wind speed. Obviously this method is not so suitable for continuous recording. With an anemometer it is a straightforward electronics problem to cube the signal and record v^3, and hence wind power, as well as v. Thus wind power can be monitored on a continuous basis. Only when this is done will an accurate figure for wind-power potential in a given locality be known.

Table 13.3 lists average wind speeds for various locations in the British Isles. For a wind speed of 10 mph the potential power availability is 55 Wm^{-2}. It is interesting to note that for localities with average speeds of less than this, winds exceeding 10 mph only occur for less than 50 per cent of the time.

Table 13.3 Mean wind speeds at various locations in the United Kingdom

Meteorological station	Location		Mean wind speed (mph)	Per cent calm	Per cent < 10 mph
Lerwick	68°08′N	1°11′W	16·7	3·5	29
Stornoway	58°12′N	6°22′W	16·4	5·8	29
Dyce	57°12′N	2°12′W	8·6	13·6	60
Bellrock	56°26′N	2°24′W	17·4	1·9	25
Leuchars	56°23′N	2°52′W	9·1	2·7	60
Tiree	56°30′N	6°53′W	15·2	3·3	30
Prestwick	55°30′N	4°35′W	9·3	6·1	55
Cranwell	53°02′N	0°30′W	10·3	4·6	48
Mildenhall	52°22′N	0°28′W	8·8	2·7	61
Felixstowe	51°57′N	1°20′W	11·1	3·3	58
Cardington	52°07′N	0°25′W	10·5	5·3	48
Shoeburyness	51°32′N	0°49′W	12·9	2·7	36
Birmingham	52°28′N	1°56′W	9·7	0·8	55
Keele	53°00′N	2°16′W	7·3	4·0	69
London, Kingsway	51°07′N	0°07′W	8·6	2·1	60
South Farnborough	51°17′N	0°45′W	6·9	8·9	71
Boscombe Down	51°10′N	1°45′W	9·8	3·2	54
Sellafield	54°25′N	3°30′W	10·4	7·1	48
Fleetwood	53°56′N	3°01′W	12·1	1·9	41
Speke	53°21′N	2°53′W	11·9	4·8	42
Manchester	53°21′N	2°16′W	11·3	3·3	45
Lizard	49°57′N	5°12′W	14·3	5·4	35
Scilly	49°56′N	6°18′W	15·7	1·8	27
Holyhead	53°19′N	4°37′W	13·1	5·1	40
Aberporth	52°08′N	4°34′W	12·6	5·6	41
Aldergrove	54°39 N	6°13′W	11·0	3·7	46

(Calculations based on data from Meteorological Office Publication No. 792, *Tables of Surface Wind Speed and Direction over the United Kingdom*. Reproduced by permission of H.M.S.O., London.)

There are many possible techniques for extracting power from the wind and converting it into mechanical energy. The simplest type of wind generator is simply a propellor mounted on a horizontal shaft. In theory the efficiency of such a propellor is highest when only one blade is used, but this can lead to balancing problems in design. Consequently, a two- or three-bladed propellor is the most likely candidate. It is found experimentally that power extraction is most efficient when such a propellor is facing directly into the wind stream so it is best to arrange that this should be so by means of a tail-vane assembly. This problem can be overcome if the propellor is replaced by a fan with a vertical axis of rotation. Since this is at right angles to the wind direction at all times steering is unnecessary. Further, there is no necessity for complicated gearing at the top of the supporting structure.

No matter which system is employed, there is still the problem of converting the mechanical energy of the propellor blades or the fan into electrical energy. There are a number of ways in which this can be achieved, for example:

(1) The rotating shaft of the wind generator can be used to drive a DC generator.

(2) The rotating shaft can be used to drive an alternator and thus produce AC electrical power whose frequency varies with the speed of rotation. The AC output can be rectified and then passed through an inverter to give an accurately stabilized AC supply. The main drawback with this approach is its expense.

(3) The rotating shaft can be used to drive an induction generator (Chapter 7) connected to the mains electricity supply. In theory, if the mains network is infinite (the National Grid is effectively so), then the induction generator will extract power from or donate power to the mains supply according to whether the speed of rotation of the shaft is less than or greater than the frequency of the mains supply. It is, of course, possible to arrange for a cut-out in the electrical circuit to disconnect the wind generator when the speed of rotation is less than mains frequency.

(4) The output from the alternator in (2) can be rectified and stored in batteries or used to electrolyse water and produce hydrogen (Chapter 14). Storage will always be necessary in isolated wind power systems to enable the consumer to cope with periods of calm weather.

Wind-powered electricity generation is certainly an attractive concept. Apart from its lack of reliability, which must be overcome by the use of some storage medium, it is attractive both technologically and from an environmental viewpoint. Thus it produces no waste heat and does not lead to noxious emissions. (It might, however, be argued that large-scale adoption of wind generation could lead to a perturbation in the air-flow system across the continents.) Nevertheless, it is unlikely that wind generators would allow large numbers of consumers to become independent of the mains electricity supply and it is likely that the use of such generators as auxiliary supply units would be uneconomic.

13.3 Geothermal Power

Part of the energy balance of the Earth's surface is made up of the heat transfer from its interior. Some of this heat is transferred by conduction and maintains a moderate ground temperature within a few metres of the surface. However, the low rate of heat flow and the low temperatures involved preclude large-scale energy conversion. Heat is also transported by convection. Occasionally hot springs or geysers and molten lava streams manage to reach the surface. The thermal energy of these hot springs can be used directly for heating purposes as is common in Iceland or the steam can be used to generate electrical power as has been the practice in Italy for most of this century. Similarly, much of Northern Europe, for example, lies over a large mass of ground water heated to a moderate temperature (60–100°C) by the Earth's internal heat. This has already been used directly to heat the studios and offices of RTF, the French television network, in Paris.

Hot springs were used for bathing and heating by the ancient Romans throughout the extent of the Empire, from the Mediterranean coast to Bath in England. Indeed there are still many flourishing Central European health resorts based on such springs. The first geothermal electric power station was completed in 1904 at the Larderello field in Northern Italy which today has a capacity of 370 MW. The first plant in the United States, with a capacity of 12·5 MW, was opened at the Geysers field, California, in 1960 and ten years later the use of geothermal power received Government support in that country with the passing of the Geothermal Steam Act. Compared to other power sources, the exploitation of geothermal power has certainly been slow, only reaching 1000 MW after nearly three-quarters of a century. This is partly due to the difficulty of finding suitable geothermal sources for, until recently, prospecting has relied purely on surface indications.

Figure 13.8 illustrates a typical geological structure in the locality of a geothermal power source. Molten rock, which is pushed up into the Earth's crust by stresses in the interior, heats the rocks in the crust close to the surface. Water in fissures or in porous rock formations at depths of, say, 10 km is heated to as much as 250°C but because of the high pressure at this depth it remains a liquid. If this water escapes through a fissure, it boils and flashes off into steam. It is possible to tap this geothermal energy either by using this emitted steam or by driving a well down into the porous layers and drawing off the super-heated water.

Hot springs and steam fields are no longer regarded as isolated natural phenomena and it is believed that there are extensive reservoirs of hot water and steam throughout the Earth's crust. In California alone the Geysers field is estimated to have a potential capacity of 3000 MW and the Imperial Valley field 10–100 GW. (This field also illustrates the major difficulty in exploiting geothermal power. The hot water contains up to 25 per cent dissolved salt and such a concentrated brine is highly corrosive.) Table 13.4 lists the geothermal power installations already in operation throughout the World together with their developed capacities.

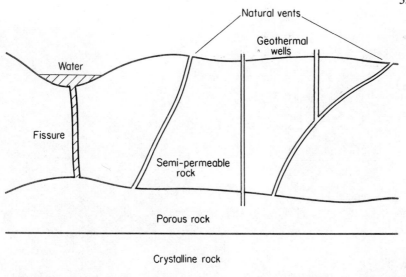

Figure 13.8. Geological structure in the locality of a geothermal field

Table 13.4 Geothermal power stations

	Installed capacity 1969 (MW)	Planned Additional capacity (MW)	Earliest Installation
Italy			
Larderello	370		1904
Monte Amiata	19		1962
United States of America			
Geysers, California	82	100	1960
New Zealand			
Wairakei	290		1958
Mexico			
Pathe	3·5		1958
Cerro Prieto	—	75	1971
Japan			
Matsukawa	20	40	1966
Otake	13	47	1967
Goshogate	—	10	
Iceland			
Hveragerdi	(used for direct heating)	17	1960
U.S.S.R.			
Kamchatka	30·75	7·5	1966

(Reference: M. King Hubbert in *Resources and Man*, Publication 1703, Committee on Resources, National Academy of Sciences—National Research Council, W. H. Freeman and Co., San Francisco, 1969. Reproduced by permission of the National Academy of Sciences.)

Figure 13.9. The Larderello geothermal steam fields, Italy. (By permission of Societa' Chimica Larderello. Ente Nazionale Idrocarburi)

There are essentially three different types of geothermal field. The first are dry-steam fields such as those at Larderello (Figure 13.9), Geysers and Valle Caldera (New Mexico). The steam can be piped direct to turbines but as its pressure is low (ranging from 1–15 bars compared with 30–200 bars in a fossil-fuel or nuclear plant) large amounts must be used and the effective size of the turbines is limited to approximately 55 MW. Similarly, as the temperature of the steam is low (250°C compared with 500°C in a fossil-fuel plant) the efficiency of power production is low. Moreover, the low steam temperature means that the steam rapidly becomes wet as it passes through the turbine. This necessitates the use of special wet-steam turbines, as described in Chapter 7 and illustrated in Figure 13.10. The second category, wet-steam fields, appear to be much more abundant than the first. Once again the steam can be used in wet-steam turbines to produce electrical power while the hot water can be used for heating or desalination, or is simply discarded. Finally, there are the low temperature (50–80°C) water fields, most of which have been discovered in Hungary and the U.S.S.R. Apart from using this water for direct heating, it has been suggested that it could be used to generate electrical power via a heat engine employing Freon (Chapter 16) as the working fluid. Such a plant is actually in existence at Kamchatka, U.S.S.R. There may also be other uses for geothermal deposits. Many fields are rich in minerals which it would be technically feasible to extract. A typical mineral-rich wet-steam field might be

Figure 13.10. Three of the wet-steam turbines installed at Wairakei Geothermal Power Station, New Zealand. Each of these units has a capacity of 30 MW and operates at a steam pressure of 50 psi. (By permission of General Electric Company Ltd.)

exploited as follows: the steam is used to produce electricity, the hot water is used in a desalination plant and the effluent from the hot water feed is concentrated into a mineral-rich brine from which the minerals can be extracted.

Another possible method for tapping geothermal power is the technique of hydraulic faulting. This involves drilling into the upthrusts of hot rock and pumping water at high pressure down the well. Large underground fracture zones result. If another well is now drilled close to the first, water can be circulated through the system, heated and then used to produce power. The feasibility of this method was demonstrated by the Los Alamos Laboratory of the United States Atomic Energy Commission when they drilled a 780 m deep well in New Mexico and succeeded in producing hydraulic faulting of even crystalline rock using pressures of $8-12 \times 10^6$ N m^{-2} (about 100 atmospheres). By extracting heat directly from hot rocks instead of being limited to rocks which are permeated with water, the usable deposits are increased substantially.

The exploitation of geothermal power certainly does not cause pollution on the scale of fossil fuel or nuclear power production but small amounts of chemicals, such as boron, which are harmful if discharged are brought to the surface. This can be overcome simply by injecting the waste material back into the wells. A more serious problem is land subsidence. Some has already occurred at the Wairakei field in New Zealand, hardly surprising since 70 million tons of water are extracted from this field each year, so much that the field has partly changed character from a wet-steam to a dry-steam one. Similarly, in Cerro Prieto, Mexico, which may be part of the Imperial Valley field, five inches of subsidence was reported in 1972 (Henahan, 1973). This problem can be solved by limiting the rate at which water is removed from the field and by returning the spent water.

From the data of Table 13.4 it is clear that more geothermal fields will have to be discovered if the resource is to have a significant impact on man's power requirements. Exploration for new fields is proceeding by way of geological studies and, more recently, by infra-red photography of the Earth's surface. At the present time prospects for new discoveries appear to be good. It is thought that large geothermal deposits occur right down the west coast of the American continent, from Alaska to Chile. Similarly, large fields have been located in the Far East, close to the ring of volcanic activity around the Pacific, in Turkey and along the African rift valley. Moreover, Europe has many spas so it is reasonable to expect that substantial fields exist underneath the whole continent, including the British Isles.

It is possible to estimate an upper limit to the extraction of power from geothermal sources. The conduction of heat to the Earth's surface proceeds at an average rate of about 0.06 W m^{-2} which represents a total heat flow to the surface of 3.2×10^{13} W. Hot springs and geysers account for only about one per cent of this flow and it has been estimated by White (1965) that the usable stored energy in the major fields is 4×10^{20} J. Allowing for a 25 per cent conversion efficiency to electricity this is equivalent to an output of 60,000 MW over a period of 50 years. This power rating is equivalent to the potential for tidal power (see later), but is only a few per cent of the potential hydropower capacity. However, if the whole geothermal reservoir below the continents can be tapped (by hydraulic faulting or otherwise) then the resource has a capacity of approximately 10^6 MW for an indefinite period of time.

13.4 Tidal Power

It is a matter of everyday experience that the water level around our coasts rises and falls twice each day. Some insight into the mechanism behind this phenomenon can be achieved by the study of a simple model for the Earth–Moon system in which the Earth is treated as a rigid sphere covered with a uniform layer of water. The Earth and the Moon rotate about their common centre of mass C (Figure 13.11) with a period of 27.3 days. The gravitational attraction between the two bodies is balanced by the centrifugal force resulting from the rotational motion in this centre of mass system of coordinates:

$$GME/R^2 = EL\omega^2 \qquad (13.1)$$

This is only true at the point O, the centre of the Earth. A unit mass at the point P on the surface of the Earth opposite to the Moon experiences a greater centrifugal repulsion and a smaller gravitational attraction than does unit

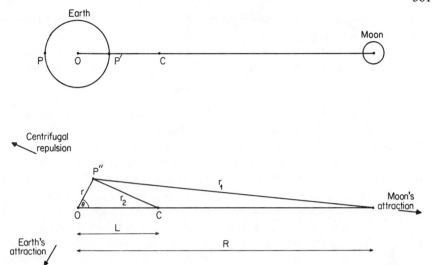

Figure 13.11. The Earth–Moon system. (For clarity the centre of mass C is shown outside the Earth. In fact, it lies just within the surface of the Earth)

mass at O. Consequently matter at P is subject to a net outward force. Similarly, matter at P′ is subject to a net force towards the Moon, again away from the Earth's surface. Thus the spherical water surface is distorted by two bulges located along the Earth–Moon direction. As the Earth rotates once every 24 hours about its own axis, a given point on the surface of the sea passes through the bulges twice each day and experiences two high tides. (It should be noted that the Earth's daily rotation, in the absence of the Moon's perturbing influence, would not cause tides but just a flattening effect and, indeed, the polar radius is less than the equatorial radius by 21 km.)

A more quantitative approach enables us to estimate the magnitude of these tides. In a reference frame rotating with the Earth–Moon system, unit mass at a point $P''(r,\theta)$ on the surface (Figure 13.11) is subjected to three forces: the gravitational attraction of the Earth, the gravitational attraction of the Moon and the centrifugal force. It is convenient to work in terms of potentials rather than forces as the surface of the sea may best be defined by an equipotential. The potential at the point P'' is

$$V(r,\theta) = -GE/r - GM/r_1 - \tfrac{1}{2}\omega^2 r_2^2$$
$$= -GE/r - GM(R^2 + r^2 - 2rR \cos \theta)^{-1/2} - \tfrac{1}{2}\omega^2(L^2 + r^2 - 2Lr \cos \theta)$$

As r/R is small (approximately 1/60), this expression can be expanded using the binomial theorem and, with the aid of equation (13.1), may be approximated to

$$V(r,\theta) = -GE/r[1 - 3/2\,(M/E)\,(r/R)^4 \cos \theta + 3/2\,(M/E)\,(r/R)^3 \cos^2 \theta]$$

Now r/R is small so the term in $\cos^2 \theta$ is dominant. Thus the main tide generating force can be described by a $\cos^2 \theta$ potential which again concurs with the evidence of two tides per day.

The surface of the sea must be a gravitational equipotential so as θ varies, r must change slightly from its equilibrium value of 6360 km in order that

$$V(r,\theta) = - GE/r[1 + 3/2(M/E)(r/R)^3 \cos^2 \theta]$$

remains constant. As θ varies from 0 to 2π, the fractional change in this expression is $3/2 (M/E)(r/R)^3$ or 1 in 12×10^6. This leads to an estimate for the change in r of about $\frac{1}{2}m$ which is in reasonable accord with observation.

The tides do have a noticeable effect on the orbital parameters of the Earth–Moon system. The motion of the moon would not be affected if the tidal bulges could follow its motion precisely. However, tidal energy is dissipated in the frictional drag between the seas and the solid earth. Because of the Earth's rotation the bulges are dragged along in advance of the Moon's position. At the same time the Earth's rotation is slowed and it would indeed appear that the length of the day is increasing by approximately 0·001 second every century. From this figure Munk and MacDonald (1960) have estimated the rate of tidal dissipation of energy on the Earth to be 3×10^{12} W.

The behaviour of real tides is, of course, much more complicated than our simple model suggests. The tidal range varies considerably from place to place. The above estimate of one half metre is certainly in good agreement with the tides observed around oceanic islands, but in the Mediterranean the range is virtually zero while in the Severn estuary it can be as large as 15 m. The most obvious limitation of our model is the assumption of a World completely covered with water. Apart from this, the tidal movements of the seas depend on their ability to undergo natural oscillations with a period of half a day. The depth of the oceans is not necessarily suited to the propogation of a surface wave whose wavelength is half the Earth's circumference and whose velocity is 450 m s^{-1} at the Equator, and this is indeed how the tides may be interpreted. There are some cases of enclosed seas capable of resonating with the tide-generating force, for example the Bay of Fundy. However, this is not the reason for the large tides at the heads of many estuaries. These result from the amplification of the tidal wave as it sweeps up the restrictive channel, concentrating its energy into a smaller volume of water.

From the point of view of utilizing tidal power, the amount of energy dissipated in the shallow seas and estuaries around the Earth's coastline is of prime importance. Recent estimates, given by Munk and MacDonald (1960), suggest that the average rate is, at most, 10^{12} W. Consequently, this is an upper limit to the amount of power that could be extracted from tidal sources although, as we shall see, the true potential is probably much less than this.

The simplest form of tidal-power installation involves impounding water in an artificial basin at high tide and then allowing it to escape into the sea at and near low tide. While it is running out of the basin it can be used to drive hydraulic turbines and generate electrical power. If all the water is stored until close to low tide and is then allowed to escape within a short period of time the turbines will be operating under a variable head of water. It may be preferable to avoid this by adjusting the rate of outflow into the sea so as to maintain a constant head. This extends the time of operation and increases the efficiency of the turbines but does not extract as much energy as the former scheme. The overall efficiency of tidal power generation can, of course, be improved by employing a duplicate set of turbines which operate while the basin is filling.

Artificial basins for these schemes can best be prepared by damming the narrow mouths of estuaries. If the average tidal range R is known, and if A is the enclosed surface area of the basin, then we can calculate the maximum amount of energy, E_{max}, that could be obtained from each cycle as follows. (E_{max} is the energy dissipated when the dam gates are closed at low tide and only reopened at high tides, and, similarly, closed at high tide and only reopened at the next low tide.) In emptying the full dam the energy dissipated is

$$\rho g A \int_0^R x \, \mathrm{d}x = \tfrac{1}{2} \rho g A R^2$$

where ρ is the water density and g is the acceleration due to gravity. Consequently,

$$E_{max} = \rho g R^2 A$$

This leads to an average power P equal to E_{max}/T where T is the period of the tides (12 h 24·4 m or $4·46 \times 10^4$ s). Table 13.5 lists the potential annual energy outputs and average power outputs for a number of the most promising sites around the World.

The first major tidal-electric power plant was installed at La Rance estuary in France and began operation in 1966. From Table 13.5 it can be seen that the average tidal range here is 8·4 m and that the enclosed basin has an area of 22 km². Electric power is generated by 24 units each producing 10 MW and giving an annual energy production of 550,000 MWh. The efficiency of the scheme is, therefore, only 18 per cent (Table 13.5) but this is quite high for tidal-power installations. (Most design computations predict efficiencies in the range 8–20 per cent.) The output of La Rance has recently been upgraded to 320 MW which is a power efficiency of approximately 25 per cent. The achievement of this figure has necessitated the use of turbines of extremely advanced design. They are slow speed, horizontal, axial-flow turbines with adjustable blades so as to permit their use during both the filling and emptying operations. In addition they can be used as pumps if necessary.

In estimating the potential of tidal power, then, it is clear that we must reduce the data of Table 13.5 by 80–90 per cent. In addition, the dissipation in suitable estuaries is only a small fraction of the total dissipation around our coastlines. Hubbert (1969) has estimated that the true potential of tidal power may be as little as 13,000 MW, approximately one per cent of the total potential hydro-power. Nevertheless, tidal-power schemes produce no pollutants and have minimal effects on the local scenery and sea life, and may therefore be considered as useful supplements to more conventional power sources. Tidal power is

Table 13.5 Tidal power sites and their potential output

	Mean range (m)	Basin area (km²)	Potential Mean power (MW)	Annual production (1000 MWh)
North America				
Passamaquoddy	5·5	262	1800	15800
Cobscook	5·5	106	722	6330
Annapolis	6·4	83	765	6710
Minas-Cobequid	10·7	777	19900	175000
Amherst Point	10·7	10	256	2250
Shepody	9·8	117	520	22100
Cumberland	10·1	73	1680	14700
Petitcodiac	10·7	31	794	6960
Memramcook	10·7	23	590	5170
South America				
San Jose, Argentina	5·9	750	5870	51500
England				
Severn	9·8	70	1680	14700
France				
Aber-Benoit	5·2	2·9	18	158
Aber-Wrac'h	5·0	1·1	6	53
Arguenon	8·4	28	446	3910
Frenaye	7·4	12	148	1300
La Rance	8·4	22	349	3060
Rotheneuf	8·0	1·1	16	140
Mont St. Michel	8·4	610	9700	85100
Somme	6·5	49	466	4090
Ireland				
Strangford Lough	3·6	125	350	3070
U.S.S.R.				
Kislaya	2·4	2	2	22
Lumbouskii Bay	4·2	70	277	2430
White Sea	5·65	2000	14400	126000
Mezen Estuary	6·6	140	1370	12000
			64125	562553

(Reference: M. King Hubbert in *Resources and Man*, Publication 1703, Committee on Resources and man, National Academy of Sciences—National Research Council, W. H. Freeman and Co., San Francisco, 1969. Reproduced by permission of the National Academy of Sciences. Strangford Lough data calculated by the authors.)

also a permanent resource to all intents and purposes. In using it we are drawing upon the kinetic energy stored in the Earth's rotation, or, more precisely, the kinetic energy stored in the Earth–Moon system. We can continue to draw upon this reserve until the Earth's rotation has been slowed to the stage where it always presents the same face to the moon. This prospect of a 1128-hour day is a distant one, indeed many thousands of millions of years will elapse before this occurs and such is the longevity of tidal power as an energy resource.

13.5 Waves and Ocean Currents

All the natural energy resources discussed so far have, to some extent, been used for power generation. There are two other schemes, however, which have yet to be evaluated practically—wave power and power from ocean currents.

In theory wave power has much to recommend it. For example, in the United Kingdom with at least 1450 km of usable coastline, it has been estimated that wave power could provide a capacity of 30,000 MW (CPRS, 1974), about half of the installed capacity of the Central Electricity Generating Board. One suggested method for tapping this resource involves a system of floating tanks tethered to each other and to the sea bed. The oscillatory motion of the tanks could be made to drive a high-pressure water pump and the high-pressure water used to drive a shore-based turbine. It is estimated that a 100 MW installation would require a line of tanks 4·5 km long.

There are, of course, many drawbacks to such a system, not the least of which is that it has yet to be tested in practice. Moreover, the line of tanks could be a shipping hazard, especially if one were to break loose, and might provoke considerable objections on aesthetic grounds. Finally, the scheme would almost certainly be more expensive than conventional power generation (Chapter 18), and, to allow for periods of calm seas, standby generating capacity would still have to be provided.

It has been suggested that power could be extracted from ocean currents by immersing a large turbine in the flow path. In theory this would be a feasible proposition. However, one can imagine the furore in the British Isles and Northern Europe if someone were to extract the energy in the Gulf Stream! Clearly, such suggestions must be treated with considerable caution.

In this and the previous chapter we have described the natural sources of power which are available to us. We need not worry about what to do when they cease to be available for when that comes to pass we shall have ceased to exist on this planet. The main question to be asked is: can the natural sources of power meet the requirements of an industrialized society such as ours? The answer, in principle, is yes, but in practice we are a long way from this goal. Nevertheless, solar power and wind power are already amenable to exploitation on a small scale and this may well be where their future application will lie. However, tidal power, geothermal power and hydropower can only be regarded in the long term as useful supplements in the large-scale generation of electricity.

References

Henahan, J., *New Scientist*, **57**, 16 (1973).
Hubbert, M. K., *Resources and Man*, National Academy of Sciences, Freeman, San Francisco, 1969.
Munk, W. H. and G. J. F. MacDonald, *The Rotation of the Earth, a geophysical discussion*, Cambridge University Press, Cambridge, 1960.
Oliver, J. W., *A History of American Technology*, Ronald Press, New York, 1956.

366

White, D. E., *Geothermal Energy*, U. S. Geological Survey Circ. 519, 1965.
CPRS, *Energy Conservation*, a Study by the Central Policy Review Staff, H.M.S.O., 1974.

Further Reading

Hydrolectric Engineering Practice, Vol. II, J. Guthrie-Brown, (Ed.), Blackie, London, 1970.
Power Generation and Environmental Change, Chaps. 10 and 11, Berkowitz D. A. and A. M. Squires, (Eds.), MIT Press, Cambridge, Mass., 1971.

14 Energy Storage

To date the theme of this book has been the production of power, whether it is by the combustion of fossil fuels, by the use of nuclear reactors or by the tapping of the natural supplies of solar power, wind power, geothermal power and so on. A number of these sources suffer from a common disadvantage in that they are intermittent. For example, this is particularly true of sources such as solar power, wind power and tidal power. If these sources are ever to be used as primary power units, then they must be linked to efficient systems for storing energy during periods of peak production in order to make provision for barren times ahead. We shall see, however, that the concept of energy storage is of universal importance in the utilization of power, however that power is generated.

Perhaps the most important example of the need for energy storage is the large-scale generation of electricity. It is the aim of the companies or generating boards who supply electricity to be in a position to provide sufficient power to meet demand. Unfortunately, demand is by no means constant; it fluctuates throughout the year as a result of climatic variations and it fluctuates during any 24-hour period with, for example, most domestic consumption occurring at meal times and in the evening. (This is illustrated by the load curve of Figure 14.1) There must be enough generating capacity to match peak demand and, ideally, production and demand ought to be matched at all times. It is true that certain types of generating equipment can be closed down or restarted quickly; for example, small gas turbines have a low response time and the sluice gates of hydroelectric power schemes can be operated fairly rapidly. However, large fossil fuel or nuclear power stations are not amenable to frequent shut-downs and, moreover, they are at their most efficient when run at a fairly constant power output. It is, therefore, advantageous for the generating boards to run their large stations continuously and to find some means by which the excess power that they generate when demand is low can be saved for use under peak load conditions.

Energy storage is also of relevance in the field of transport. Diesel, petrol and kerosene fuels enable us to live in a highly mobile society. It may be that these fuels will become increasingly scarce in the future and electricity generated by nuclear stations could be our main source of power. It is a relatively simple matter to electrify railways either by using conductors between the rails or by using overhead cables, and cables can also be used for trolley buses or tramcars. On the other hand, the private car does not lend itself easily to such solutions.

367

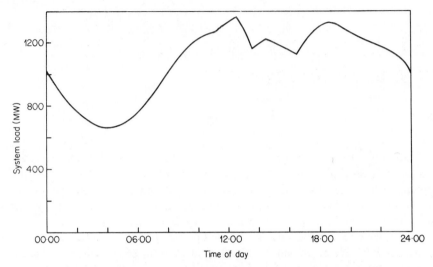

Figure 14.1. Typical winter daily load curve for an electricity supply network. (Reproduced by permission of the Electricity Supply Board, Ireland)

An electric car requires a portable source of electrical energy, the most familiar example being the battery. Even if petrol supplies remain abundant, the electric car may have an important role to play in an urban scenario by virtue of its cleanness and silence, so research into the storage of electrical energy for transport is well justified.

Energy storage usually involves energy conversion. For a given power source there are usually several possibilities for converting the primary energy into secondary forms which may be more manageably stored. A particular example, that of wind power, will serve to illustrate this. Energy storage must attempt to smooth out the differences between the rapidly fluctuating power source and the consumer's demand for power, which may also be extremely variable. One possibility is to use the wind power to pump water to a higher altitude. The water is then a potential source of electrical power via, say, a turbine. The wind power can, of course, be converted directly into electricity. This can be converted again into electrochemical energy stored in batteries; into thermal energy stored, say, in a hot water tank; or into chemical energy stored in the hydrogen and oxygen produced by the electrolysis of water. When the wind is insufficiently strong to meet demand directly, the hot water can supply space heating requirements, the batteries can supply electricity and the hydrogen can be used as fuel directly or recombined with oxygen in a fuel cell to produce electricity.

It should be obvious from this single example that a complete presentation of all the possible techniques for energy conversion and storage would be most unwieldy. We shall content ourselves for the most part with a discussion of the principal techniques.

14.1 Off-peak Electrical Power

The first major application of energy storage is in overcoming the demand fluctuations in the large-scale generation of electrical power. There are two principal methods for dealing with this problem already in existence. The first does not attempt to store the surplus electricity directly but instead strives to achieve a more uniform demand level throughout the day. Consumers are encouraged to use power when demand is low, for example during the night, by the incentive of reduced tariffs.

For space heating applications, the relatively cheap 'off-peak' power must be stored as thermal energy by the consumer for use during his own peak demand period. There are essentially two ways in which this can be done: by latent-heat storage which involves a phase change in the storage material, or by sensible heat storage. In a purely latent-heat process, storage occurs at constant temperature whereas in a sensible heat process the storage material experiences a rise in temperature.

Acceptable latent heat storage materials should have a high latent heat per unit volume and should undergo the phase change within predictable limits and without subcooling. They must also be non-corrosive and chemically stable. Many inorganic salts are being actively considered for this role. Indeed, sodium nitrite for one has an extremely high volumetric heat capacity (Table 14.1).

Table 14.1 Storage capacity of various materials. (This is illustrated by the amount of material required to store 10 kWh of energy.)

	Operating range (°C)	Weight (kg)	Volume (l)	Volumetric capacity (kWh/l)
(a) Fusion materials				
Sodium hydroxide	135–500	32	19	0·53
Sodium nitrite	135–310	65	32	0·31
Paraffin wax	45–320	40	62	0·16
(b) Materials using sensible heat only				
Iron-steel	65–500	170	21	0·48
Concrete	65–500	70	31	0·32
Brick	65–500	95	49	0·20
Water	45–100	155	155	0·065

Water has been extensively used as a sensible heat storage medium but it is clear from Table 14.1 that its volumetric thermal capacity is considerably poorer than that of solids such as concrete, brick and iron. The use of high-density materials such as alumino-silicate bricks in storage heaters is common in the United Kingdom. Eastern European countries, on the other hand, have tended to favour iron storage heaters. The main problem with storage heaters is regulation. By their very nature they are sluggish in responding to sudden changes in weather conditions with the result that much electrical power may be wasted on warm winter days.

An interesting variant on storage heaters is underfloor heating which employs a concrete floor as the storage medium. This system has a number of advantages over the standard storage heater. Because of the large mass of the concrete floor, low-temperature storage can be used. This is less likely to cause draughts than is a smaller, high-temperature heater. Moreover, underfloor heating leads to a more uniform and smaller vertical temperature gradient in the space being heated which is preferable on grounds of comfort. It does, of course, suffer from the same slow response time as the other storage systems.

The utilization of off-peak power is by no means restricted to space heating. By a suitable rearrangement of work schedules, any automated industrial process can be carried out using off-peak power. Ironically, the United Kingdom Central Electricity Generating Board has sold off-peak power to its consumers so successfully that new peaks are appearing in the load curves when all the off-peak appliances are activated within a short space of time.

14.2 Pumped Storage Schemes

As an alternative to the storage of electrical energy in the form of heat by the consumer, the generating authorities can store it prior to transmission by the technique of *pumped storage*. Surplus electrical power is in effect used to imitate the role of solar energy in hydroelectric power schemes. During off-peak periods water is pumped from a lower to an upper reservoir in which it is stored, the electrical energy being converted into the gravitational potential energy of the water. Then, during a period of peak demand, the system is run in reverse as a straightforward hydroelectric power station. The pumps used to drive the water uphill and the turbines used to generate electricity on its return journey are actually one and the same. They are so designed that by rotating in opposite senses they act either as pumps or turbines and accordingly consume or produce electrical power. Typically, modern pumped storage schemes can achieve efficiencies of close to 80 per cent, twice the efficiency of the best fossil fuel power stations.

A suitable site for such a pumped storage scheme is a natural lake in hilly countryside. The natural lake can act as the lower reservoir while the top of a nearby hill can be excavated so as to form an artificial upper reservoir. (Of course, there is no reason in principle why the natural lake should not be used as the storage reservoir with the lower reservoir excavated underground, a design which might be expected to find more favour with the environmental lobby.) As an example of a pumped storage scheme, Figure 14.2 shows the ESB Turlough Hill installation near to completion together with a schematic view of the pipeline system. The upper reservoir was made partly by excavation and partly by using the excavated material to construct an embankment. Its capacity is $2 \cdot 3 \times 10^6$ m^3 with a maximum depth of 19·4 m. The power station as such is actually sited underground and consists of four pump turbines capable of producing a total of 292 MW of power at 500 rpm. It is connected to the upper reservoir by a 548 m long, 4·8 m diameter steel-lined pressure tunnel, and to the lower reservoir by a 106 m, 7·2 m diameter concrete-lined tailrace tunnel. The flow rates in the pressure and tailrace tunnels at nominal turbine output are 6·25 m s^{-1} and 2·86 m s^{-1}, respectively; these figures are reduced to 4·92 m s^{-1} and 2·10 m s^{-1} at nominal pump output. Each power unit has a mean flow of 28·3 m^3 s^{-1} at the nominal output of 73 MW, and a mean flow of 22·1 m^3 s^{-1} when pumping at a load of 68·2 MW. The pumps are started using a shaft mounted pony motor of 5·5 MW.

The Turlough Hill scheme, with a maximum stored energy of 1860 MWh per cycle, pales into insignificance compared with the planned Dinorwig scheme in Wales, the largest of its type in Europe. This has an 1800 MW input and a storage capacity of up to 7·8 GWh per cycle. Apart from its storage role, it is planned that the Dinorwig turbines will be driven continuously throughout the day by 15 per cent of the peak water flow. This will allow the turbines to spin synchronously with the National Grid and to respond within ten seconds of a major failure in the rest of the production and distribution network.

The use of pumped storage schemes is expanding throughout the World. An even larger installation than Dinorwig is planned for Storm King, on the Hudson River. This project, devised by the Consolidated Edison Company of New York, will have an initial capacity of 2000 MW. By 1971 more than 7000 MW of pumped storage capacity had been or was being installed in the United states with planning permission pending for another 10,000 MW (Berkowitz, 1971).

The electricity producers' faith in pumped storage is not difficult to comprehend as these schemes have much to recommend them. Firstly, they can contribute to peak load generating capacity. Secondly, as we have already noted with respect to the Dinorwig installation, they are capable of a rapid response to a sudden load adjustment. Moreover, when the turbines are running as pumps they are in effect a load which can be shed rapidly to release generating capacity for unexpected peaks in demand. Thirdly, they can be used to smooth the daily load curve. (For example, the effect of the Turlough Hill installation on the ESB load curve is obvious from Figure 14.3) This smoothing effect means that steam plant, relieved from load following, can be used more efficiently. Consequently the overall efficiency of electricity production is raised.

In some respects pumped storage is environmentally more desirable than the more straightforward hydroelectric installations. Fairly small stream flows can be utilized to provide large generating capacities. The construction of large dams is avoided. Land with a steep profile and with the provision of space for the construction of an artificial storage reservoir is all that is necessary. On the other hand, the water level in the reservoirs is subject to very rapid and large fluctuations which may cause difficulties for aquatic life.

It may safely be assumed that the concept of pumped storage will continue to attract power engineers by its elegance as a solution to the problem of energy storage. Its high efficiency and versatility enhance this attractiveness. Nevertheless, it must be remembered that pumped storage seldom saves fuel as such. The savings resulting from the use of high efficiency thermal stations by night

(a)

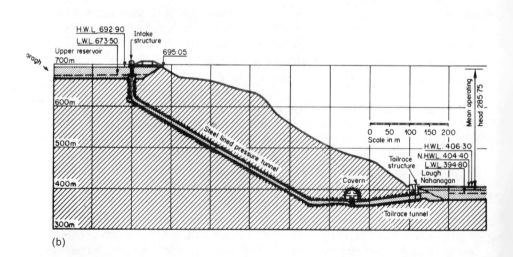

(b)

(c)

Figure 14.2. Turlough Hill pumped storage scheme near Dublin, Ireland. (a) Aerial view of the artificial upper reservoir and the natural lower reservoir. (b) Schematic view of the pipeline linking the two reservoirs. (c) Interior of the underground power station nearing completion. (Reproduced by permission of the Electricity Supply Board, Ireland.)

and the displacement of inefficient plant by day are offset by losses incurred at the pumped storage station itself.

14.3 Chemical Storage

If it is accepted that the present trend towards an all-electric economy is to continue, especially so if more electricity is generated by nuclear power stations and if fossil fuels become scarce or more expensive, then it is clear that there will be a need for a secondary fuel. Similarly, installations capitalizing

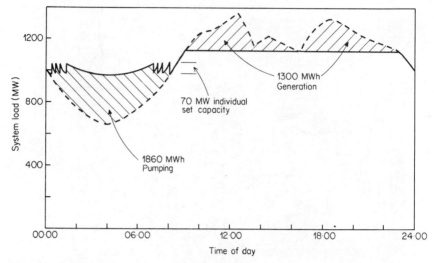

Figure 14.3: Effect of the Turlough Hill pumped storage scheme on the ESB daily load curve of Figure 14.1. (Reproduced by permission of the Electricity Supply Board, Ireland)

on natural power sources such as tidal and geothermal power are likely to be sited in remote areas whence it may be more convenient to distribute secondary fuel rather than electricity.

A good secondary fuel should have a high energy content per unit volume and per unit weight, it should not be too expensive to produce and it should be safe to store and transport. In these respects gasoline is an admirable choice despite the danger of explosion and the sophisticated handling and storage techniques that are necessary. Taking octane as representative of gasoline, its performance is compared to that of other fuels in Figure 14.4 which indicates the weight and volume of fuel required to yield 1 MWh of energy on combustion. It can be seen that the small-volume fuels are solids which are difficult to handle while the low-weight fuels are cryogenic liquids. The ideal fuel would be liquid and would be represented in Figure 14.4 by a point closer to the origin than that for octane. In this respect hydrogen, a cryogenic liquid, is not the best choice but we shall see that there are many good reasons for its selection as a secondary fuel.

Hydrogen is of interest because of its availability in ordinary water and for its ease of manufacture. It burns with an extremely high flame temperature (2500°C) and so is a near ideal working fluid for a heat engine, and, moreover, it burns cleanly. The only combustion product, apart from small traces of nitrogen oxides formed from the air near the flame, is water. Consequently its use as a domestic fuel is attractive. It is also attractive ecologically in that the atmosphere is a very efficient and rapid recycler of water. For these reasons hydrogen is today accepted as the main claimant to the secondary fuel throne.

The use of hydrogen for energy storage was advocated as long ago as 1933

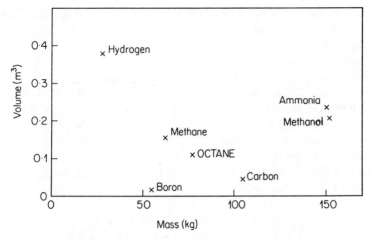

Figure 14.4. Comparative performances of secondary fuels. The diagram shows the volume and mass of fuel required to provide 1 MWh storage

by R. A. Erren, a German resident in England, who proposed that off-peak electrical power could be used to produce hydrogen by the electrolysis of water. Since that time the cause of hydrogen has been championed by F. T. Bacon, the designer of the hydrogen–oxygen fuel cells used by N.A.S.A. in the American space programme (see Section 14.5). More recently the idea of an overall 'hydrogen economy' has gained credence (Gregory and coworkers, 1972). Its proponents suggest that nuclear power stations should be used to produce hydrogen either by electrolysis or by using the waste heat which they produce to thermally decompose water. The hydrogen could be stored and then transported for use as fuel or as chemical feedstock. To begin with, though, we must examine the possible methods for hydrogen production. In all cases the basic raw material is water.

Electrolysis

At the present time most hydrogen is manufactured from petroleum feedstock but some is produced by electrolysis. The efficiencies of the best plants are only in the range 65–70 per cent and it would appear that with existing design larger plant will have even poorer performance. Schenk and coworkers (1973) have given a detailed costing of the electrolytic process based on an electricity tariff of £9·30 per 10^{10} J and a theoretical consumption of 2·8 kWh per m³ (at N.T.P.) of hydrogen produced (or, with 60 per cent efficiency, a consumption of 4·6 kWh m^{-3}). They find that for plants with an output of 10,000 m³ h^{-1} of hydrogen (7000 tons per annum), the cost would be approximately £20 per 10^{10} J. As this is twice the cost of the electricity consumed, it is clear that conventional electrolysis does not produce inexpensive hydrogen. The above figure is perhaps a little high as some allowance should be made for the sale of the oxygen produced; this could be used by the steel industry or in fuel cells. More optimistically, it should be noted that electrolytic efficiencies as high as 85 per cent have been achieved under laboratory conditions. Consequently, a considerable reduction in cost must be possible through improvements in electrolytic cell design. One such design, described by Mrochek (1968), is the Allis–Chalmers bipolar porous electrode cell. It has been demonstrated that this cell has a power consumption of only 80 per cent of the conventional bipolar cell. Similarly, the General Electric Company in the United States has adapted the technology of the advanced fuel cells used in the space programme to design a cell which has a solid perfluorinated sulphonic acid polymer electrolyte (Titterington, 1973). This has already achieved performances comparable with the Allis–Chalmers cell.

Fusion Water Blanket

Eastland and Gough suggested in 1972 that the plasma leakage or exhaust from a nuclear fusion reactor could be used to generate an intense ultra-violet photon source. The photons could then be absorbed by a water blanket surrounding the reactor thus causing the water to decompose. This futuristic scheme is not without its problems in that the hydrogen is produced in a mixture with oxygen and no operational fusion reactor exists (Chapter 11).

Thermal Decomposition

Straightforward thermal dissociation of water molecules is certainly possible but it requires a temperature of 2500°C!

Thermochemical Cycles

Considerable effort is being expanded in circumventing this problem by the use of multi-step chemical processes operating below 1000°C (Marchetti, 1973). Research by the Euratom organization has identified a very promising cycle, driven by process heat from a nuclear reactor, having an overall efficiency for hydrogen production of 55 per cent. The process involves $CaBr_2$, $HgBr_2$ and water, and naturally the materials other than water are recycled. However, this involves the handling of mercury compounds which are dangerous pollutants. A more general criticism of thermochemical cycles is that many of the reactions are isothermal and require a specific temperature which may present problems of control. In addition, with nuclear reactors as the heat source, an intermediate heat exchange medium would have to be used on grounds of safety. This means that a reaction which runs at 800°C would require a significantly higher reactor temperature. At the moment this is at the limit of nuclear technology: the hot gas from a high temperature gas cooled reactor has a maximum temperature of just over 800°C. Consequently lower temperature cycles must be found. One possibility identified by the Euratom group involves ferrous chloride and requires only three steps:

$$6 \, FeCl_2 + 8 \, H_2O \longrightarrow 2 \, Fe_3O_4 + 12 \, HCl + 2 \, H_2 \; (650°C)$$

$$2 \, Fe_3O_4 + 3 \, Cl_2 + 12 \, HCl \longrightarrow 6 \, FeCl_3 + 6 \, H_2O + O_2 \quad (200°C)$$

$$6 \, FeCl_3 \longrightarrow 6 \, FeCl_2 + 3 \, Cl_2$$

$$2 \, H_2O \longrightarrow 2 \, H_2 + O_2$$

Once again, though, temperature control is of paramount importance. At only a slightly lower temperature the first reaction step produces ferrous hydroxide and no hydrogen.

Once manufacured, the large-scale distribution of hydrogen should be a relatively simple matter. Natural gas is already transported by pipeline in many American and European countries (Chapter 17) and this method of distribution could equally well be used for hydrogen. Although three times as much hydrogen as natural gas would have to be pumped for the same power requirements (hydrogen has approximately one-third the energy content per unit volume of natural gas), the lower density and viscosity of hydrogen allow the same pipeline to handle a flow rate three times greater than for natural gas, albeit using slightly higher compressor power. The conversion of existing natural gas pipeline networks to carry hydrogen would, therefore, be quite straightforward. Hydrogen could be stored in bulk at major production and distribution centres and then piped or transported by cryogenic tankers to outlying districts.

The storage of hydrogen in cryogenic tanks is already a standard technique. Tanks with a capacity of up to 5000 m^3 of liquid hydrogen have been used in the N.A.S.A. space programme and it is estimated (Bartlit and coworkers, 1972) that 20,000 m^3 (5 million gallon) vacuum-jacketed tanks could be built using existing technology. There is no denying, though, that there is an ever present risk of an explosion when there is a mixture of hydrogen and oxygen (or air). Leakage from storage tanks is a problem with gaseous hydrogen because of the small molecular size but at the same time this provides a partial cure as the gas quickly diffuses away from the neighbourhood of the container. With sensible precautions gaseous or liquified hydrogen can be handled safely — at least as safely as gasoline.

Storage safety can be further improved by the use of metal hydride 'sponges'. Hydrogen will form chemical compounds with many powdered metals and with a variety of organic compounds. Slight warming of the hydride is usually sufficient to cause the release of the hydrogen once again. The capacity of these 'sponges' is extremely high. For example, magnesium can absorb over 1000 times its own volume of hydrogen. This may be of particular importance in transportation. Schoeppel (1972) has shown that a magnesium hydride fuel tank would only be four times as large as a gasoline tank for the same energy content. Unfortunately, magnesium hydride has a somewhat high operating temperature (287°C for 1 atmosphere dissociation pressure) and much research is needed to find a hydride with equivalent capacity and cheapness but with a lower operating temperature.

The potential uses for hydrogen are indeed multifarious. The following is by no means an exhaustive list but should give some indication of hydrogen's versatility as a secondary fuel.

Chemical Feedstock

Hydrogen is already used as a basic raw material in the fertilizer industry. (Ammonia synthesis accounts for some 55 per cent of hydrogen production.) In addition, hydrogen could eventually become a basic feedstock for what is now called the petrochemicals industry.

Domestic Heating and Cooking

Hydrogen as a central heating fuel has the advantage of requiring no flue and therefore giving a better efficiency. In addition it provides its own humidification system! It is an extremely clean fuel and can be made more so if catalytic combustion is used: catalytic bed temperatures as low as 100°C can be employed which prevents the formation of noxious nitrogen oxides in the flame. As additional safeguards for domestic use an odorant could be used to facilitate detection (as has already been done for natural gas) and, similarly, an illuminant could be added to make visible the somewhat faint hydrogen flame, and improve heat transfer.

Transport

A conventional internal combustion engine can easily be converted to run on hydrogen instead of gasoline. All that is required is a modification to the carburettor to enable it to handle gaseous fuel and to prevent pre-ignition. The major problem at the moment is fuel storage. Safety problems may well encourage the use of hydrogen in fuel cells (Section 14.5) rather than by direct combustion. Hydrogen could also prove to be an ideal aircraft fuel. In this application its very lightness compared to an energetically equivalent amount of kerosene would be a positive advantage, although the larger fuel tanks required would involve some redesign of airframes. Furthermore, it would appear that hydrogen is an ideal fuel for gas turbine engines. As an additional bonus, it would appear that by using hydrogen as a fuel, airspeeds of Mach 10 can theoretically be achieved, compared with a limit of Mach 5 for liquid hydrocarbon fuels. Considering the present furore over supersonic airliners, however, it may well be prudent to avoid over-emphasizing this particular point!

Local Power Generation

Hydrogen, like natural gas, can be used as a fuel for gas turbine generators.

In addition, electricity can be generated locally using hydrogen–oxygen fuel cells (Section 14.5).

Standby Fuel

Gas turbine generators are already used to some extent as standby generators for national electricity networks. As hydrogen can conveniently be stored in large cryogenic tanks and can be used instantly when required, it is an ideal standby fuel.

It is ironic that after spending many years ravaging the Earth for its fossil fuels, mankind may well turn in the future to the abundant supplies of sea water for his primary *and* secondary power sources. Deuterium could be the fuel for his nuclear fusion reactors and hydrogen his secondary fuel. One wonders if this is what Jules Verne had in mind when he wrote:

"...and what will men burn when there is no coal? Water. Yes, my friends, I believe that one day water will be employed as a fuel, that hydrogen and oxygen which constitute it, used singly or together, will furnish an inexhaustible source of heat and light."

(*The Mysterious Island*)

14.4 Batteries

Instead of converting electrical energy directly into the chemical energy of compounds which are later burnt as fuels, it is possible to employ a cyclic process in which such energy is stored chemically but is later reconverted to electrical energy. This is the principle of electrochemical storage. The traditional electrochemical power unit is the battery. There are two distinct types. Primary batteries convert some of the chemical energy stored in certain compounds into electrical energy via reactions in an electrolyte. Once these reactions have proceeded to completion the battery is defunct as a power source and is discarded. In secondary batteries the electrolyte reactions can be conveniently reversed. Once the reactions are complete electrical power can be applied to the battery terminals in such a way as to cause the reverse reaction to occur, that is to charge the battery. This charge–discharge cycle can be repeated many times before the useful life of the battery is complete.

The current United Kingdom market for batteries is of the order of £35 million per annum for primary units (in torches, portable radios, etc.) and £60 million for secondary units of which approximately 80 per cent is spent on vehicle starter batteries. Indeed, some 40 per cent of the western world's consumption of lead is accounted for by such batteries, which are invariably of the lead–acid type. The only other secondary battery to have claimed a sizable proportion of the market, with United Kingdom sales of £5 million, is the nickel–cadmium cell which has the twin advantages of lightness and durability. Primary batteries, such as zinc–carbon dry cells will give a low power output of up to 10 W for between ten and twenty hours. This is quite acceptable for some applications but, as they are not suited to repeated electrochemical storage and are discarded after use, we shall neglect them in the discussion which follows.

The most familiar secondary battery is, without doubt, the combination of lead–acid cells used in cars to power the starter motor and the electrical system. The electrodes of the cell consist of porous lead impregnated with sparingly soluble lead sulphate. The electrolyte is sulphuric acid saturated with lead sulphate. During the charging process the cathode donates electrons to the Pb^{2+} ions forming lead atoms which are deposited on that electrode. At the anode Pb^{2+} ions lose electrons to become Pb^{4+} ions which then form lead oxide, PbO_2, by hydrolysis. Thus the charging reactions are

$$Pb^{2+} + 2e \longrightarrow Pb \qquad \text{(cathode)}$$

$$Pb^{2+} + 2\,H_2O \longrightarrow PbO_2 + 4\,H^+ + 2e \quad \text{(anode)}$$

The reverse process, discharge, converts the electrochemically stored energy back into electrical energy which can drive an external load connected between the electrodes (Figure 14.5). Now the role of the electrodes is reversed: the plate which was the anode during charging becomes the cathode and vice versa. The electrode reactions are now

$$PbO_2 + 4\,H^+ + 2e \longrightarrow Pb^{2+} + 2\,H_2O \text{ (cathode)}$$

$$Pb \longrightarrow Pb^{2+} + 2e \qquad \text{(anode)}$$

and the overall reversible cell reaction is

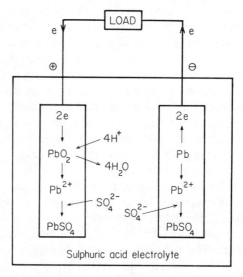

Figure 14.5. Discharging processes in a lead–acid cell

$$\text{Pb} + \text{PbO}_2 + 4\,\text{H}^+ \underset{\text{charge}}{\overset{\text{discharge}}{\rightleftarrows}} 2\,\text{Pb}^{2+} + 2\,\text{H}_2\text{O} \qquad (14.1)$$

Equation (14.1) implies that there should be a decreasing acidity of the electrolyte as the battery discharges (H^+ ions are being used up). It is this reduction in sulphuric acid concentration which causes the density of the electrolyte to decrease. When fully charged the lead–acid cell has an open cell voltage of approximately 2 V—the common 12 V car battery contains six lead–acid cells connected in series.

The electrochemical processes in such a cell are governed by Faraday's laws of electrolysis:

(1) The mass of a substance produced or consumed at an electrode is proportional to the electrical charge which has passed through the cell in the time concerned.

(2) The masses of different substances produced or consumed by a given charge are proportional to their respective chemical equivalents. Faraday's constant, F, is defined as the charge carried by 1 mole (6.023×10^{23}) of electrons and has the value 96,847 C. Consequently, Faraday's laws can be reduced to the statement that a charge F will produce or consume one gramme equivalent of a substance at each electrode of a cell. The masses of materials involved in electrode reactions are related to the charge transferred, $\mathrm{d}Q$, by

$$\mathrm{d}Q = nZF \qquad (14.2)$$

where n is the number of moles which react and Z is the number of electrons transferred.

It is possible to obtain a value for the open-cell voltage E (the emf) from thermodynamic considerations. The Gibbs free energy G is defined as

$$G = H - TS$$

$$= U + pV - TS$$

using the notation of Appendix 2. The change in free energy at constant pressure and temperature, $(\mathrm{d}G)_{TP}$, is, therefore, given by

$$(\mathrm{d}G)_{TP} = \mathrm{d}U + p\,\mathrm{d}V - T\,\mathrm{d}S \qquad (14.3)$$

For an electrolytic cell the change in internal energy, $\mathrm{d}U$, is not simply given by the sum of the heat supplied and the mechanical work, it also includes a term describing the electrical work involved. This electrical work is equal to the product of the charge passed and the cell emf E, $E\,\mathrm{d}Q$ or, using equation (14.2), $nZFE$. Thus the change in internal energy is given by

$$\mathrm{d}U = T\,\mathrm{d}S - p\,\mathrm{d}V - nZFE$$

As a result equation (14.3) is simplified to

$$(\mathrm{d}G)_{TP} = nZFE \qquad (14.4)$$

This analysis has assumed equilibrium processes. In a real cell irreversible

382

processes will occur which alter equation (14.4) to

$$(dG)_{TP} < -nZFE$$

$$= -nZFV$$

where V is the actual cell voltage when a current is flowing.

For the lead–acid cell described above the change in free energy for the electrolyte reaction of equation (14.1) is 394 kJ per mole. Substituting this into equation (14.4) with $n = 1$ and $Z = 2$ yields a value of 2·05 V for the open-cell voltage. This is indeed close to the measured open-cell voltage for a lead–acid cell.

The most important characteristics of any battery are its storage capacity, its energy density and its power density. Storage capacity, the amount of charge available without recharging, tends to decrease with the age of the battery. This is a result of changes in the microstructure of the electrodes. For example, in a lead–acid battery the lead sulphate reaction products may block part of the lead electrode which is then rendered ineffective in the charging process. Consequently, the capacity of the battery is reduced. Under operating conditions, the voltage produced by a battery decreases as the cell becomes more and more discharged (Figure 14.6). It is often the case that the voltage–current relationship for the cell also depends on the degree of discharge as illustrated schematially in Figure 14.7.

Energy density and power density refer to how much total electrical energy or power is available from a given size of battery in one cycle of operation. Both of these quentities are of obvious importance in portable applications. The lead–acid cell has a typical energy density of 15 Wh kg^{-1} and a power

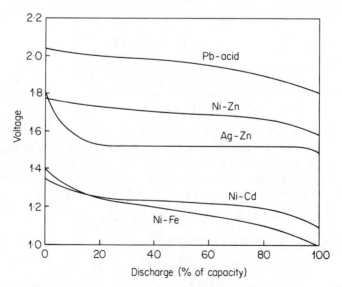

Figure 14.6. Variation of cell voltages with degree of discharge

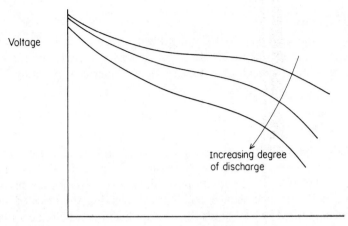

Figure 14.7. Voltage–current curves for a lead–acid cell for different degrees of discharge

density of about 150 W kg^{-1}. It is these parameters which must be improved upon if batteries are to play a major role in transportation and the like as we shall discuss later. Table 14.2 lists the performance data for various batteries

Table 14.2 Performance of batteries and fuel cells

	Equilibrium potential (V)	Typical energy densities[a] (Wh kg^{-1})	Typical power densities[a] (W kg^{-1})
Lead–acid	2·04	15	150
Nickel–cadmium	1·48	40	150
Nickel–iron	1·4	30	
Silver–zinc	1·70	110	
Sodium–sulphur		200	
Lithium–chlorine		300	
Zinc–air		30	
Sodium–amalgam–air		55	
Magnesium–air		120	
Hydrogen–oxygen	1·20	1000–2000	100
Hydrazine–oxygen	1·56	750	

[a] These figures must be treated with some caution. Battery performance will vary significantly according to the conditions of use.

which are either commercially available at the present time or are being extensively developed. It is clear that significant improvements in battery performance are to be expected but we must first examine the various designs to appreciate the technological problems involved.

Lead–Acid

Although the lead–acid battery has been in use for many years it would be a mistake to think that no further improvement in its performance is possible. Indeed, its storage capacity has undergone a 35 per cent uprating in the last decade.

Nickel–Cadmium

Nickel–cadmium batteries are already commercially available. Those with sintered nickel electrodes have energy densities as high as 40 Wh kg^{-1} and power densities of 150 W kg^{-1}. In addition to this improved performance over lead–acid batteries, they also require less maintenance but the price of nickel and cadmium compared to lead limits them to applications where the initial capital investment is not the main consideration.

Silver–Zinc

These have particularly high energy densities, and could already be used in electric cars. However, silver is very expensive and there is a tendency for the cell plates to change shape after only a few hundred charge–discharge cycles.

Zinc–Air

Non-rechargeable zinc–air batteries are already on the market. They use concentrated potassium hydroxide as the electrolyte with an amalgamated zinc powder anode. The 'air' cathode is made up of a number of layers of cylindrical plastic. The outer layer is a hydrophobic PTFE film which allows oxygen to pass through into the electrolyte but prevents the electrolyte from passing in the opposite direction. The inner face of the PTFE has a layer of catalyst in contact with the electrolyte. This encourages the breakdown of the oxygen molecules into negative hydroxyl ions. Rechargeable versions with an energy density of 130 Wh kg^{-1} have been developed. Here the fine particles of zinc oxide (the reaction product) are precipitated into the circulating electrolyte fluid and are then stored in a spent fuel tank. Recharging is accomplished in the usual way; passing a current causes the fluid to decompose into oxygen and zinc and the metal is then electroplated back onto the anode until the battery is fully charged. An alternative method of recharging is simply to draw off the spent slurry and replace it by a refill of zinc powder slurry.

Sodium–Sulphur

This battery, developed by the Ford Motor Company of America and in the United Kingdom at the Capenhurst Research Laboratory of the Electricity

Council, is expected to have a tenfold storage advantage over lead–acid batteries. It differs from all batteries described previously in that it has a solid electrolyte and liquid electrodes. The electrolyte is a ceramic test-tube-shaped piece of β-alumina (permeable to sodium ions) containing the liquid sodium anode and surrounded by liquid sulphur (the cathode) contained in a stainless steel vessel. The liquid sodium is replenished from another stainless steel reservoir. The use of liquid sodium and sulphur does, of course, necessitate a high operating temperature of between 250 and 400°C. Interest in this battery was stimulated when Capenhurst demonstrated an 18 cwt van which was powered by 960 sodium–sulphur cells with a total capacity of 50 kWh.

Lithium–Chlorine

This is being developed by the United States Atomic Energy Commission at their Argonne National Laboratory, near Chicago. Chlorine is a good electron acceptor, and lithium is an excellent electron donor and is also light in weight, one-twentieth the density of lead. High energy densities have been obtained but the cell operates at even higher temperatures (approximately 700°C) than the sodium–sulphur one. Moreover, chlorine is difficult to handle.

There are many problems still to be solved in battery technology. In many cases the electrode reactions are not yet fully understood, the important rate-determining step has not been identified. Corrosion is a major difficulty as are side effects which reduce battery performance. Similarly, electrocatalysis is a most important factor but is little understood. These problems also apply to the second type of electrochemical power source, the fuel cell.

14.5 Fuel Cells

In a secondary battery the materials which take part in the electrochemical reaction are all contained in one unit. The reaction itself can be reversed at will. But in a fuel cell the reactants are supplied from outside fuel tanks and, often, the reaction is not easily reversible. A fuel cell may be defined as a device which converts chemical energy into electrical energy by electrochemical reactions with a continuous supply of reactants in a galvanic cell. Fuel is supplied to one electrode and an oxidant, usually oxygen, is supplied to the other, fuel and oxidant combining in the form of ions.

The first fuel cell was produced in 1839 by Grove. A current density of 1 mA cm^{-2} at a potential of 1 V was generated when hydrogen and oxygen were supplied to two separate electrodes, platinum black strips, immersed in sulphuric acid. In 1889 Mond and Langer demonstrated the reduction in the reactivity of the platinum black with time and managed to extend the cell's life by storing the electrolyte in a porous non-conducting container. They also introduced electrodes of platinum pierced with tiny holes and covered with platinum black. This was probably the first fuel cell with a measurable power output which

386

was truly invariant in that the electrodes and electrolyte were unchanged by the reactions. Siegl (1913) was able to effect a reduction in the cost of this cell by supporting the platinum on carbon particles. Becquerel in 1855 had been the first to attempt the direct oxidation of carbon in a fuel cell; he used a fused nitre electrolyte in a platinum vessel and obtained a current between the platinum and a carbon rod placed in the fused salt. From that time direct oxidation was studied by a number of workers without much success until Davtyan in 1946 published his results on solid electrolytes for high temperature cells using carbonaceous fuels.

Meanwhile Bacon and his colleagues in Cambridge, England, had focussed their attention since 1932 on hydrogen–oxygen fuel cells operating at high temperatures and pressures with alkaline electrolytes. By 1959 they were in a position to demonstrate a 5 kW fuel cell battery and it is a tribute to their work that the American Gemini space probes and Apollo moon probes used fuel cells of this type to provide electrical power for their missions (and to supplement the drinking water by the reaction products). Nevertheless, it must be admitted that the development and manufacturing costs of these cells was not of prime concern. That no commercially viable fuel cell has as yet appeared is an indication of the technical and financial problems associated with their design.

Figure 14.8 illustrates the simple hydrogen–oxygen fuel cell. Hydrogen gas is diffused through a porous metal electrode, nickel in this example. A catalyst embedded in the electrode permits the hydrogen gas to be adsorbed on the electrode surface as hydrogen ions which then react with hydroxyl ions in the electrolyte to form water:

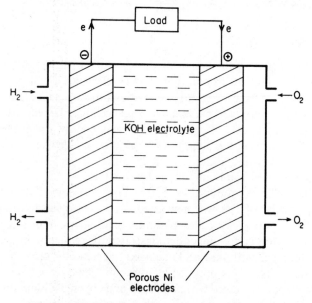

Figure 14.8. The simple hydrogen–oxygen fuel cell

$$H_2 \longrightarrow 2\,H^+ + 2e \quad \text{(catalyst)}$$

$$2\,H^+ + 2\,OH^- \longrightarrow 2\,H_2O$$

At the other electrode oxygen diffuses through, is adsorbed on the surface and is reduced to form hydroxyl ions which then migrate through the electrolyte to the hydrogen electrode:

$$O_2 \longrightarrow 2\,O \quad \text{(catalyst)}$$

$$O + H_2O + 2e \longrightarrow 2\,OH^-$$

Thus hydroxyl ions produced at one electrode are used in the reaction at the other electrode and electrons donated by the oxygen electrode are accepted by the hydrogen electrode. The overall reaction is the production of water from hydrogen and oxygen. The motivation for the whole process is the reduction in free energy of the components during the reaction. This energy difference will appear as the energy of the electrons which flow in an external closed circuit from the hydrogen to the oxygen electrode. On open circuit, electrical double layers are built up at the electrodes and these inhibit further reactions between the electrolyte and the gases.

Any ionic redox reaction can in theory be used in a fuel cell. However, in practice the reaction must proceed at a reasonable rate and the free-energy change involved should be high in order that the voltage developed is adequate (equation (14.4)). When these kinetic and thermodynamic considerations have been satisfied the ease of handling and cost of the fuel must also be taken into account. Finally, it is desirable that the fuel used should lead to manageable reaction products and, while ionizing rapidly at the anode, should be inert at the cathode. Some possible reactions are listed in Table 14.3.

Table 14.3 Some possible fuel cell reactions

Fuel	Reaction	Theoretical e.m.f. (V)	Temperature (°C)
Hydrogen	$2\,H_2 + O_2 \longrightarrow 2\,H_2O$	1·20	25
		1·00	700
Hydrazine	$N_2H_4 + O_2 \longrightarrow N_2 + 2\,H_2O$	1·56	25
Carbon monoxide	$2\,CO + O_2 \longrightarrow 2\,CO_2$	1·33	25
		1·00	700
Methanol	$2\,CH_3OH + 3\,O_2 \longrightarrow 2\,CO_2 + 4\,H_2O$	1·21	25
Ammonia	$4\,NH_3 + 3\,O_2 \longrightarrow 2\,N_2 + 6\,H_2O$	1·13	25
Carbon	$C + O_2 \longrightarrow CO_2$	1·02	25
		1·02	70
Methane	$CH_4 + 2\,O_2 \longrightarrow CO_2 + 2\,H_2O$	1·05	25
		1·04	70

The main attraction of fuel cells is the prospect of high energy conversion efficiencies, a point first appreciated by Ostwald in 1894. Assuming perfect electrodes and an electrolyte with a negligible internal resistance, all of the free-energy change for the net chemical reaction taking place in the fuel cell can be made available to do useful work; the intrinsic efficiency is 100 per cent. Strictly speaking we ought to compare the energy output of a fuel cell with the amount of heat produced by the combustion of the fuel, that is the enthalpy change ΔH in the reaction. It is on this figure that the efficiency of a heat engine is based. For the hydrogen–oxygen cell the enthalpy change is 242 kJ mole^{-1} while the free-energy change is 229 kJ mole^{-1}. Thus the thermal efficiency, η_T, of the cell is 229/242 or 95 per cent.

Of course, electrodes are never perfect and electrolytes do have some resistance to the flow of an electric current. The overall efficiency η of a cell is in fact the product of three terms: the thermal efficiency η_T, the Faradic efficiency η_F and the electrical efficiency η_E. The Faradic efficiency refers to the degree of utilization of the fuel at the electrodes. For various reasons, such as the involvement of some of the reactant in side reactions, not all of the reactant is used in the main fuel cell reaction. This failure to utilize all the fuel correctly is characterized by η_F, defined as

$$\eta_F = \frac{\text{Coulomb output}}{ZF \text{ (moles consumed)}}$$

For most simple loss processes the under-utilization can be taken as being proportional to electrode area, and equal to λ times the area where λ is a constant independent of current. If it is assumed that only one electrode is inefficient in this respect, then it is possible to derive a simple expression for η_F:

$$\eta_F = \frac{J/ZF}{J/ZF + \lambda}$$

where J is the current density. Thus η_F increases with current output. The electrical efficiency is the ratio of the on-load voltage of the cell to the open-cell voltage, V/E, and the thermal efficiency is given by

$$\eta_T = ZFE/\Delta H$$

It follows that the overall efficiency is

$$\eta = \frac{VJ}{\Delta H(J/ZF + \lambda)}$$

In practical cells overall efficiencies close to 80 per cent have been achieved.

The structure of the electrodes is a major factor in determining the efficiency of a cell. Normally gas diffusion electrodes are used and the reaction can only proceed at an adequate rate in the presence of a catalyst embedded in the electrode or coated on its surface. Thus the electrode reactions can take place only at certain regions where the three phases of gas, electrolyte and catalyst are in mutual contact. Obviously such regions are limited and much of the design effort in new cells goes into increasing the contact area. One method is to use porous electrodes. Three-phase contact regions may be formed within the pores of the electrode; capillary forces will force electrolyte into the pores while the gas pressure will force gas into the pores. Only in pores where these forces are finely balanced will there be a genuine three-phase interface. High gas pressures or small pore sizes will lead to gas escaping directly into the electrolyte with a consequent reduction in the Faradic efficiency. On the other hand, low gas pressures or large pore sizes will allow the pores to become flooded with electrolyte in which case no reaction can take place. In a practical porous electrode fuel cell the pores will not be of uniform size, some will be ineffective through being flooded with gas or electrolyte and the efficiency will be impaired.

Figure 14.9 illustrates the three-phase region inside an electrode pore. Only in the reaction zone AB do optimum conditions exist. To the left of this region, nearer to the pore head, the resistance to diffusion of water molecules and ions is high because of the remoteness of the electrolyte from the bulk. To the right of AB the larger amount of electrolyte poses difficulties for the diffusion of gas molecules. In the reaction zone, conditions are such that resistance to the diffusion of ions is reasonably low and that the diffusion of gas molecules to the catalyst surface is not inhibited. One way of enhancing the reaction zone is to prepare electrodes with double porosity having larger holes on the gas side than on the electrolyte side.

In addition to these difficulties the output voltage of a fuel cell is always less than the theoretical value due to polarization or overpotential effects. This polarization may take a number of forms. Chemical polarization refers to the activation energy which must be supplied in overcoming (in particular) the rate limiting electrode reactions. It is normally only important at low current densities. Similarly, concentration polarization can reduce the efficiency of a cell. In a hydrox (hydrogen–oxygen) cell hydroxyl ions diffuse through the electrolyte to the reaction zone of the hydrogen electrode. For a given current

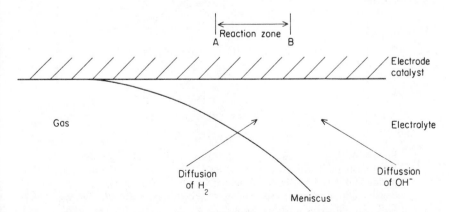

Figure 14.9. Reaction zone of an electrode pore in a hydrox fuel cell

390

density a certain rate of supply of these ions must be maintained but their concentration is limited by diffusion rates through the electrolyte. (In addition convection currents in the cell can disturb this diffusion.) This also applies to the supply of hydrogen and oxygen leading to gas-side concentration polarization. As an additional point, if air is supplied instead of pure oxygen the oxygen has to diffuse through the inert nitrogen as well. To achieve high current densities the concentration polarization must be reduced by minimizing the diffusion path lengths for the gases and the reactants. Finally, there is resistance polarization. All cells have an internal resistance r which reduces the output voltage by Ir when a current I flows. As the output emf of a fuel cell is usually about one volt, and as high current densities are desirable, it is imperative that this internal resistance be reduced to as low a value as possible.

The effects of polarization and the voltage–current relationship for a cell under load are shown in Figure 14.10. Liebhafsky and coworkers (1965) have given a semiempirical expression for this curve:

$$V = E - V_P$$
$$V_P = a \log J + b J - c \log ((J_L - J)/J_L) + d \qquad (14.5)$$

where V_P is the polarization, J is the current density, J_L is the limiting current density for the cell and a, b, c, d are parameters for the particular cell. For a hydrox cell at 25°C these are

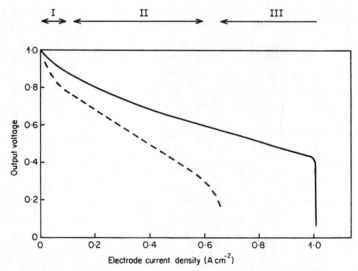

Figure 14.10. Polarization in a hydrox cell. The solid curve is the theoretical prediction while the dashed curve is a typical set of experimental results. (H. A. Liebhafsky and coworkers, 'Current density and electrode structure in fuel cells', in *Fuel Cell Systems*, R. F. Gould, (Ed.), *Advances in Chemistry Series No. 47*, 1965. Reproduced by permission of the American Chemical Society)

$$a = 0\cdot03 - 0\cdot12 \text{ V}$$

$$b \sim 0\cdot3 \, \Omega \, \text{cm}^{-2} \text{ (the internal resistance)}$$

$$c = 2\cdot3 \, RT/ZF$$

$$d \sim 0\cdot2 \text{ V (the polarization at 1 mA cm}^{-2}\text{)}$$

The diagram shows that the curve defined by equation (14.5) is in good qualitative agreement with experimental results.

The polarization curves can be divided into three regions (I, II, III) in which the rate of decrease of output voltage with current has particular characteristics. From equation (14.5)

$$-\,\mathrm{d}V/\mathrm{d}J = \mathrm{d}V_\mathrm{P}/\mathrm{d}J = \underset{\text{(I)}}{a/2\cdot3J} + \underset{\text{(II)}}{b} \; \underset{\text{(III)}}{-\,c/2\cdot3(J_\mathrm{L} - J)}$$

In region I the rate of decrease is inversely proportional to current. The main effect here is chemical polarization. In region II the rate is just equal to the internal resistance. Electrode structure has only a minimal effect in both these regions. However, the electrode structure does play a dominant role in determining the limiting current J_L and in region III the rate of decrease varies rapidly as J approaches J_L.

Fuel-cell development today centres on improvements in electrode design and in the electrochemical reaction rate. The rate can be increased by raising the temperature but this also hastens the corrosion reactions. If an aqueous electrolyte is to be used at temperatures above 100°C, the cell must be pressurized which in turn leads to construction problems. The effect of increasing pressure and temperature is to reduce the chemical and concentration polarizations while raising the temperature alone causes the internal resistance to be reduced.

To conclude this section a brief discussion of the main types of fuel cell design follows.

High Temperature Hydrox Cell

This cell, originally developed by Bacon, has become famous through its use in the Apollo moon flights (Figure 14.11). It has been demonstrated (Bacon and Fry, 1973) that higher operating temperatures lead to improved performance although corrosion problems do become severe if the temperature exceeds 200°C. The main danger of corrosion is, not unexpectedly, at the oxygen electrode but this is overcome by the use of nickel in constructing the electrodes and other cell parts. Potassium hydroxide, preferred to the less expensive sodium hydroxide because of its lower electrical resistance, is used as the electrolyte at dilutions of about 50 per cent, although the Apollo cells used 80 per cent solutions and thus required pressurization to only 4 atmospheres instead of 20. Nickel is normally used as the catalyst at the hydrogen electrode while lithiated nickel oxide has been used by Bacon at the oxygen electrode. The open-cell

Figure 14.11. Fuel cell power plants used in the NASA Apollo missions. (Reproduced by permission of the Pratt and Whitney Aircraft Corporation)

voltage with this design is in the range 1·0–1·1 V, lower than the theoretical voltage of 1·23 V because perhydroxyl ions as well as hydroxyl ions are formed at the oxygen electrode. The removal of the reaction product in a pressurized cell is fairly simple in that the water evaporates into the circulating gas.

The cells used in the Apollo lunar missions were assembled into three units, of 31 cells each, weighing 102 kg. They were capable of supplying between 563 and 1420 W of power at 27–31 V with an overall efficiency of 72 per cent and with energy densities of 880 Wh kg^{-1}. As Bacon and Fry (1973) have pointed out, these cells were designed in 1962 and considerable improvements have been made since then. Even so, the energy density of the Apollo cell is at least 100 times higher than that of a lead–acid battery. Moreover, power to weight and power to volume ratios of 88 W kg^{-1} and 50 kW m^{-3} have been achieved (Grevstad, 1969).

Low Temperature Hydrox Cell

This type of cell may possibly have more commercial potential than the high-temperature cell because of its relative simplicity. In the range 20–100°C carbon electrodes have been used allowing high current densities to be passed. The hydrogen electrode is prepared by depositing a noble metal catalyst on the carbon surface and treating it with a water-repellant to prevent 'drowning' of the electrode pores. A lifetime of 2000 hours at a current density of 1000 mA cm^{-2} has been reported for such an electrode (Litz and Kordesch, 1965).

As in the high-temperature cell, potassium hydroxide is again the electrolyte but now at a concentration of 30–40 per cent. The main problem with this type of cell is water removal. One method of doing so is to circulate the diluted electrolyte and to remove the water externally. Alternatively, a high gas flow rate can be employed so that the water is taken up by the gas stream which is then passed through a condenser.

Ion-Exchange Hydrox Cell

Ion-exchange cells utilize an ion-exchange membrane to divide the cell with electrodes in contact with it on either side. The membrane, in the form of a continuous sheet typically less than 1 mm thick, is a polymer with covalently bonded ion-exchange sites I^-. A typical ion-exchange reaction might be

$$I^- Na^+ \text{ (solid)} + K^+ Cl^- \text{ (solution)} \longrightarrow I^- K^+ \text{ (solid)} + Na^+ Cl^- \text{ (solution)}$$

This type of cell has a number of advantages over the more conventional designs. The membrane is light and compact, there is no loss of gaseous reactants to the electrolyte, and the deposition of the catalyst is less critical as are the requirements for pore size and waterproofing. On the other hand, heat removal is difficult as the electrolyte is not circulating. This is important because the organic polymer membranes are extremely sensitive to heat. The problem of heat removal can be lessened by using a dual membrane cell in which two membranes are separated by sulphuric acid, the acid acting as a heat-transfer fluid.

Methanol Cell

Electrolyte-soluble liquid fuels have a number of advantages over gaseous fuels: they are more easily stored and transported and, through the elimination of heavy gas cylinders, they should enable cells with better power and energy densities to be produced. The ideal liquid fuel would have to satisfy a long list of requirements—cheapness, availability, high calorific value, stability and solubility in strong acids or bases, and a wide temperature range for the liquid phase. In addition, it should ionize readily at a suitable anode with the minimum polarization. Methanol is by no means capable of meeting all these requirements but it can be produced in bulk and can be electrochemically oxidized to harmless reaction products, carbon dioxide and water. Consequently, methanol fuel cells have been attracting much attention.

Most experimental methanol fuel cells have employed a dissolved oxidant as well as a soluble fuel. The electrolyte in the first such cell (Grimes and coworkers, 1961) was potassium hydroxide with dissolved methanol and hydrogen peroxide. When the electrodes, nickel sheets coated with platinum on one side and silver on the other, were assembled in series in a battery, the silvered sides acted as cathodes and the platinized sides as anodes. Boies and Dravnieks (1965) have reported a methanol cell employing dissolved sodium chlorite as the oxidant in a sodium hydroxide electrolyte and they predict that energy densities of up to 400 Wh kg^{-1} may be attainable.

Hydrazine Fuel Cell

Hydrazine is another potential electrolyte-soluble fuel. The hydrazine molecule provides us with an ideal storer and transporter of hydrogen—attached to nitrogen. (Hydrazine in aqueous solution is quite safe although the crystalline hydrate is readily detonated.) Unfortunately, it is extremely expensive. The main reason for its high cost is the last stage in the production of the crystalline hydrate from aqueous solution in the Raschig process. It may be, though, that this last step could be eliminated and the solution used directly in fuel cells. Hydrazine also has the disadvantage of being extremely toxic and with a molecular weight of 50 and a yield of only four electrons per molecule it compares unfavourably with methanol (molecular weight 32 and six available electrons per molecule).

In theory the hydrazine–oxygen cell should have an open-circuit voltage of 1·56 V (Latimer, 1952) but it is generally found that the voltage in practical cells is approximately equal to that of a hydrox cell with the same electrolyte. It may be that the equilibrium

$$N_2H_4 + 4\,OH^- \rightleftharpoons N_2 + 4\,H_2O + 4e$$

is not readily established and that the hydrazine simply breaks down at the catalyst into nitrogen and hydrogen:

$$N_2H_4 \rightleftharpoons N_2 + 2\,H_2$$

Moreover, it is possible that ammonia is produced from the absorbed hydrogen and hydrazine.

It is found that the hydrazine electrode in such cells has a low voltage efficiency but has a Faradic efficiency of virtually 100 per cent (Gillibrand and Lomax, 1963). Major improvements in the performance of the oxygen electrodes will be necessary if full advantage is to be taken of the hydrazine electrode's potentialities. The other main line of design development is the prevention of hydrazine from reaching the cathode where wasteful direct oxidation of the fuel can occur.

Ammonia Fuel Cell

Ammonia is an inexpensive, readily available material with a high hydrogen content but it is gaseous, toxic and difficult to oxidize electrochemically. However, it does have a higher energy content per unit weight than hydrazine. The theoretical energy contents of the three soluble fuels discussed are: methanol, 5·9 Wh g^{-1}; ammonia, 5·7 Wh g^{-1}; hydrazine, 3·5 Wh g^{-1} (for the crystalline hydrate).

Hydrocarbon Fuel Cells

The main drawback to the use of hydrocarbons as fuels is the high operating temperature necessary to achieve good reactivity. In cells with molten carbonate electrolytes, temperatures in the range 500–650°C are required and in those with solid oxide electrolytes, a temperature of greater than 1000°C is necessary. Under these conditions thermal cracking of the hydrocarbon may occur. The hydrogen released can be utilized as fuel but carbon may be deposited in the fuel lines and in the anode compartment. The efficiency of the cell will be maximized if the hydrocarbon is completely oxidized electrochemically to carbon dioxide and water. One way of ensuring this is to add steam and operate the cell at 750°C with a reforming catalyst (nickel). This produces a hydrogen rich gas suitable for electrochemical oxidation. For example, if propane is the hydrocarbon concerned a 5:1 ratio of steam to fuel is required (Chambers and Tantram, 1960). It is possible to operate a low temperature hydrocarbon fuel cell but this necessitates the use of a phosphoric acid electrolyte at 150–200°C. Phosphoric acid is extremely corrosive at these temperatures and suitable catalysts which are corrosion-resistant are very expensive. Of all these types of cell the molten carbonate one is probably nearest to commercial application but this is still a long way off.

Regenerative Fuel Cells

In a regenerative cell the reactants are regenerated from the reaction products and recycled. In a thermally regenerative cell the products are fed into a high-temperature regenerator where they are dissociated into the original fuel and oxidant. One example is the lithium–hydrogen cell which has a lithium cathode and a hydrogen anode with a lithium chloride/lithium fluoride eutectic electrolyte fused at 600°C. The product, lithium hydride, is pumped into a regenerator at 900°C and the metallic lithium and gaseous hydrogen separated from the molten electrolyte and recycled. Open-circuit voltages of 0·45 V and efficiencies of 6–10 per cent have been reported (Taylor and coworkers, 1963). An intriguing possibility is that of photochemical regeneration in which solar energy may be used to decompose the reaction product. One example is the nitric oxide–chlorine cell (McKee and coworkers, 1960) in which the reaction product, nitrosyl chloride (NOCl), is decomposed by light with a quantum efficiency of two. The most attractive method of regeneration from the point of view of storing electrical energy is, of course, direct electrical regeneration. In one design (Lee and Handelwich, 1961) electrolysis produces gases which are removed for storage. Alternatively, the simple ion-exchange cell can be used and efficiencies of 50 per cent for energy storage have already been reported (Bone and coworkers, 1961).

It is obvious from this short discussion that fuel-cell development is still at a very early stage. Nevertheless, the advantages of fuel cells as energy storers and energy converters—efficiency, reliability, robustness, lack of moving parts, quietness and cleanness—are sufficient to warrant the intensive research effort in which many laboratories throughout the World are already engaged.

14.6 Electrochemical Transport

One of the most important applications of batteries and fuel cells in the future may well be as power units for vehicles. The United Kingdom is already an experienced user of battery-powered vehicles, with more than 80,000 fork lift trucks and 40,000 milk delivery vans in service powered by lead–acid batteries. The range of these vehicles is, of course, limited by the low energy density of lead–acid cells but this is not a serious drawback for the type of application quoted. In fact, battery operated delivery vans are more desirable than their petrol or diesel engined relatives when the delivery 'round' is short. Under stop–start driving conditions an electric motor can be cut-out with the minimum of bother whereas the internal combustion engine is left to idle.

Urban bus transport is another particularly good example of the potential for electric vehicles. During peak period traffic congestion the average speed of buses is low and their mileage is severely curtailed. In fact, as many as 90 per cent of buses on routes through city centres in the United Kingdom do not cover as much as 40 miles during this period. Similarly, almost half of a typical urban bus fleet is in use for less than seven hours per day (Figure 14.12). This is an ideal situation for electric buses. Such a bus can operate during the morning peak, then have its batteries charged and return to the streets for the evening peak. The main problem is the achievement of sufficiently rapid battery re-charging. Recently, Chloride Technical Limited have overcome this problem with their 'Programmed Rate of Rise of Voltage' charger. This can recharge lead–acid batteries in $3\frac{1}{2}$ hours instead of the usual eight hours. Moreover,

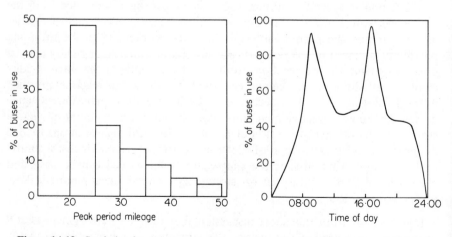

Figure 14.12. Statistics for urban bus routes. (Reproduced by permission of Chloride Technical Ltd.)

Chloride have gone further than this and now have an experimental 'Silent Rider' bus in service in Northern England.

A brief description of this experimental bus may well serve to illustrate the salient features of electric vehicle technology at the present time. The basic power unit is an assembly of 165 lead–acid cells each of 330 ampere hours capacity under five hour discharge conditions. The energy density of these batteries, 24 Wh kg^{-1}, is actually much less than that of lead–acid starter batteries (35 Wh kg^{-1}) but it is found that the longer life of the former more than compensates for this deficiency and gives a lower running cost per mile. The 72 kW (2100 rpm) series-wound DC motor is thyristor controlled. The voltage can be switched on and off at up to 500 times per second which permits accurate power control either by pulse width or pulse rate. (The current remains smooth because of the inductance of the motor's field winding and armature.) Regenerative braking is employed; that is, the engine acts as a generator and charges the batteries. Under heavy braking, a back-up compressed air system is employed. The bus's specification includes a maximum speed of 64 km h^{-1}, a range of 65 km, an acceleration from rest of 1 m s^{-2} and a hill climbing ability of one in eight. This is achieved at an overall energy consumption of 104 Wh for every tonne km.

The applications mentioned so far are fairly limited in scope. Moreover, the advantages offered by electrically powered vehicles—reduced noise, reduced pollution—are such that it is worth considering more general applications. Indeed, what is the likelihood of electrochemical private transport? A typical small saloon car weighing one tonne requires 20 kWh of energy to travel 35 miles (55 km). It can do this on one gallon (4 kg) of petrol but its electrical counterpart today would require 1500 kg of batteries occupying $\frac{1}{2}$ m^3 of space. Consequently, today's electric car is a two-seater with a range of 40 miles (65 km). It is obvious that wholesale application of electric vehicles will have to await developments in battery technology.

The DC series-wound traction motor is favoured in most 'stop–start' electric vehicles because of its high torque at low speeds. But high speed driving (greater than 25 mph) necessitates the use of sophisticated electronic controls (as in the Chloride bus) which would add considerably to the cost of the vehicle. Similarly, the requirements of an electric car are for a top speed of 60 mph (100 km h^{-1}), a hill climbing ability of one in four, and sufficient power reserve for quick overtaking in suburban conditions. This would imply the use of an extremely large series motor just to overcome the high thermal losses in the motor windings when the car is at top speed. Alternatively, the motor can be linked to the drive wheels by a clutch and a gearbox but this removes one of the virtues of electric cars, simplicity. A better solution is to employ a torque converter to couple the motor to the driving axle, or series-shunt hybrid winding with switching—in effect simulating a gearbox.

There is another possible solution to the electric car problem. Storage batteries, as we have seen, have good power densities but poor energy densities. Fuel cells, on the other hand, have their virtues and handicaps interchanged relative to batteries. A hybrid power source consisting of both batteries and fuel cells may well be the answer. During steady cruising, the fuel cells would supply power to the electric motor as well as charging the batteries. On steep gradients, or when rapid acceleration is necessary (when starting from rest or when overtaking) the batteries could be used to provide bursts of power.

14.7 Summary

Energy storage is, of course, a secondary problem to power production. Nevertheless, if power is to be consumed without wastage, the techniques of energy storage must be perfected and used. Inducements for the use of off-peak electrical power and pumped storage schemes are already an established feature of the energy scenario. On the other hand, the technology of hydrogen production and of batteries and fuel cells must be improved enormously if the hydrogen economy and the electrochemical economy are to establish themselves. Until recently there has been little incentive to develop these concepts but in the aftermath of the fuel price crisis renewed effort may well bring about more rapid progress.

References

Bacon, F. T. and T. M. Fry, *Proc. Roy. Soc. (London)*, **A334**, 427–452 (1973).

Bartlit, J. R., F. J. Edeskuty and K. D. Williamson, *7th U.S. Intersociety Energy Conversion Conf.*, 1972.

Becquerel, A. C., *Traite d'electricité*, Paris, 1855.

Berkowitz, D. A., in *Power Generation and Environmental Change*, D. A. Berkowitz and A. M. Squires, (Eds.), MIT Press, Cambridge, Mass. 1971.

Boies, D. B. and A. Dravnieks, *Fuel Cell Systems*, p. 262, Am. Chem. Soc., 1965.

Bone, J. S., L. W. Niedrach and M. D. Read, *Proc. Ann. Power Sources Conf.*, **15**, 47 (1961).

Chambers, H. H. and A. D. S. Tantram, *Proc. 2nd Int. Symp. on Batteries*, Bournemouth, England, 1960.

Davtyan, O. K., *Izvest. Akad. Nauk S.S.S.R., Otdel. Tekh. Nauk*, **107**, 215 (1946).

Eastland, B. J. and W. C. Gough, *Paper 41, Symp. on Non-fossil Chemical Fuels*, Am. Chem. Soc., Boston, 1972.

Gillibrand, M. I. and G. R. Lomax, *Proc. 3rd Int. Symp. on Batteries*, p. 221, D. H. Collins, (Ed.), Pergamon, Oxford, 1963.

Gregory, D. P., D. Y. C. Ng and G. M. Long, *Electrochemistry of Cleaner Environments*, Bockris, (Ed.), Chap. 8, Plenum Press, 1972.

Grevstad, P. E., *Paper 176, Proc. 4th Intersociety Energy Conversion Engineering Conf.*, 1969.

Grimes, P. G., B. Fidler and J. Adams, *Proc. Ann. Power Sources Conf.*, **15**, 29, 1961.

Grove, W. R., *Phil. Mag.*, **14**, 127 (1839).

Latimer, W. M., *The Oxidation States of the Elements and their Potentials in Aqueous Solutions*, 2nd ed., p. 128, Prentice-Hall, New York, 1952.

Lee, J. M. and R. M. Handelwich, *Proc. Ann. Power Sources Conf.*, **15**, 43 (1961).

Liebhafsky, H. A., E. J. Cairns, W. T. Grubb and L. W. Niedrach, 'Fuel cell systems, *Am. Chem. Soc.*, 116 (1965).

Litz, L. M. and K. V. Kordesch, *Fuel Cell Systems*, Am. Chem. Soc, 166 (1965).

McKee, W. E., E. Findl, J. D. Margenum and W. B. Lee, *Proc. Ann. Power Sources Conf.*, **14**, 68 (1960).
Marchetti, C., Chemical Economy and Engineering Review, Japan, 1973.
Mond, L. and C. Langer, *Proc. Roy. Soc.*, **46**, 1889.
Mrochek, J. E., in USAEC Symp. Series No. 14, Abundant Nuclear Energy, 1968.
Schenk, H., W. Wenzel, F. R. Block and E. Wortberg, Euratom Rpt. EUR 416d, 1973.
Schoeppel, R. J., *Chemtech*, 476, (1972).
Siegl, K., *Elektrotech. Z.*, **34**, 1317 (1913).
Taylor, J. E., N. Fatica and G. H. Rohrback, *Fuel Cells*, p. 13, Chem. Eng. Progr. Tech. Manual, Am. Inst. Chem. Eng., N. Y., 1963.
Titterington, W. A., *8th International Energy Conversion Conf.*, Philadelphia, 1973.

Further Reading

Baker, B. S. (Ed.), *Hydrocarbon Fuel Cell Technology*, Academic Press, New York, 1965.
Bockris, J. O'M. and D. M. Drazic *Electrochemical Science*, Taylor and Francis Ltd., London, 1972.
Central Electricity Generating Board, Annual Report and Accounts, 1973–1974.
Electricity Supply Board Ireland, 47th Annual Report, 1974.
Fuel Cell Systems, Symposia sponsored by Division of Fuel Chemistry at 145th and 146th meetings of the American Chemical Society *(Advances in Chem. Series*, No. 47, R. F. Gould, (Ed.), Am. Chem. Soc.), 1965.
Guthrie-Brown, J., (Ed.), *Hydroelectric Engineering Practice*, Chapter 18, Pumped Storage, Blackie, London, 1970.
Liebhafsky, H. A. and E. J. Cairns, *Fuel Cells and Fuel Batteries*, John Wiley and Sons Inc., New York, 1968.
Proceedings Hydrogen Economy Miami Conference, National Science Foundation and University of Miami, 1974.
Williams, K. R., (Ed.), *An Introduction to Fuel Cells*, Elsevier, Amsterdam, 1966.

15 Energy Waste

Energy production is a major industry in all the developed countries. As consumers, we are all accustomed to readily available and reasonably inexpensive electrical power and to the benefits which it brings into our homes and factories. It would appear, however, that it is only in times of emergency, for example the 1974 oil embargo by the Middle East producing countries, that we give any serious thoughts to economies in the use of energy or to the elimination of wasteful practices. It is precisely this aspect of energy production and utilization that we shall be examining in this chapter.

Energy waste has many connotations. Inefficiency in the production of power has already been dealt with, to some extent, in earlier chapters. A large proportion of the heat produced by the combustion of fossil fuels in a power station, or by a nuclear reactor, is discharged into the cooling water. This waste heat cannot be eliminated completely without violating the Second Law of Thermodynamics, but uses can be found for it. Similarly, we can view energy waste in terms of inefficiency in the use of energy, for example in space heating or in motor vehicles. Finally, society produces many waste products, both refuse and sewage, and we shall discuss how the wasted energy represented by these products might be recovered, at least in part.

To appreciate the potential for savings in energy consumption, it is interesting to examine the breakdown between the various 'end-uses' for energy. Table 15.1 gives the percentage of annual consumption in the United Kingdom and in the United States for each group of consumers. In the United Kingdom, transportation and domestic consumption alone account for approximately half of the nation's energy requirements. These also happen to be sectors in which large savings can be made with relatively little effort, as we shall soon see.

There is, in addition, a more sinister aspect to energy waste, namely pollution. Power production leads to the emission of gases and particles into the atmosphere and to the discharge of vast quantities of heat into the environment. Moreover, there are many environmental hazards involved in the extraction and transportation of fuels. In Chapter 2 we saw how power production and consumption could influence climate, and in Chapter 10 we looked at the problems of radioactive waste disposal in the nuclear power industry. We begin this chapter by analysing those remaining aspects of pollution which are specifically brought about by man's activities in the field of energy and power.

Table 15.1 Energy consumption by final users in the United Kingdom and United States of America on the basis of heat supplied. (These figures are quoted as percentages of gross national consumption.)

United Kingdom (1972)			United States of America (1968)	
Domestic		25	Domestic	19·2
Industrial		43·5	Industrial	41·2
iron and steel	11			
foodstuffs	3·5			
engineering	6			
agriculture	1·5			
chemical industries	6			
Transport		20·5	Transport	24·7
road	16			
rail	1			
air	3			
water	0·5			
Public services		6		
			Commercial	14·4
Others		5		
		100		100

(References: United Kingdom Data—CPRS, 1974. (Reproduced by permission of H.M.S.O.) United States of America Data—SRI, 1971.)

15.1 Pollution

Broadly speaking, there are two aspects to pollution caused by power production and consumption. The first is the pollution associated with extracting fuel and then transporting it. Thus coal mining can lead to land erosion and subsidence, while off-shore oil drilling and the transportation of oil in tankers can pollute the World's oceans with crude oil. Secondly, the production of power from fossil fuels leads to the discharge of gaseous and particulate emissions into the atmosphere, and, in common with nuclear power production, to the discharge of vast quantities of waste heat into the environment.

Coal is mined by underground and open-cast or strip techniques (Chapter 5). Underground mining is by far the more important in the United Kingdom, but in the United States strip mining accounts for one-third of all coal produced, that is $1·8 \times 10^8$ tonnes per annum. Strip mining is particularly attractive in that the costs involved are significantly lower than for underground mining while the productivity of such open-cast mines is roughly twice as high. However, the damage caused to the surface is much greater in the case of strip mining.

Area mining (strip mining in flat regions) involves removing a stretch of overburden to expose the coal seam which is then extracted. Then the overburden parallel to the first cut is removed and deposited in the region of the first cut. After a period of time the land is transformed into a series of ridges and valleys which are particularly susceptible to erosion by wind and rain.

(a)

(b)

Figure 15.1. Parkhouse Colliery tip (a) before and (b) after removal by opencast mining. (Reproduced by permission of the National Coal Board)

Contour strip mining is carried out when there is a coal seam outcropping in a hillside. Here the overburden is removed by a first cut directly above the coal. The overburden is discarded down the mountainside and the coal extracted. Successive cuts involve the removal of more and more overburden. The mining follows the contour of the hill and can continue until the ratio of overburden to coal seam is so high as to render the process uneconomical. Once again the result is an eyesore and erosion is a serious problem.

By 1965 some 1·3 million acres of the United States had been affected, with roughly equal amounts of area and contour mining (Perry, 1971). Worse still, 70 per cent of the affected land is in the five states of Pennsylvania, Ohio, West Virginia, Illinois and Kentucky. However, land disturbed by strip mining can be reclaimed, and this is especially simple if the reclamation is carried out soon after mining (Figure 15.1). Reclamation essentially involves the construction of drainage control dams, diversion ditches and stream channels, and planting in order to achieve soil stabilization against erosion and slides. Once reclamation is complete, the land can be rehabilitated for recreational, commercial, industrial and agricultural use.

Underground mining poses the problem of subsidence when insufficient coal is left in place to support the overburden. This is particularly important in urban areas where the disturbance can cause damage to streets and buildings, but even in rural areas it can damage drainage channels and for example, prevent the construction of dams or reservoirs in affected regions.

In addition, all coal mining produces refuse banks or slag heaps of the waste material left behind after the coal is cleaned. The amount of waste in each tonne of coal produced in the United States is approximately 0·2 tonnes (Perry, 1971). These refuse banks are unsightly, have a tendency to burn and cause local air pollution, and contaminate streams in their vicinity by seepage of water through the bank. One solution is to landscape the banks and conceal them with vegetation but a more desirable approach may be to find other uses for the waste material. For example, it may be used in the manufacture of lightweight aggregates, in road construction, or for the backfilling of both active and abandoned mines.

Pollution from oil production and transportation is a more serious and more widespread problem. In 1969, when world oil production was 2700 Mt and when 1200 Mt was transported by tanker, the total loss into the oceans was an estimated 2·1 Mt (SCEP, 1970). Three-quarters of this figure were equally accounted for by losses from ships, losses from tankers and accidental spills from ships and refineries. A mere 5 per cent results from off-shore oil production and the remainder may be attributed to industrial waste carried to the sea in rivers. Incidentally, it is worth noting that sea-bed pipelines from off-shore oil drilling rigs should not be pollution hazards even in the event of fracture because of the higher hydrostatic pressure outside the pipelines which would tend to force water into, rather than oil out of, the pipe. Although the most conspicuous oil pollution is localized in the form of accidental spills in off-shore waters, only 10 per cent of the total contamination is produced in this way. Nevertheless, with the trend towards larger and larger tankers, the effects of a single accident will become more pronounced (Chapter 17).

Crude oil and oil fractions can kill marine organisms directly by coating, asphyxiation or contact poisoning, or through exposure to the dissolved toxic components of oil. This latter effect may occur at some distance from the original spill. Moreover, one of the principal causes of death in birds which survive the immediate effects of oil exposure is the reduced resistance to infection brought about by the uptake of sublethal amounts of oil. Petroleum hydrocarbons can further disrupt the ecosystem by the mass destruction of the food sources of higher species. There is also possible aesthetic degradation of the environment.

There are a number of possible approaches to dealing with this problem. Firstly, it is clear that there should be stringent controls on the discharge of waste oil into rivers by industry. Improved navigational controls and monitoring of tanker movements could reduce the risk of collisions and strandings. Similarly, new methods must be found for cleaning out the bilges and fuel tanks of ships without the discharge of oil into the sea. Finally, the techniques for the removal of accidental spills must be reappraised. The use of non-toxic dispersants does succeed in breaking down the oil slick into fine droplets but these are potential poisons for filter-feeding organisms such as mussels and clams. Burning the oil, if it can be done, causes severe air pollution. The best technique is probably containment followed by physical removal which at the moment is most effectively achieved by absorbing the oil in chopped straw!

We now turn to the emission of pollutants during fuel combustion. Table 15.2 lists the particulate and gaseous emissions from all sources in the United States during 1968. It is clear that there is a high proportion of emission from fuel combustion and this is especially true for carbon monoxide, sulphur oxides

Table 15.2 Pollutant emissions in the United States of America (1968)

	Particles		SO_x		Hydro-carbons		NO_x		CO	
	(Mt)	(%)	(Mt)	(%)	(Mt)	(%)	(Mt)	(%)	(Mt)	(%)
Fuel combustion	9·2	35·7	22·9	75·9	15·7	54·1	16·5	87·8	60	65·7
Transport		4·3		2·4		51·9		39·3		63·8
Coal		4·3		60·5		0·6		19·4		0·8
Oil		1·0		13·0		0·3		4·8		0·1
Gas		0·7						23·3		
Wood		0·7				1·3		1·0		1·0
Other sources	16·5	64·3	7·3	24·1	13·4	45·9	2·3	12·2	31·3	34·3
Industrial processes		26·5		22·0		14·4		9·7		9·6
Solid waste disposal		3·9		0·3		5·0		7·8		7·8
Miscellaneous (forest fires, etc.)		33·9		1·8		26·5		16·9		16·9
Total	25·7		30·2		29·1		18·8		91·3	

(Reference: National Air Pollution Control Administration, 1970. 'Nationwide Inventory of Air Pollution Emissions'.)

and hydrocarbon emissions. Thus the control of this source of emissions would go a long way towards reducing the overall problem. The techniques for dealing with particulate emissions are already well understood and are incorporated as standard in most modern fossil fuel burning power stations (Chapter 5). We shall therefore concentrate on gaseous emissions.

Carbon monoxide is the major atmospheric pollutant after carbon dioxide. In fact, apart from carbon dioxide, it is normally emitted in greater quantities than all the other gaseous pollutants put together, especially in urban areas. The main technological source of the gas is the incomplete combustion of carboniferous fuels in transportation, heating, industrial processes, refuse burning and so on. It is also produced naturally in volcanoes, marsh gases and gas in coal mines (Flury and Zernik, 1931), during electrical storms (White, 1932), and by forest fires. However, the amounts so produced are small compared with man's production. The total global emission of carbon monoxide has been estimated to be 200 Mt yr^{-1} (Robinson and Robbins, 1968) and, by analogy with Table 15.2, transport is expected to be the main villain of the piece. Indeed, a diurnal variation in carbon monoxide concentration in urban areas has been observed (Jaffe, 1970) which can be correlated with the use of motor vehicles. The concentration is low in the early morning and then peaks during the morning rush-hour. It remains fairly high during the day and peaks again in the evening rush hour.

Carbon monoxide is long-lived in the lower atmosphere with a residence time of between 0·1 and 5 years (Robbins and coworkers, 1968; Weinstock, 1969). Given the level of emissions, this suggests that the background levels of the gas should rise by 0·04 ppm per year, whereas they have remained steady. This implies that there is a fairly effective scavenging process in the atmosphere. The mechanism is not understood but a number of potential removal processes have been suggested. The first is migration of the gas to the upper atmosphere where, in the presence of molecules such as NO_2 or O_3, it is converted into carbon dioxide by the high intensity ultra-violet radiation (Harteck and Reeves, 1967). Secondly, it is known that some micro-organisms can metabolize carbon monoxide (Schnellen, 1947). Finally, it is well known that carbon monoxide combines with biological porphyrin compounds such as the hemes but this is not a permanent sink for the gas as it is later released. In any case, it would appear that there is no cause for alarm as far as a permanent build-up in the carbon monoxide concentration is concerned, but measures should be taken, and indeed are being taken, to limit the emissions in order to prevent localized pollution. The most effective measures are obviously those which limit emissions from automobile engines.

Hydrogen sulphide is the main sulphur based gas produced by natural processes. Fuel combustion does produce some hydrogen sulphide but by far the most important sulphur bearing pollutant is sulphur dioxide. Total world emissions in 1965 were 130 Mt (Robinson and Robbins, 1970) of which 70 per cent was due to coal burning and 16 per cent to oil burning. The remainder was accounted for by smelting operations and petroleum refining.

Sulphur dioxide in the atmosphere is scavenged fairly effectively by a number of mechanisms. It is known that vegetation will take in the gas at a rate of about 2.5 g m^{-2} day^{-1} (Katz and Ledingham, 1939) when the atmospheric concentration is 10^{-3} ppm. Alternatively, and this is the principal mechanism in daylight, sulphur dioxide is photochemically oxidized to sulphuric acid in the presence of nitrogen dioxide and hydrocarbons. Direct absorption by water droplets can also occur. The solubility is low in droplets of low pH, but the presence of ammonia or carbonate promotes the solubility by neutralizing the acid formed in the droplets by the dissolved gas (Junge and Ryan, 1958). Once the sulphur is in the form of a sulphate aerosol it is rapidly removed by rainfall so that the upper limit to its residence is probably a week or so. This short residence time is confirmed by the large amplitude of climatic fluctuations of sulphur dioxide concentration (Ericksson, 1963). As in the case of carbon monoxide, there is no evidence for an accumulation of sulphur dioxide in the atmosphere.

Oxides of nitrogen are produced in fuel combustion when this takes place at a high enough temperature for the nitrogen in the air to be fixed and when the combustion gases are quenched sufficiently rapidly to reduce their subsequent decomposition. Global emissions of nitrogen dioxide from industrial processes were estimated by Mayer in 1965 to total 48·1 Mt, equivalent to 14·6 Mt of nitrogen. This should be compared with emissions from biological processes of 455 Mt nitrogen dioxide (135 Mt nitrogen), 540 Mt of nitrous oxide (370 Mt nitrogen) and 5365 Mt ammonia (4450 Mt nitrogen). Typical atmospheric concentrations of these gases over the continents are 0·006 ppm ammonia, 0·25 ppm nitrous oxide and 0·0004 ppm nitrogen dioxide. Over the oceans the concentration of nitrogen dioxide appears to be smaller still. Thus O'Connor (1962) has measured the concentration in average North Atlantic air (off the west coast of Ireland) and finds it to be as low as 0·0002 ppm. This low level, coupled with the emission data given above, suggests that there is a very effective scavenging process for nitrogen dioxide in the Earth's atmosphere; in fact, the residence time for the gas is only three days (Robinson and Robbins, 1970). The probable removal process is vapour-phase absorption by water to form nitric acid. The nitric acid vapour is rapidly removed in the formation of nitrate with, for example, atmospheric ammonia.

Water is also a major pollutant produced during electricity generation. Thermal power stations, fossil fuel and nuclear, use water to cool their condensers. The most common way of doing this is by once-through cooling. Water is taken in from a river, is heated and is then discharged downstream from the power station. The discharged effluent spreads in plumes on the surface of the receiving water and is carried off by surface currents. Heat dispersal depends on many factors such as current speed, turbulence, humidity and the temperature difference between the water and the air.

Power station cooling water is the major source of thermal pollution. In the United States some 80 per cent of all cooling water is used by power plants (Federal Water Pollution Control Administration, 1968), 95 per cent of this by fossil-fuel plants and the rest by nuclear installations. These latter produce more waste heat in the cooling water than do fossil-fuel stations of equivalent size— for one thing, some heat is lost up the stack in a fossil-fuel plant. Consequently, as more nuclear power stations are commissioned, the waste-heat problem will become relatively greater. It has been estimated (Belter, 1971) that the 1980 United States consumption of cooling water will be 7.5×10^9 litres per day, one-sixth of the total daily run-off. Moreover, as two-thirds of the yearly run-off occurs in one-third of the year, during the rest of the year electricity generation will account for half of the daily run-off.

This waste heat is a potential pollutant for a number of reasons. Increasing a stream's temperature will reduce its oxygen content by accelerating the oxygen consuming decay processes and so degrade its attractiveness to aquatic organisms. But at the same time, higher temperatures increase the organisms' metabolic rates and raise their demand for oxygen. Higher temperatures can also be lethal for fish as it is known to reduce the affinity of fish blood haemoglobin for oxygen (Clark, 1969). Under such conditions fish tend to grow faster and have a shorter life span. Moreover, reproduction and embryonic development will only occur if the water temperature is within certain limits.

These harmful effects of the once-through cooling process can be minimized by a number of techniques. The use of large capacity pumps allows more water to be used for a given degree of cooling and hence reduces the discharge temperature. This can also be achieved by mixing the heated cooling water prior to discharge with large quantities of ambient water. Another possibility involves taking the inlet water from deep in the stream where it is cooler. This means that the outlet water will be cooler.

There are alternative cooling technologies. Artificial cooling ponds may be used instead of rivers but this depends on the availability of land (a 1000 MW nuclear plant requires 2000–3000 acres of lake surface) and is about twice as expensive initially as once-through cooling. The heated cooling water may be sprayed into the air to promote heat transfer, caught in a reservoir (spray pond) and recirculated to the condenser. By combining a spray pond with an artificial cooling pond a smaller cooling pond can suffice. Wet cooling towers, four times as expensive to install as once-through schemes, are essentially spray ponds in a building. The purpose of the tower is to channel the water and air flows and to prevent excessive loss of water from the system by wind-carry of droplets as opposed to useful loss by evaporation. Inside the tower, the water falls over partitions which break up and delay the falling drops and hence improve the efficiency of heat transfer. Forced draught towers use fans to achieve the air flow: these have a higher operating cost but are cheaper to install. Finally, there are dry cooling towers in which the coolant is sealed in the system and heat transfer takes place only by conduction and convection.

In using these alternative methods an assessment must be made of the relative harm caused by discharging the waste heat into streams or directly into the atmosphere near the cooling towers (Chapter 2). In addition, chemicals such as chlorine are utilized in wet cooling towers to hinder the growth of organic matter. These chemicals are discharged into the atmosphere along with the water vapour. All things considered, it would appear that the dry cooling towers are the best solution but their initial cost is an order of magnitude greater than once-through systems. A more radical approach to the problem is to find a use for the waste heat, for example in agriculture to extend the growing season, in fish-farming of shrimps and such like, or in district heating, but this latter application merits separate consideration in Section 15.4.

15.2 Efficiency of Electricity Generation

Since steam turbines are the prime movers for most electricity production, the efficiency of electricity generation is limited ultimately by the efficiency of the heat engine. Most of this inefficiency is of an intrinsic nature. However, even small improvements in efficiency can lead to huge savings in fuel consumption so it is prudent to examine the possibilities.

There is little that can be done to improve the efficiency of a given power station. Modern power station design is such that the various components are closely matched, a small change in one component entailing modification to the whole station, and the cost of such modifications would be of similar magnitude to any gains which might be accrued during the working life of the station. Therefore, with the existing types of power station, an increased overall efficiency will only come about in time with the introduction of new plant. As we saw in Chapter 6, the most modern fossil fuel installations have efficiencies of about 40 per cent compared with an overall efficiency of approximately 30 per cent for the British Isles as a whole.

There is considerable interest at present in the use of *combined cycle generators* involving a gas turbine and a conventional steam turbine. The gas turbine has the advantage of working at higher temperatures than steam plant. It is used to drive a generator while its exhaust gases are used to raise steam for the steam turbine. In the United States prototypes of 250–400 MW output have already achieved efficiencies of just over 40 per cent. Unfortunately, gas turbines require the more expensive light fractions of petroleum. Admittedly, they do not require 100 per cent distillate fuel, but they are not at present competitive with conventional plant burning residual fuel oil. Even at efficiencies of 55 per cent they would not be competitive with nuclear stations for base load duty (CPRS, 1974). Nevertheless, their low cost could render them attractive as intermittent stations, and, moreover, they could in future be linked with coal gasification schemes (Chapter 5).

The overall efficiency of electricity generation can be increased by the incorporation of more pumped storage installation in the network although it must be remembered that such schemes rarely save energy (Chapter 14). Similarly, the use of waste heat for district heating (Section 15.4) may improve the overall efficiency of power production but not that of electricity alone. The remaining possibilities are the direct energy conversion devices.

In a conventional power station, the chemical potential energy of the fuel is first converted into thermal energy, then into the mechanical energy of the turbines and finally into electrical energy. It is the thermal stage that places a thermodynamic limitation on the efficiency of power production. In Chapter 14 we saw that the efficiency of fuel cells was not Carnot limited because the potential energy of the fuel is converted directly into electrical energy. Other techniques which do involve a conversion into thermal energy but which do not involve a mechanical stage have also been dubbed, somewhat erroneously, direct conversion devices. These are magnetohydrodynamic (MHD), thermoelectric and thermionic generation. Although Carnot limited, the lack of a mechanical stage enables much higher temperatures to be employed than in conventional stations with a consequent increase in efficiency. Thermoelectric and thermionic generators have already been mentioned in connection with solar power (Chapter 12) and we shall restrict ourselves to a brief discussion of MHD generators. A theoretical treatment of MHD generation has already been given in Chapter 7.

In a conventional generator a voltage is induced in an electrical conductor— a copper strip—by moving it across a magnetic field. This is also the principle behind MHD generation except that the copper strip is replaced by a gaseous conductor, a partially ionized gas. The gas is passed at high velocity through a high magnetic field and the generated current is extracted by placing suitable electrodes in the gas stream. To achieve good power outputs a gas velocity of 10^3 m s^{-1} is required (Chapter 7). The major problem is the achievement of adequate conductivity in the gas stream by ionization. Only a 0·1 per cent degree of ionization is necessary but to achieve this simply by heating the gas would necessitate temperatures of many thousands of degrees. This can be overcome by *seeding* the gas with elements of low ionization potential in which case temperatures of around 2000 K can be utilized. This is just within the limits of existing materials technology and can be achieved in combustion reactions.

Conventional steam plant is actually quite efficient (allowing for Carnot inefficiency) in the temperature range up to 600°C. MHD generators, on the other hand, are restricted, for the immediate future at least, to much higher operating ranges. Consequently, a composite plant, in which an MHD generator works at the highest temperature and sends its exhaust, slightly cooled by the extraction of useful energy, into conventional plant, is an attractive concept. The overall efficiency of such a plant would be high but it is by no means clear that the extra efficiency would justify the cost of the MHD 'topper'. It is worth noting that the theoretical upper limit to the temperature in a nuclear reactor is exceptionally high, and MHD might well provide a means of utilizing this high temperature. However, as we saw in Chapter 10, present day reactor technology is unable to make use of the high temperatures which are in principle available. Indeed, most nuclear reactors operate at temperatures which are low by the standards of conventional fossil fuel power stations.

Many problems still remain to be solved before MHD 'toppers' can become a practical proposition. The theoretical power output from an MHD generator (see Chapter 7) varies roughly as $\sigma B^2 u^2$, where σ is the conductivity of the gas,

B is the applied magnetic field and u is the gas velocity. Improved performance will result from increases in any of these parameters.

Conductivity

It can be shown that there is little to be gained by increasing the degree of ionization of the gas beyond 0·1 per cent. Reasonable values of conductivity, 10–100 mhos m^{-1} (compare 10^7 mhos m^{-1} for a typical metal), must be achieved. To do this by means of thermal ionization would require a temperature greater than 10,000 K. By a simple intuitive argument we should expect the conductivity to vary as $\exp(-eV_I/2kT)$, where V_I is the ionization potential of the gas atoms. Thus halving V_I improves the conductivity by a factor of 7·5. This is the basis of the seeding technique in which atoms of lower ionization potential than the gas itself are added to the gas stream. The seed ionizes more readily than the gas and so enhances the electrical conductivity. Caesium has a very low ionization potential of 3·9 V but is very expensive, so potassium, at 4·35 V, is taken as the next best. According to Frost (1961) conductivities between 10 and 100 mhos m^{-1} can be achieved in this way at temperatures of 2000–2400 K.

Magnetic Field

As in other applications demanding high magnetic fields, significant improvements can only be made by employing superconducting magnets. Flux densities of 10 Wb m^{-2} have already been achieved with such magnets but only over small volumes. Nevertheless, there is considerable cause for optimism in this field as new superconductors such as NbZr and Nb$_3$Sb begin to be used.

Gas Velocity

Thermal energy in a gas is, of course, another way of speaking about the kinetic energy of the gas molecules. In MHD it is desirable that this kinetic energy should be highly directional, that is the gas stream should have a high velocity in a well-defined direction. This can be brought about by standard nozzle expansion techniques by which means velocities of 10^3 m s^{-1} can be achieved. In theory there is no upper limit to the gas velocity which can be used for MHD, but in practice subsonic velocities are desirable to avoid shock phenomena which would affect the MHD process and also place heavy demands on the design and construction of equipment. It can be shown (Ralph, 1964) that the generator design can be optimized by ensuring that the temperature, pressure and velocity of the gas fall off so as to maintain a constant Mach number throughout the generator region. However, the mass flow through the generator must also be constant so the ducting which channels the gas through the generator region must be flared. The conditions of constant mass flow and Mach number lead to an analysis specifying the best profile for the duct.

Many experimental MHD generators have been built with a variety of geometrical configurations but they can be broadly classified into those which operate on an open cycle and those which operate on a closed cycle. In an open-cycle system the gas passes once through the generator to the exhaust. The use of seed materials in such systems poses severe problems. Firstly, the seed, potassium, is highly reactive and great care must be taken to prevent corrosion. Secondly, although only 0·1 atomic per cent of seed is required in the gas stream, this may involve the addition of as much as 30 per cent by weight of seed to the fuel. It is, therefore, necessary on economic grounds to recover the seed and this is a major unsolved problem in MHD generation. Other difficulties to be overcome include the development of suitable duct materials to withstand the high-temperature impact of the gas stream and the development of non-corrosive electrodes. Closed-cycle generators are being developed for eventual use as 'toppers' in conjunction with nuclear reactors.

As a final point in this section, it is worth noting that MHD can also be applied using working fluids other than gases. The power output of an MHD generator is proportional to σu^2 and suitable values of this quantity can be achieved with liquid metals. Assuming a conductivity of 10^6 mhos m^{-1} for a liquid metal, four orders of magnitude higher than that of an ionized gas, and a velocity of the order of 100 m s^{-1}, the power output would be 100 times better than for gaseous generators. Moreover, the working fluid is at a much lower temperature in this case.

15.3 Space Heating and Insulation

Space heating accounts for a sizeable proportion of national energy consumption in the United Kingdom. Exact figures are not available at present but it is reasonable to assume that the proportion is comparable to that in the United States where some 18 per cent of the total energy budget is spent in this way (SRI, 1971). Traditionally fuels such as coal, oil and gas have been cheap in both countries—so, in reality, has been electrical power—and there has been little incentive to improve the efficiency of space heating. Since the fuel price crisis all this has changed and it is unlikely that energy sources will ever be drastically underpriced again.

The efficiency of space heating and the associated fuel consumption depend on a number of factors—climatic conditions, internal temperature setting, efficiency of the heating unit, controls, degree of ventilation and insulation. Obviously the efficiency of the heating unit is important but this is already very high in most installations, and only requires adequate maintenance of the unit to ensure that there is no unnecessary wastage. Huge savings can be made, however, if traditional units are replaced by heat pumps. As this will be discussed in detail in the following chapter we shall restrict our treatment in this section to the remaining factors.

The internal temperature setting is clearly a matter of personal choice. However, there has been a trend towards unnecessarily high room temperatures.

During the 1974 energy crisis, when restrictions were placed on the temperatures in offices, factories and shops, many of us discovered that lower temperatures were not necessarily a bad thing. Excessively high temperatures in centrally heated buildings can lead to extreme dryness of the air and hence irritation of the throat and nasal passages. However, educating the public to accept room temperatures of not more than 18°C (65°F) will not lead to immediate fuel savings for there are still many houses which are inadequately heated, and just as overheating can be unhealthy, cold houses are positively health hazards.

The chosen internal temperature and the ventilation requirements are two parameters which affect the necessary heat input to a building. The others are external temperature and wind speed; climatic factors. The necessary heating capacity must be sufficient to match the sum of the total rate of heat loss from the building and the heat required to warm the fresh air inlet (both ventilation and leaks). In fact a higher capacity is essential to allow for warm-up after overnight switch-off and even with continuous operation thermostatic control will be more efficient if the heating unit is slightly oversized.

Heating engineers can estimate the annual heating load from meteorological records by using the concept of *degree days*. It is a matter of experience that the fuel consumption on a given day is proportional to the difference between the average temperature on that day and some critical temperature T_c. (Note that this is not equal to the design internal temperature as heating systems do not switch on until the external temperature is somewhat less than this.) Unfortunately, various definitions of the degree day are used throughout the World. In the United Kingdom, where an internal temperature of 65°F is thought to be adequate, T_c is taken to be 60°F. In the United States room temperatures of 70°F are deemed to be necessary so T_c is taken to be 65°F. Finally, on the continent of Europe T_c is taken as 18°C. No direct conversion between these various methods is possible without consulting complete meteorological records.

To illustrate the concept, we shall take the example of the British system. If T is the mean temperature for the day in question (usually calculated only for the duration of the heating period), then $(60 - T)$ is the number of degree days for that day. By taking values of T which are averages over many years, we can compile a degree day plot for a given locality. The number of days, N, with positive degree day values determines the length of the heating season. The total number of degree days, D, is obtained by summing the daily values, but neglecting negative values. The total heat input, H, required for the heating season can be written as

$$H = nQ_0 N(5 + D/N)/(65 - T_0)$$

where Q_0 is the rate of heat loss under design conditions, T_0 is the external design temperature and n is the average number of hours per day when the heating system is operative (Diamant and McGarry, 1968).

We shall now turn our attention to a consideration of the possible methods for reducing the heat losses represented by Q_0. Heat is lost from buildings by conduction through the roof, walls and floor, by conduction and radiation through windows, and by ventilation—warm air escapes via cracks around doors and windows, and via chimneys and spaces between floorboards, and is replaced by cold air from outside. To combat these losses, various insulation materials are available of the following types:

(1) fibrous or cellular mineral matter (asbestos, glass, silica),
(2) fibrous or cellular organic matter, naturally occurring (cane, hair, rubber, wood, cork),
(3) cellular organic plastics (polystyrene, polyurethane),
(4) heat reflecting metals (aluminium foil).

The thermal conductivities of such materials are listed in Table 15.3.

Table 15.3 Typical thermal conductivities for building materials at 20°C

	Btu h^{-1} ft^{-1} °F^{-1}	W m^{-1} °C^{-1}
Aluminium	140	240
Copper	220	390
Lead	20	34
Steel	28	48
Aerated concrete	0·1	0·17
Breeze block	0·1–0·2	0·17–0·34
Brick	0·4	0·7
Concrete	0·4–1·1	0·7–1·9
Cork	0·025	0·04
Glass	0·5	0·85
Plaster	0·6	1·0
Timber	0·1	0·17
Vermiculite concrete	0·12	0·2
Expanded polystyrene	0·019	0·03
Glass fibre	0·025	0·04
Hair	0·021	0·035
Polyurethane foam	0·02–0·025	0·035–0·04
Urea formaldehyde foam	0·021	0·035
Wool	0·03	0·05
Air	0·016	0·03
Water	0·35	0·6

The low conductivity of air and the relatively high conductivity of water are significant. There is little to be gained from using a material of low conductivity in building construction if it will tend to absorb water and hence become a poor insulator. On the other hand, the thermal conductivity of air is much lower than that of any solid (this is the basis of foam insulation). If we are considering heat transfer across a wall, for example, the layer of air next to the wall is virtually static and, because of the insulating nature of this layer, convection and radiation are the main mechanisms for heat transfer from the surface of the wall to the turbulent air beyond.

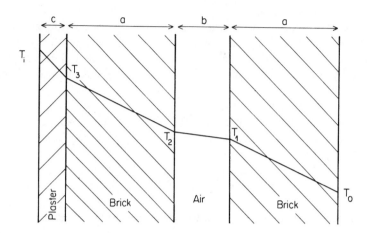

Figure 15.2. Temperature distribution across a cavity brick wall.
Typically, $a = 11{\cdot}3$, $b = 5{\cdot}0$, $c = 1{\cdot}3$ cm

Insulation techniques are already highly developed but it is true to say that their implementation is lagging. The problem of reducing wasteful practices in houses and other buildings is more one of educating the public than one requiring major technological innovations. Nevertheless, it is relevant that we should, at this point, give a brief account of existing methods of building insulation. To illustrate this, let us examine the insulation of a typical European cavity-wall brick house.

We begin by examining wall insulation. Consider a cavity wall as depicted in Figure 15.2. The inner brick leaf is coated with a layer of plaster. Assuming that we know the thermal conductivities of brick and plaster (Table 15.3), and assuming that conduction is the only mechanism of heat transfer, then we can write down an overall heat transfer coefficient, U, defined in terms of the rate of heat flow, Q, across unit area of the wall.

$$Q = U(T_i - T_0)$$

As this rate of heat transfer must be constant throughout the thickness of the wall,

$$Q = (k_3/c)(T_i - T_3)$$

$$Q = (k_1/a)(T_3 - T_2)$$

$$Q = (k_2/b)(T_2 - T_1)$$

$$Q = (k_1/a)(T_1 - T_0)$$

where k_1, k_2 and k_3 are the thermal conductivities of brick, air and plaster, respectively. By adding these four equations we find that

$$(T_i - T_0) = Q(c/k_3 + a/k_1 + b/k_2 + a/k_1)$$

Consequently,

$$1/U = c/k_3 + 2a/k_1 + b/k_2$$

Using the values of k_1, k_2 and k_3 given in Table 15.3, the value of U for the wall of Figure 15.2 is found to be 0·49 W m^{-2} °C^{-1}. It is interesting to compare this with the losses across a single leaf of bricks with plaster on the inside. In this case U has the larger value of 5·6 W m^{-2} °C^{-1}.

This analysis has been considerably oversimplified. In practice, radiation and convection effects are important, and only the boundary layers of air at either side of the cavity and at the inner and outer surfaces of the wall limit the flow of heat by conduction. The influence of these boundary layers can be seen in the experimental results for convection losses across a cavity (Figure 15.3). The losses are virtually independent of the cavity width as long as this is greater than about 5 cm. For smaller widths, less than the sum of two boundary layers, conduction dominates and the heat losses vary inversely as the cavity width. We can define a total heat transfer coefficient h_{cav} in terms of the rate of flow of heat across unit area of the cavity per degree temperature difference. For the cavity of Figure 15.3 it is found that $h_{cav} = 7$ W m^{-2} °C^{-1} approximately. Similarly, we can define heat-transfer coefficients, h_i and h_0, for the boundary layers at the inner and outer surfaces of the wall. In still air conditions a typical value for these coefficients would be 8·5 W m^{-2} °C^{-1} (ASHRAE handbook 1967). The total heat loss across the cavity wall is now given by U_T where

$$1/U_T = 1/h_i + 1/h_{cav} + 1/h_0 + c/k_3 + 2a/k_1$$

This leads to a heat loss of 1·4 W m^{-2} °C^{-1}.

The boundary layer on the outside surface of the wall will, of course, be thinned if there is a movement of air across the surface. With a wind of 13 m s^{-1} (30 mph) it is estimated (ASHRAE) that h_0 is increased to something like

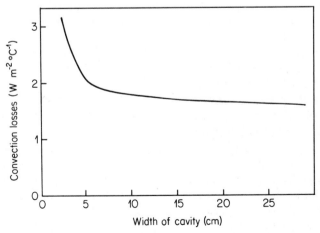

Figure 15.3. Convection losses across the cavity of Figure 15.2

70 W m^{-2} °C^{-1}. In addition, ventilators and imperfections in building construction will permit draughts to be induced in the cavity itself. It is difficult
to estimate the effect of this on h_{cav} but this might well be raised to 20–25 W m^{-2}
°C^{-1}. Now the value of U_T is increased to 1.8 W m^{-2} °C^{-1}. Rainy weather will
increase this still further and if we estimate the effect of moisture by eliminating
the external boundary layer in our calculations, U_T becomes 1.9 W m^{-2} °C^{-1}.
As a final modification, let us suppose that the cavity is now filled with urea
formaldehyde foam. Under still air conditions the value of U_T is now 0.5 W
m^{-2} °C^{-1}, and with a 30 mph wind it is 0.53 W m^{-2} °C^{-1}. For comparison,
the outcome of all these calculations is reproduced in Table 15.4.

Table 15.4 Heat losses across cavities

	Heat Loss (W m^{-2} °C^{-1})	
	(a) Still air	(b) 30 mph Wind
(i) Single-leaf brick wall plastered on inside [a]	3·5	5·7
(ii) Double-leaf brick wall plastered on inside	2·2	3·0
(iii) Cavity brick wall (5 cm cavity) plastered on inside	1·4	1·8
(iv) As in (iii) but with urea formaldehyde foam in cavity	0·5	0·53
(v) Single-pane glass 0·3 cm thick	4·1	7·7
(vi) Double-glazing: two 0·3 cm panes separated by 2·5 cm air gap	2·35	3·25

[a] Standard British $4\frac{1}{2}''$ (11·3 cm) brick with $\frac{1}{2}''$ (1·3 cm) of plaster.

The advantages of having a cavity wall are immediately obvious but so
also should be the value of cavity-wall insulation. This technique is already
being employed and is becoming more common in the aftermath of the fuel
price crisis of 1974. Urea formaldehyde resin is produced in the 'resol' stage
by mixing 30 parts urea with 100 parts aqueous (30 per cent) formaldehyde,
neutralized with sodium hydroxide. The mixture is boiled for some minutes,
a small amount of acetic acid added, and is further heated for a few hours to
produce a viscous liquid. This liquid can be stored and transported to the
required site in a cylinder. Immediately prior to use, a thin stream of this liquid
is mixed with a thin stream of liquid from a second cylinder containing a
catalyst (cinnamic acid) and a detergent dissolved in water. On mixing, the
liquids are whipped into a foam which is then injected through the outer leaf
of the brick wall under pressure. The foam sets rapidly and forms a porous,
water repellent solid of low thermal conductivity. Other foams such as the
polyurethanes can also be used to advantage—the only restriction is that the
foam should be of the closed-cell variety and, therefore, impervious to water.
If this is not the case then moisture penetration is likely to occur under conditions of driving rain.

Until recently most of the effort in house insulation was expended in reducing heat losses through the roof. Most new buildings today are adequately insulated in this respect and it is a fairly simple procedure to improve matters in older property. One standard technique is to fill the spaces between the ceiling joists with polystyrene chippings and to cover these with glass fibre. The top of the roof space is usually lined with roofing felt which, in addition to its insulating action, provides some weatherproofing beneath the roofing tiles.

With the roof and walls adequately insulated, the windows provide the next main route for the passage of heat. Heat losses via this exit can, of course, be reduced by decreasing the window area of the house but this is not necessarily a sensible approach (and obviously cannot be applied in existing buildings). In south-facing rooms, windows produce a greenhouse effect during the day and can actually lead to a net gain of heat during a 24-hour period. Heat losses at night can be further reduced by drawing heavy curtains across the windows, or, to be slightly old-fashioned, by closing wooden shutters in front of them. By analogy with cavity walls, we can also reduce losses through windows by constructing cavity windows, that is by the use of double glazing. The calculations are similar to those given above for cavity walls and the results are also to be found in Table 15.4. The improvement brought about by the installation of double glazing should be clear from these data.

The expense of cavity-wall insulation or double glazing is to some extent wasted if the building is prone to draughts. Fortunately, this is a fairly simple, though often tedious, problem to correct. Window frames and door frames can be lined with a material such as sponge rubber to ensure an air-tight fit. Similarly, the flue of an open fire can be blocked off when not in use.

Once a house has been effectively insulated in this way many unexpected heat sources begin to play a noticeable part in space heating. For example, an individual sitting at rest is losing some 200 W of heat to his environment so that five people sitting in a room generate as much heat as a 1 kW electric heater! Electrical appliances such as television receivers are another example. This is especially so when one considers the present trend towards colour receivers which contribute 250 W of heating power compared with less than 100 W for a monochrome receiver. Indeed, even the lights become an important source of heat when the insulation is adequate. Lamps are actually extremely inefficient sources of light energy in that most of the input of electrical energy is converted into thermal energy. The most up to date building designs attempt to incorporate this into the overall heating supply.

This is especially true in large buildings such as office blocks. The interiors of such buildings require continuous artificial lighting, and air conditioning is also necessary in order to maintain the temperature at an acceptable level in these regions. An alternative to air conditioning is IED—integrated environmental design—in which lighting, heating and ventilation are all incorporated into one scheme. For example, this can be done by the use of ventilation slots in the light fittings. Air passed through these slots extracts the heat from the interior light fittings and distributes it around the whole building, thus avoiding the need for much conventional heating. On the other hand, the extreme outside rooms of the building can capitalize to some extent on solar heating. The building can be warmed simply by switching on the lights and occupying it. This can, however, lead to some snags, for such buildings tend to be cold in the morning, soon after the occupants have arrived for work, and then to become warmer as the day progresses. Nevertheless, many improvements have already been made and in a relatively short space of time, so these minor drawbacks should not be allowed to hinder further developments in this direction.

15.4 District Heating and Total Energy Schemes

It is common practice for space heating to be achieved by the generation of heat within the building concerned. An alternative technique, *district heating*, delivers heat to the building from a central generator in the form of hot water or steam. Large units, such as housing estates or even small towns, can be supplied from the one generator and, in addition to space heating, hot water for industrial process heat or for domestic consumption can also be supplied. Clearly, the technique is particularly applicable to large scale housing schemes such as those which exist in Sweden, France, Germany and Eastern Europe, and it is worth noting that Denmark, with similar haphazard housing development to the United Kingdom, has many district heating networks. The malfunctioning of the early British schemes, together with the individualism of the average Briton, has inhibited development in the United Kingdom, but, even so, approximately 14 per cent of present local authority house building in the United Kingdom involves district heating (CPRS, 1974).

Simple district heating schemes have much to recommend them. Fuel costs can be reduced because a large centralized plant can deal with poorer quality fuel than can furnaces installed in individual buildings (compare electricity generation). Moreover, the cost of transporting fuel can be minimized if the central unit is suitably sited. In addition, air pollution can be reduced as specialized filtering techniques can be employed in the larger unit. The case for district heating becomes stronger if it is linked with electricity generation, as we shall discuss shortly. On the debit side, it is difficult to install district heating in existing towns or housing estates, and no completely satisfactory heat meter exists which means that it is difficult to monitor and control individual consumption figures.

The first district heating scheme in the United Kingdom was completed in 1919 at Gorton and Blackley, Manchester, but failed as a result of pipeline corrosion. Many small schemes were implemented after this but real impetus to the development of district heating was only given in 1953 with the publication of the DSIR Study Group report. This advocated three new major schemes but only one was completed. This is the Pimlico scheme in London. Hot water from the back-pressure turbines of Battersea power station is passed through twin pipelines underneath the river Thames to the north bank. In the year of completion (1956) this supplied 2403 apartments spread over an area of 31 acres.

By comparison, district heating was rapidly developed in Germany during the late 19th century and the early 20th century. By 1964 the total annual heat supplied had reached 13,000 GWh with a capacity of 6000 MW. The total pipeline length of the network then exceeded 1300 km. Similar rapid development took place in Denmark and the U.S.S.R. In the United States the first scheme commenced operation in 1877. Unlike the European practice, it is customary for the American networks to supply steam instead of hot water. This is partly due to the demand for air conditioning in summer as well as heating in winter. Some 'district cooling' schemes have also been implemented which supply refrigerated water for this purpose!

Various different types of pipeline network are in use. Single pipe systems, common in the U.S.S.R., do not, obviously, involve return of the water or steam condensate, but they can be economic when the heat is being transported over large distances. Two-pipe systems incorporating a return line are common in most European installations. A further sophistication, of which one example is the Berlin system, is the three-pipe network. The third pipe is simply used to provide the supply of hot water for domestic purposes during the summer months. Finally, four-pipe networks can be utilized which separate the hot water supplies for consumption and for space heating. The main advantage of incorporating a return line in the case of a steam system is to save on the water purifying costs (hydrazine is used to remove oxygen). Steam is easier to pump uphill and to meter, but hot water systems can make use of low temperature heat from thermoelectric power stations (that is, power stations combining electricity generation with the supply of heat). In this case the heat flow rate can be varied simply by altering the water temperature whereas in steam systems the flow must be increased by pumping.

Consumer connections depend on the supply temperature. If high supply temperatures are involved, heat exchangers must be used at substations or, perhaps, in the basements of large blocks of flats. This type of system has the advantage of operating at low pressure (separate pumps are used for circulating water through the building) and permits narrow mains to be used. Alternatively, low-temperature water can be supplied directly which simplifies the network design. Naturally, steam systems must employ the former approach.

It is perfectly feasible to charge for the heat supplied to individual consumers merely by applying a fixed tariff for a given floor area of premises. However, this method is undesirable in that there is no incentive on the part of an individual consumer to eliminate wasteful practices. On the other hand, exact monitoring of each heat supply presents its own problems. Metering of steam is fairly straightforward. It is either achieved by the use of an integrating pressure flow meter, a common technique in the United States, or by the use of a condensate meter. This comprises a rotating drum with a number of chambers and a geared counting chain. As steam condenses after passing through the heat exchanger, the water produced runs through the meter, alternately filling and emptying the chambers and rotating the drum. Small instruments of this type can easily monitor flow rates in the range 0–100 litres per hour. They are usually constructed of Silumin alloy as protection against attack by water.

Hot-water metering is much more difficult. Where consumption is large a mechanical heat flow meter is used, but these are too expensive to install in small buildings. Therefore, it is customary to combine heat meters with simple quantity meters which give some indication of the relative consumption of the various buildings supplied via the one heat meter. Heat meters of the Pollux, Pegus and similar types (Cole and Jones, 1964) monitor the temperature difference between water in the inlet and return lines. In one variety, mercury in steel thermometers in the inlet and return lines actuate a spring which in turn operates a cam via a system of levers. A simple vane water meter is included in the return line and is coordinated with the cam so that the water flow and temperature are integrated and displayed as heat supplied.

Many different sources of heat have been utilized in district heating schemes. Apart from good quality coal and residual fuel oil, one can burn coal dust, peat and poor quality lignite. Similarly, crude oil can be burnt directly in large installations, although this might be deemed wasteful as far as the valuable light fractions are concerned. Many existing schemes involve the incineration of domestic and industrial refuse (Section 15.6). Moreover, as we saw in Chapter 13, geothermal heat in the form of steam or hot water from geysers can be used directly. The most important source of heat, however, is that produced as a by-product of electricity generation. The combined production of heat and electrical power in a thermoelectric power station can achieve efficiencies of 70–75 per cent compared with 30–40 per cent for electricity generation alone.

The majority of today's high output power stations are sited in isolated areas, so it might be thought that the cost of distributing hot water or steam to towns would be prohibitive. This is not necessarily the case. Wilk (1965) has shown that, for a given amount of energy transported, the relative costs of distributing it in the form of oil, gas, district heat in a twin-pipe system and electricity (50 kV cable) are in the ratio 0·4:0·45:1:1·5 (although see Chapter 17). Given that there is a case for thermoelectric power stations, how is the heat to be generated? The temperature of the warm water effluent from the condensers of a conventional power station is only 30°C, low for district heating purposes, but more appropriate temperatures can be attained through the use of back-pressure turbines or intermediate take-off condensing turbines (ITOC's).

As shown in Figure 15.4, district heating can be supplied simply by raising the temperature and pressure on the backpressure side of a normal condensing turbine. Instead of leaving the turbine at 30°C and 0·6 psi, the water vapour is discharged at between 90 and 150°C and at a pressure of 14–30 psi. To run a large turbine on pure backpressure operation can be quite impractical as the heat discharged will be large compared to the electricity generated. Conse-quently, during periods of low heating demand, it would be necessary to shut down the turbine and switch to alternative plant. Moreover, the use of low power backpressure turbines is uneconomical as low feed pressures lead to a deterioration in turbine efficiency.

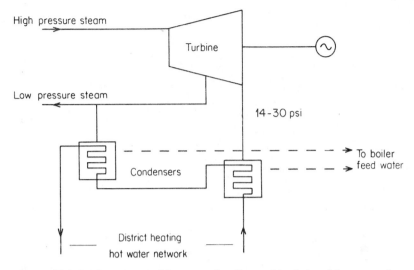

High pressure steam

Turbine

Low pressure steam

14-30 psi

To boiler
feed water

Condensers

District heating
hot water network

Figure 15.4. Backpressure turbine operation for combined electricity generation
and district heat production

Pure backpressure turbines are only really suitable under the following
demand conditions.

(1) If the heat demand is constant throughout the year. This is the case,
for example, in the New York Consolidated Edison system where the steam
supplied in used in winter for heating and in summer to power air-conditioning
units.

(2) If electricity generation is a minor objective compared with heat supply.
This is often the case with industrial consumers who require large amounts
of process heat. It is more economical to produce this steam at high pressure
and allow it to do some work (electricity generation) before supplying it to
industry as low pressure steam.

(3) If periods of low electricity demand are coincident with periods of low
heat demand, that is the plant is merely acting in a standby capacity. Such
circumstances exist, for example, in Scandinavian countries where the bulk
of the demand for electricity during the summer is met by hydroelectric installations.

In all other circumstances the ITOC turbine is to be preferred for combined
heat and electricity production. The turbine is divided into at least two parts and
steam can be bled from between the turbines and at a number of positions on
the various turbine units (Figure 15.5). The maintenance of a high vacuum
on the low pressure side of the turbine set ensures that the whole set can be
compared with a normal condensing turbine. In fact, the overall efficiency
when operating in a completely condensing cycle (100 per cent electricity
generation) is only one per cent or so less than that of a custom-built condensing

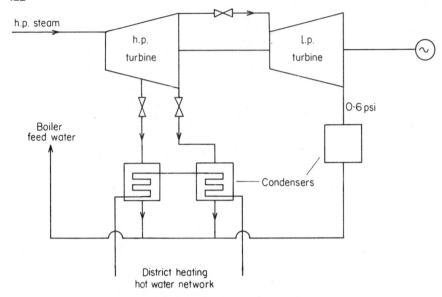

Figure 15.5. Intermediate take-off condensing turbine (ITOC) for combined electricity generation and district heat production

turbine. The difference arises from steam head losses caused by dividing the set, but even this has been overcome in the most recent designs.

When heat is required, steam from the end of the high-pressure turbine (Figure 15.5) is used directly for district heating or is condensed and the heat so obtained used for warming the water in the district heating mains. The temperature of the heating water can also be boosted with the aid of additional condensers fed with bled steam. As heating demand falls, the amount of by-pass steam is reduced and the whole system begins to resemble a straightforward condensing station. This flexibility of the ITOC turbine is further emphasized during periods of peak electricity demand in winter. Thus all by-pass steam lines can be shut and the station run on 100 per cent electricity generation until the peak period has passed. During this time the district heating network can be maintained at a reasonable temperature by utilizing stored hot water. (Even without such storage, it is found that an interruption in the supply for a couple of hours has little effect on space heating.) The flattening of demand by ITOC turbine operation is illustrated in Figure 15.6. When heating demand dominates electricity demand, the ITOC turbine can be used at heating factors of up to 50 per cent—a higher fraction renders the turbine uneconomic. Thus it is best to design such turbines to cater for basic demand and use standby stations to deal with peak demand.

In conclusion, it is apposite to reiterate the optimum conditions for the economic operation of district heating schemes:

(1) High density housing—in particular, it is much less expensive to connect

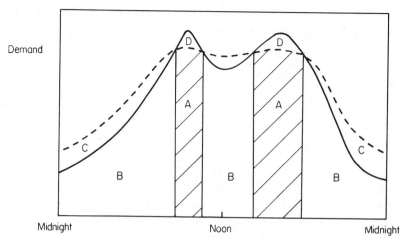

Figure 15.6. Effect of ITOC turbine on the daily power load curve. The full curve represents demand while the dotted curve represents the total output from the turbine. During ITOC operation, regions B, the excess heat produced, C, is stored against peak demand. During peak demand the turbine is run on a purely condensing cycle, A, and the excess demand D is met by the stored heat

large apartment blocks rather than individual houses to district heating networks.

(2) New developments where pipeline laying is cheaper.

(3) Fairly steady heat demand—in this sense industrial consumers are to be preferred to domestic ones.

(4) Large networks.

(5) Operation for most of the year at full load without short, severe peaks in demand—the British climate, for example, with its long but moderate winters, is ideally suited.

(6) A high proportion of buildings within the network area actually connected to the network—this is difficult to achieve without centralized planning but can be done, for example, in municipal housing estates.

Before leaving this section we should note that district heating based on combined heat and electricity production is just one illustration of the so-called *total energy concept*. The diesel engine and the gas turbine also offer possibilities for small-scale combined schemes. The diesel engine can burn both heavy and light oils with an efficiency of up to 40 per cent. Some 60 per cent of the waste heat can be recovered from the high temperature exhaust gases and the cooling water, which gives a total efficiency of 75 per cent. The gas turbine is only 20 per cent efficient for electricity generation, but, assuming the same figure of 60 per cent for the fraction of waste heat recoverable, we still end up with an overall efficiency of 70 per cent. The diesel engine is to be preferred if electricity generation is more important than heat production, and the gas turbine if the converse is true. Using this total energy approach it is possible for large industrial consumers to become independent of the national electricity supply.

15.5 Transport

It is clear from Table 15.1 that the transport sector accounts for more than a fifth of total energy consumption in the United Kingdom and the United States. Indeed, 25 per cent of all oil consumed in the United Kingdom is used for transport, 77 per cent of it by road vehicles, 15 per cent by aircraft, 4 per cent by the railways and 4 per cent by ships. It seems certain that road transport will continue to be dominant. There are two basic approaches to savings in this sector: technical improvements in the efficiency of vehicles and measures which attempt to alter the travelling habits of individuals.

The most obvious place for technical improvements is in the car engine—the petrol engine is only 25–30 per cent efficient (Chapter 6). Considerable interest is currently expressed in weak-mixture petrol engines such as the 'stratified charge' engine being developed in Japan and the United States. Unfortunately, this gives a much poorer performance than the conventional petrol engine. The diesel engine (Chapter 6) is approximately 30 per cent more efficient than the petrol engine and is already used on many commercial vehicles. However, further acceptance of the diesel engine will probably only occur if a number of problems can be overcome. Thus it is heavier and noisier than petrol engines, suffers more from vibration, and is somewhat 'smoky' when badly tuned. Moreover, cold starting is difficult and can involve some delay. Perhaps the best hope for significant savings is the introduction of electric vehicles, at least for 'stop–start' driving in congested urban conditions. This has already been discussed in Chapter 14.

Efficiency can also be improved by changes in vehicle body design. Acceleration and hill climbing create the greatest thirst in a car engine. Consequently, the mass of the vehicle directly affects fuel consumption. However, it is wrong to argue for lighter cars if this entails less safe construction. Most motorists will opt for more protection in the event of an accident instead of fuel economy at the expense of safety. The shape of the car body does affect wind resistance at high speeds, but at low speeds (less than 40 mph) other losses such as rolling resistance dominate. The use of steel-belted radial tyres can reduce rolling resistance by as much as 50 per cent and, more as a result of their better handling characteristics and longer life, these are already being adopted as standard by many manufacturers.

The second approach to fuel savings in road transport, namely measures which limit personal choice, is a much more emotive subject. Increased use of public transport is urged in many quarters but this is extremely difficult in rural areas. The attraction of private transport in such areas is emphasized by the extremely high per capita car ownership in East Anglia and Northern Ireland compared with the more developed areas of the United Kingdom. It is also argued that lower speed limits would achieve considerable fuel savings. Indeed a reduction in speed from 70 to 50 mph in motorway driving produces a 25 per cent economy in petrol consumption. When all types of driving conditions are considered, however, it is found that such a reduction in the speed limit would only lead to 2·5 per cent fuel saving in the United Kingdom (CPRS, 1974).

It is probably true to say that significant economies could be made in the fuel consumed in transport without the introduction of any new technical developments. This would entail many restrictive measures on the part of government. Whether or not these would be acceptable is not a matter for this book.

15.6 Refuse Incineration and Recycling

The problem of refuse disposal is an enormous one. In the United Kingdom more than 15 million tonnes of domestic waste are produced each year, not to mention the vast and varied amounts of industrial waste. Traditionally, much of this refuse has been discarded on tips until it can be burnt or buried, and these tips have become familiar eyesores around urban areas. There is a limit to how much land can be given over to this purpose, and already many of those entrusted with the task of handling the nation's rubbish are looking seriously at two alternatives to tipping—recycling and incineration.

A large fraction of the refuse produced may be thought of as a debit on the country's energy budget. For one thing, many of the waste products, now assumed to have exhausted their useful life, were originally produced by energy intensive industrial processes. If it were possible to recycle these products, if only to a point beyond the first stage in a manufacturing process, a considerable energy saving could be effected. An alternative point of view, and one which is practicable on the basis of existing technology, is that the rubbish dump fails to capitalize on the considerable calorific value of the refuse (see later) and that this should instead be burnt. We begin by examining this approach.

Since the war, the density of refuse has been decreasing steadily in all industrial countries. In the United Kingdom this has been partly due to a reduction in the ash content, a trend which began with the introduction of the Clean Air Act in 1956, since when domestic solid fuel consumption has more than halved (Chapter 5). The most important world-wide factor, however, has been the rapid increase in the use of packaging and non-returnable containers; the paper content of refuse in Britain is now more than 70 per cent by volume. The poorer ash content of refuse has tended to reduce its calorific value but this has been more than compensated for by the increased paper and plastic content.

The actual calorific value will have an important bearing on incinerator design for a number of reasons. The size of the combustion chamber will be determined by the heat content of the gases produced by combustion and by the resulting volume increase of the flue gases. Similarly, the refractory brick-work of the combustion chamber must be built to withstand the heat generated, and the amount of air which must be supplied for complete combustion is also related to the calorific value. Typical figures for the *net* calorific value of domestic refuse in the London area are 8000–12500 kJ kg^{-1} (Skitt, 1972), compared with 25000–30000 kJ kg^{-1} for coal (Chapter 4). The net value refers to the gross calorific value less the latent heat of the moisture contained in the refuse and is a measure of the maximum usable heat that can be obtained. The moisture content is extremely variable and can be a considerable problem in incineration.

The heat produced by refuse incineration can be exploited either by using it directly for district heating (Section 15.4) or to raise steam for electricity generation. For most of the existing installations of this type it must be admitted that power is not produced economically. To supply a large incineration plant entails the collection of refuse over a wide area and its transport to the central plant. The costs of this, together with the capital investment in the incinerator, are not balanced by the receipts from the sale of district heat or electricity. However, it must be remembered that refuse incineration has other benefits which are not so easily costed. It is almost completely hygienic and leads to a volume reduction of 10:1 or more. This is better than can be achieved by any other disposal technique and the residue is well suited to land reclamation and similar purposes in that it is almost completely inorganic.

Many local authorities are now deciding that, on balance, it is worth investing in refuse incinerators which are linked to electricity production. One such example is the Greater London Council's Edmonton incinerator, depicted schematically in Figure 15.7. Some 1600 tonnes of crude refuse is brought to the plant on each day of the working week and this is burnt at the rate of 1140 tonnes per day. The total bunker capacity is 3900 tonnes. The combustion chamber incorporates a roller grate which is an inclined series of parallel hollow rollers through which is fed primary air. Complete combustion at a temperature of 1700–1900°C is ensured by agitating the burning refuse as it turns and falls from roller to roller. After leaving the final roller the clinker and ash is passed through a quench bath to a residuals handling area. Residual ferrous metals are sold as scrap while the clinker can be used in road making. Electrical power is generated at 11 kV in four 12·5 MW generators and up to 25 MW may be fed to the national electricity network and elsewhere. One problem on the generation side is that the calorific value of the refuse is conti-nually changing but electronic controls are employed to govern the pressure in the steam mains.

Recycling is already a well-established technique in many industries. In the United Kingdom 40 per cent of paper and board is recycled (CPRS, 1974), as is 50 per cent of all crude steel and 70 per cent of lead. One example of lead recycling is afforded by the practice of garages in returning old crank-case oil to the refineries where the lead additives can be extracted. That the above percen-tage figures are so high simply reflects that the recovery of such materials is

427

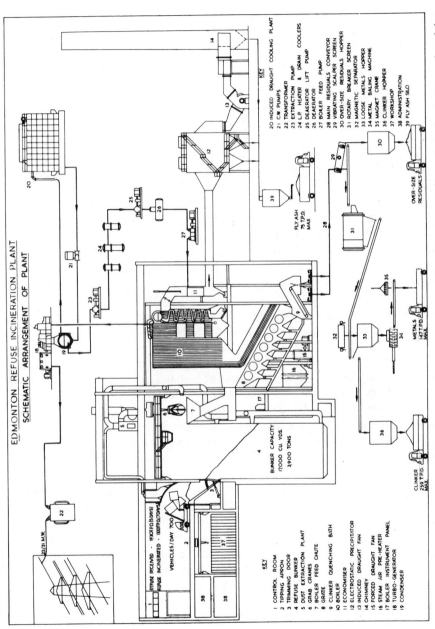

Figure 15.7. A schematic diagram of the Greater London Council's Edmonton incineration plant. (Reproduced by permission of the Greater London Council)

EDMONTON REFUSE INCINERATION PLANT
SCHEMATIC ARRANGEMENT OF PLANT

KEY
1 CONTROL ROOM
2 TIPPING APRON
3 TRIMMING DOOR
4 REFUSE BUNKER
5 DUST EXTRACTION PLANT
6 GRAB CRANES
7 BOILER FEED CHUTE
8 GRATE
9 CLINKER QUENCHING BATH
10 BOILER
11 ECONOMISER
12 ELECTROSTATIC PRECIPITATOR
13 INDUCED DRAUGHT FAN
14 CHIMNEY
15 FORCED DRAUGHT FAN
16 STEAM AIR PRE-HEATER
17 BOILER INSTRUMENT PANEL
18 TURBO-GENERATOR
19 CONDENSER

KEY
20 INDUCED DRAUGHT COOLING PLANT
21 CW PUMPS
22 TRANSFORMER
23 EXTRACTION PUMP
24 L.P. HEATER & DRAIN COOLERS
25 DEAERATOR LIFT PUMP
26 DEAERATOR
27 BOILER FEED PUMP
28 MAIN RESIDUALS CONVEYOR
29 VIBRATING SCALPER SCREEN
30 OVER-SIZE RESIDUALS HOPPER
31 ROTARY BREAKER SCREEN
32 MAGNETIC SEPARATOR
33 LOOSE METALS HOPPER
34 METAL BALING MACHINE
35 MAGNET CRANE
36 CLINKER HOPPER
37 WORKSHOP
38 ADMINISTRATION
39 FLY ASH SILO

REFUSE RECEIVED - 1900 T.P.D. (5 DAYS)
REFUSE INCINERATED - 1300 T.P.D./1700 T.P.D.
VEHICLES/DAY 700

BUNKER CAPACITY
17,000 CU. YDS
3,900 TONS

FLY ASH
75 T.P.D.
MAX.

OVER-SIZE
RESIDUALS

METALS
167 T.P.D.
MAX.

CLINKER
259 T.P.D.
MAX.

an extremely attractive economic proposition. In many other cases recycling is not at present economic but this may well change as the availability of world resources varies. Increasing fuel costs may also provide an incentive to recycle. For example, the production of aluminium by electrolysis requires 17,000 kWh per tonne but recycled aluminium can be purified at a cost of only 450 kWh per tonne.

It is perhaps worth clearing up some popular misconceptions concerning recycling. It is not true to say that most of our waste products could be recycled and that it is only market forces which preclude so doing. In many cases the waste product cannot act as a raw material for further similar products. For example, although, as we have seen, a high proportion of paper is recycled, it is not possible to recycle good quality paper as such. Each successive recycling degrades the quality of the paper until it can only be used for cardboard manufacture. Of course, paper recycling should still be encouraged, especially as there is a world-wide shortage of paper and this is likely to be the case for some time to come. Glass is another example. Many pundits argue for greater recycling of this material, but it is impossible to melt down a bottle and remake a similar bottle from the melt. Nonetheless, it is present practice to include some 25 per cent of recycled glass in the production of new glass. The emphasis with glass packaging should rather be on reuse with bottles being returned to dairies or breweries for sterilization.

The major problem with recycling is the separation of the various components of rubbish. One technique currently being studied is fluidized bed separation. A trough of sintered bronze supports a bed material of iron powder, ferrosilicon or magnetite (the density of the bed material must lie between those of the materials being sorted). Vibration causes the high density wastes to fall through the bed material and move up the trough to be discharged. Similarly, the lighter wastes which remain on the top of the bed material float along the surface and into another discharge chute. Such separators can deal with waste components ranging in size from 10 to 0·2 cm. The density difference between the components to be separated must be at least 0·2 g cm^{-3}. Already throughputs of 0·5 tonne h^{-1} have been achieved for metal–plastic mixtures.

A new potential technique for refuse handling, pyrolysis, has been developed on a laboratory scale. Refuse is fed into a retort-like chamber and heated in the absence of air to between 600 and 1000°C. A volumetric reduction of 10:1 is achieved and the end-products are gases, such as methane and hydrogen, together with (depending on the temperature) carbon char, oils and tars. This technique may well have advantages over incineration in that air pollution is avoided, the gaseous and liquid products can be used as fuels, and metal separation is easily effected.

Sewage is another waste product which, like refuse, can be an energy source. In municipal sewage disposal works, sewage sludge undergoes aerobic bacterial decomposition and yields sewage gas. This is typically 75 per cent methane and 20 per cent carbon dioxide with small amounts of nitrogen, hydrogen and hydrogen sulphide. Its calorific value is approximately 27000 kJ m^{-3}. This gas finds a use today as a fuel for the sewage works and also for motor vehicles. The carbon dioxide and hydrogen sulphide are removed and the remaining gas stored in steel containers at 350 atmospheres pressure. From there it is transferred into the storage cylinders of motor vehicles which have been modified to operate on this fuel.

References

Belter, W. G., *Power Generation and Environmental Change*, Berkowitz and Squires, (Eds.), Chap. 21, MIT Press, Cambridge, Mass., 1971.
Clark, J. R., *Scientific American*, **220**, 18 (1969).
Cole, E. A. and K. W. James, Heat Metering Study, National Coal Board, London, 1964.
CPRS, *Energy Conservation—A Study by the Central Policy Review Staff*, H.M.S.O., London, 1974.
DSIR, District Heating—A Report by the Department of Scientific and Industrial Research Study Group, H.M.S.O., London, 1953.
Diamant, R. M. E. and J. McGarry, *Space and District Heating*, Iliffe, London, 1968.
Erickson, E., *J. Geophys. Res.*, 68, 4001, 1963.
Federal Water Pollution Control Administration, Industrial Waste Guide on Thermal Pollution, U.S. Dept. of the Interior, 1968.
Flury, F. and F. Zernik, *Schadliche Gase, Dampfe, Nebel, Rauch and Staubarten*, Julius Springer, Berlin, 1931.
Frost, L. S., *J. Appl. Phys.*, **32**, 2029 (1961).
Harteck P. and R. R. Reeves, Symposium on the Chemistry of the Natural Atmosphere, 155th Annual Meeting of the Am. Chem. Soc., Chicago, 1967.
Jaffe, L. S., Conf. on Biol. Effects of Carbon Monoxide, New York Academy Sciences, 1970.
Junge, C. E. and T. Ryan, *Quart, J. Meteorol. Soc.*, **84**, 46 (1958).
Katz, M. and G. A. Ledingham, Nat. Res. Council of Canada, in Effects of Sulphur Dioxide on Vegetation, NCR No. 815, Ottawa, 1939.
Mayer, M., A Compilation of Air Pollution Emission Factor, U. S. Public Health Service, Divn. of Air Pollution, Cincinnati, Ohio, 1965.
O'Connor, T. C., Atmospheric Condensation Nuclei and Trace Gases, Final Report, Dept. of Physics, University College, Galway, Ireland, Contract No. DA–91–591–EUC–2126, 1962.
Perry, H., *Power Generation and Environmental Change*, D. A. Berkowitz and A. M. Squires, (Eds.), MIT Press, Cambridge, Mass., 1971.
Ralph, J. C., in *Magnetohydrodynamic Generation of Electrical Power*, Coombe, (Ed.), Chapman and Hall, London, 1964.
Robbins, R. C., K. M. Borg and E. Robinson, *J. Air Pollution Control Assn.*, **15**, 423 (1968).
Robinson, E. and R. C. Robbins, Stanford Research Institute, Project 6755, 1968.
Robinson, E. and R. C. Robbins, in *Global Effects of Environmental Pollution*, Singer, (Ed.), D. Reidel, Holland, 1970.
SCEP, Man's Impact on the Global Environment, a report by the Study of Critical Environmental Problems Work Group, MIT Press, Cambridge, Mass., 1970.
SRI, Patterns of Energy Consumption in the United States, Stanford Research Institute, 1971.
Schnellen, C. G., Ph.D. Thesis, Technische Wetenschap, Delft, Rotterdam, Netherlands, 1947.
Skitt, J., *Disposal of Refuse and Other Waste*, Charles Knight and Co., London, 1972.
Weinstock, B., *Science*, **166**, 224 (1969).
White, J. J., *U.S. Naval Medical Bulletin*, **30**, 151 (1932).

430

Wilk, S., Rortransport av Energi-En kostnadsjamforelse med andra transportsatt, VVS 11, p. 595, 1965.

Further Reading

ASHRAE Handbook of Fundamentals, Chaps. 18 and 19, American Society of Heating, Refrigeration and Air-conditioning Engineers, 1967.
Berkowitz, D. A. and A. M. Squires, *Power Generation and Environmental Change*, MIT Press, Cambridge, Mass., 1971.
Coombe, R. A., *An Introduction to Direct Energy Conversion*, Pitman, London, 1968.
Singer, S. F., (Ed.), *Global Effects of Environmental Pollution*, D. Reidel, Holland, 1970.
Skitt, J., *Disposal of Refuse and Other Waste*, Charles Knight and Co., London, 1972.

16 The Heat Pump*

It is a fundamental tenet of physical science that energy is always conserved. Indeed, so basic is this idea to our way of thinking that, when an apparent discrepancy arises in the energy balance of some system, we invent a new type of energy to account for the deficit. Thus from the original concepts of mechanical potential and kinetic energy, the definition of energy has been enlarged to include electromagnetic, chemical and electrochemical energy, to name but a few of the varieties. Energy conservation becomes so entrenched in the scientist's subconscious that he may find it difficult to accept at first sight the action of a heat pump or refrigerator whereby heat is supplied or removed at a faster rate than the supply of power to the machine. The scientist will quickly appreciate, however, that these machines are simply using a small amount of energy to effect the transfer of a much larger equivalent amount of heat.

The heat pump and the refrigerator are essentially one and the same device. Each transfers heat from a cold to a warm environment. When the primary purpose of the machine is to cool an enclosure, it is termed a refrigerator. Alternatively, when the aim is to warm an enclosure, the machine is called a heat pump. The energy input to the machine is used to drive the heat against the temperature gradient. As we shall see, however, this is done by a cyclic process in such a way that there is never an actual heat transfer against this gradient. Of course, the classification into heat pumps and refrigerators is somewhat arbitrary for in many applications both the heating and cooling effects can be utilized: for example, it is possible to maintain a cold room and run a hot water system using only one machine.

The heat pump, then, is by no means the weird contraption that it often appears when first encountered. It neither enables us to 'get something for nothing' nor does it at any stage of the process cause heat to travel 'in the wrong direction'. In the United Kingdom there is general familiarity with the refrigerator but scepticism still abounds where the heat pump is concerned. Not so in the United States nor in many European countries where heat pumps have been extensively used in both large-scale and domestic applications.

The concept of the heat pump is not of recent origin. Lord Kelvin first proposed a design for such a device in 1852 but this was destined to go no further than the drawing board for three-quarters of a century until Haldane

---

*(Throughout this chapter we shall follow standard refrigeration and air-conditioning practice in the United Kingdom and the United States of America by using the British system of units. In most instances SI units will be quoted in parentheses.)

built the first practical machine and used it to heat his home in Scotland. He used the atmosphere as his source of heat, backed up by the local water supply when atmospheric conditions were not favourable (Haldane, 1930).

The cyclic processes utilized by heat pumps and refrigerators are essentially the familiar power cycles of heat engines operated in reverse. In a power cycle heat is received by some working fluid at a high temperature and rejected at a low temperature while work is done by the fluid. In the reverse case, heat is received at a low temperature and rejected at a high temperature while work is done on the fluid. For the reader who is unfamiliar with the application of thermodynamics to power cycles a brief resume of the relevant theory is given in Appendix 2. As the working fluids of the devices are usually gases or vapours, a brief summary of the properties of vapour–liquid mixtures is also included. It is hoped that this segregation of background theory should allow for more continuity in the present chapter.

16.1 Operating Cycles

As has been stated previously, the refrigeration or heat pump cycle involves the transfer of heat from a low-temperature source to a high-temperature sink by the input of work to the machine. Figure 16.1 illustrates this schematically and also compares the cycle with the heat engine cycle. In the heat pump cycle an amount of heat Q_C is taken in at temperature T_C and a larger amount Q_H is rejected at the higher temperature T_H. This necessitates the input of an amount of work W. For an ideal machine with no losses such as friction, so that W represents the total work that has to be done, the difference between Q_H and Q_C is simply equal to W:

$$Q_H = Q_C + W \tag{16.1}$$

The efficiency of a heat engine is defined as the ratio of the work done by the engine to the heat taken in at the high temperature. We require a corresponding

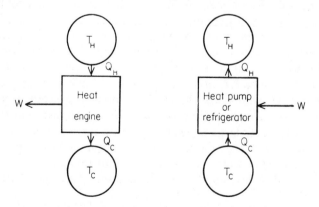

Figure 16.1. Schematic representations of a heat engine and a heat pump

measure for the performance of refrigerators and heat pumps. In the case of a refrigerator, we are concerned with the machine's performance in removing heat from the low-temperature source. A good measure of its 'efficiency' in this respect is the ratio of heat removed to the total work that has to be supplied. This ratio is known as the *coefficient of performance* for cooling, CP_c. Thus

$$CP_c = Q_C/W$$

In the case of a heat pump, on the other hand, we are more interested in the amount of heat delivered at the high temperature so it is more relevant to quote the ratio of the heat delivered to the work done. Thus the coefficient of performance for heating, CP_h, is given by

$$CP_h = Q_H/W$$

(An alternative nomenclature which may be encountered in the literature restricts the use of 'coefficient of performance' to refrigerators and describes CP_h as the 'performance-energy ratio'.) From equation (16.1) it can be seen that the coefficient of performance for cooling, CP_c, may be greater than unity whereas the coefficient of performance for heating, CP_h, must always be greater than unity. The equation also suggest the simple relationship between the two coefficients:

$$CP_h = CP_c + 1$$

There are three main types of refrigeration cycle:

(1) The vapour compression cycle,
(2) gas cycles,
(3) the absorption refrigeration cycle.

For heat pump applications only the vapour compression cycle is very practicable owing to disadvantages in cost, size or efficiency for the other designs. Before examining these cycles in some detail, we shall first consider the ideal standard cycle from the thermodynamic point of view. This is, of course, the *reversed Carnot cycle*, the refrigeration cycle which is analogous to the Carnot engine (Appendix 2).

The reversed Carnot cycle can best be illustrated using a temperature–entropy $(T - s)$ diagram. Figure 16.2 shows such a cycle together with the analogous heat engine cycle. Consider the sequence of operations on the working fluid used in a reversed Carnot cycle. Starting at point 1, the working fluid is compressed adiabatically to point 2, followed by an isothermal compression to point 3. During this latter stage an amount of heat Q_H is lost by the fluid. The working fluid is now allowed to expand adiabatically to point 4, and, to complete the cycle, expands isothermally to point 1, taking in heat Q_C in the process.

Consider the performance of such a cycle. For the Carnot heat engine

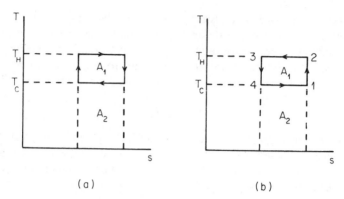

Figure 16.2. Temperature–entropy diagrams for (a) a Carnot cycle heat engine, (b) a Carnot cycle heat pump

(Figure 16.2 and Appendix 2), the efficiency is given by the work done as a fraction of the heat taken in, that is to say

$$\eta = W/Q_H = (Q_H - Q_C)/Q_H$$

We know from thermodynamics that this can be rewritten as

$$\eta = (T_H - T_C)/T_H$$

Alternatively, this can be expressed as the ratio between the areas A_1 and $A_1 + A_2$ in the T–s diagram. It follows that CP_c for such a cycle is given by the ratio A_2/A_1, and CP_h by $(A_1 + A_2)/A_1$. These may be rewritten as

$$CP_c = T_C/(T_H - T_C)$$
$$CP_h = T_H/(T_H - T_C) \tag{16.2}$$

Just as the Carnot heat engine has the highest theoretical efficiency of all heat engines operating under similar conditions, so it is true that the performance of the Carnot refrigerator and heat pump represent ultimate attainable values for practical machines.

The Carnot cycle is reversible in the thermodynamic sense (Appendix 2). This demands that there are negligible temperature differences between the working fluid and the environment at the upper and lower reservoirs. Heat transfer through a finite temperature difference, as must be the case if the transfer is to be efficient, leads to departures from this ideal and causes the cycle to become irreversible. The cycle now entails an overall increase in entropy and requires a greater energy input than before. Thus the efficiency of the engine or the performance factors of the refrigerator or heat pump will be lower than in the ideal case. The deficiencies in an actual machine are a result of the properties of the refrigerant (the working fluid in a refrigerator or heat pump), the particular cycle employed and the machine design. The cycle to be used is often dictated by the choice of refrigerant so that there is only a freedom of choice in the design considerations. As we shall see in later sections, these in turn reflect a compromise between various financial, physical and other restrictions. It should be clear, then, that there are many factors which cause the performance of real machines to fall well below that of a Carnot machine.

Nevertheless, equation (16.2) does indicate why we should be interested in the heat pump in the first place. Suppose that a heat pump is used to extract heat from the atmosphere at 32°F (0°C) and supply heat to the inside of a house at 70°F (21°C). The theoretical coefficient of performance for a Carnot heat pump under these conditions is 529/38 (on the Rankine scale, 294/21 on the Kelvin scale), approximately 14. If the heating load for the house is 14 kW when ordinary electrical resistance heating is employed, this can now be reduced to little more than 1 kW with the aid of a heat pump. Of course, the saving will not be nearly so dramatic in practice, but, as we shall see, it is still extremely attractive.

We now turn our attention to the most important of the cycles listed above, the *vapour compression cycle*, which uses a liquid–vapour mixture as its refrigerant. Figure 16.3 provides a schematic view of the process in addition to the pressure–enthalpy (P–h) and T–s digrams for the cycle. Just before entering the compressor at point 1 the refrigerant is in the form of a dry saturated vapour at low pressure and temperature. It is then compressed adiabatically and reversibly to a high temperature, high pressure superheated vapour (point 2)

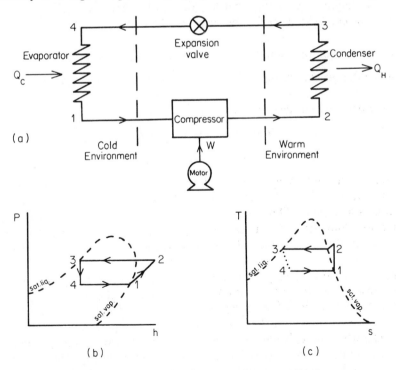

Figure 16.3. Vapour compression cycle: (a) schematic, (b) pressure–enthalpy diagram, (c) temperature–entropy diagram

and enters the condenser. There it is condensed reversibly at constant pressure (after being desuperheated) giving up heat Q_H. At point 3 in the cycle it leaves the condenser as a high pressure, medium temperature saturated liquid. It is then allowed to expand irreversibly via a throttling valve to become a low quality vapour (i.e. some of the refrigerant flashes into vapour during the throttling process) at a low temperature and pressure (point 4). The cycle is completed when the refrigerant is passed through the evaporator where it is evaporated reversibly at constant pressure to the original dry, saturated state 1. The heat Q_C required for the evaporation is taken in from the low temperature environment. The cycle is such that the heat transfers at the condenser and the evaporator occur with negligible temperature differences except during the desuperheating process in the condenser.

The nature of the various steps in the cycle is emphasized in the $P–h$ and $T–s$ diagrams. The pressure–enthalpy diagram clearly shows the isenthalpic throttling process 3 to 4. However, this cannot be displayed on the temperature–entropy diagram as it is an irreversible process. Only the start and finish points of this process can be plotted. It should be emphasized that the cycle is not ideal because of the irreversible expansion and because there is a finite temperature difference in the desuperheating process.

The performance of such a cycle can be analysed most conveniently using the $P – h$ diagram. Assuming that kinetic and potential energy terms can be ignored, the energy balance for the cycle may be written as follows:

1–2 work of compression $h_2 - h_1$ 2–3 heat of condensation $h_2 - h_3$

3–4 isenthalpic expansion   4–1 input heat of evaporation $h_1 - h_4$

Thus the performance factors are

$$CP_h = (h_2 - h_3)/(h_2 - h_1)$$
$$CP_c = (h_1 - h_4)/(h_2 - h_1)$$

(16.3)

Naturally there are many factors influencing the performance. In practice there will have to be finite temperature differences across the condenser and evaporator and there will be pressure losses in the pipes, valves, condenser and evaporator. Similarly, the compressor will be less than 100 per cent efficient. In fact, though, the performance of various refrigerants can be compared using equation (16.3).

The cycle described may be termed the 'ideal' vapour compression cycle. In practical cycles there are a number of differences. Thus the compression is often carried out in the superheat region as it is difficult to arrange for the expansion to stop at state 1. This also ensures that no liquid enters the compressor mechanism where, being incompressible, it might cause extensive damage or, less seriously, cause excess lubricant to be swept through the system. In this case, heat transfer efficiency in the evaporator could be reduced. In addition the liquid leaving the condenser may be subcooled before entering

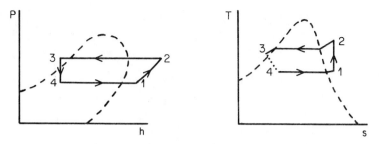

Figure 16.4. Modified vapour compression cycle

the throttling valve. This can be achieved fairly simply by making use of the significant temperature difference between the condenser and the environment. The modified cycle is shown in Figure 16.4. Superheating beyond state 1 may give a slight increase in CP_h but the liquid subcooling will always improve both the heating effect and CP_h.

It is worth considering at this juncture the requirements that must be placed on the refrigerants to be used in the vapour compression cycle. From an examination of the cycles of Figures 16.3 and 16.4, we can make the following observations. Firstly, the critical temperature of the refrigerant should be higher than the highest temperature attained in the cycle. Similarly, the saturated vapour pressure should be greater than atmospheric pressure at the lowest evaporator temperature. It is also useful if the refrigerant has a low specific heat in the liquid phase thus ensuring that the saturated liquid line is steep, as should be the saturated vapour line. From the point of view of compressor performance, the compression ratio between the vapour pressures in the evaporator and the condenser should be as small as possible. Finally, the refrigerant should have a high latent heat of vaporization in order that a small amount of refrigerant will suffice for a given power output. We shall return to such considerations in more detail in Section 16.4.

Although it involves a slight digression, it is convenient at this point to to mention a unit commonly encountered in refrigeration and air-conditioning practice: the 'ton' of refrigeration. A refrigerator or heat pump is said to have a capacity of one ton if it can remove 200 Btu min^{-1} of heat in the evaporator. (This is sufficient to freeze one ton of water at 32°F in one day.) Referring to Figure 16.3, we can write down the following relationships for a one ton machine:

Refrigerant flow rate	$= 200/(h_1 - h_4)$ lb min^{-1}
Volume of vapour entering compressor	
(specific volume v_1)	$= 200\,v_1/(h_1 - h_4)$ cf min^{-1}
Work of compression	$= 200\,(h_2 - h_1)/(h_1 - h_4)$ Btu min^{-1}
Heat rejected in condenser	$= 200\,(h_2 - h_3)/(h_1 - h_4)$ Btu min^{-1}

The second type of refrigeration cycle is that employing a gas such as air. It might be thought that the reversed Carnot cycle for a gas as the working

438

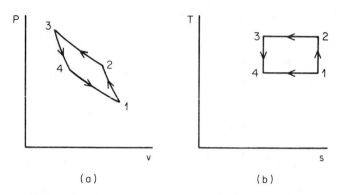

Figure 16.5. Reversed Carnot gas cycle: (a) pressure–volume
diagram, (b) temperature–entropy diagram

fluid would be the most desirable choice. However, this cycle (Figure 16.5) involves both adiabatic and isothermal steps. The adiabatic stroke requires a high speed while the isothermal stroke requires a low stroke speed. Such a variation in stroke speed is not practicable. In fact the first successful refrigerator was the Bell–Coleman machine which operated on the *reversed Joule cycle* and used air as the working fluid.

The major design difference between the vapour compression and the reversed Joule cycles is that the latter requires an expansion device rather than a throttling valve. This is because the temperature drop is very small when a gas is passed through a throttling valve, indeed it is zero for a perfect gas. In the reversed Joule cycle the gas is actually allowed to expand in such a way that it does work and thus suffers a reduction in internal energy.

The reversed Joule cycle is illustrated by means of pressure–volume and temperature–entropy diagrams in Figure 16.6. Also depicted is the arrangement of equipment. The compressor takes in low-pressure air at point 1 in the cycle and compresses it to a high pressure, high temperature state. The air passes through a heat exchanger where it gives up some of its heat. It is now allowed to expand in the expansion device doing positive work: this work is actually fed back to the compressor. The air now takes up heat in the low pressure heat exchanger and the cycle is completed. The compression and expansion are both adiabatic in an ideal cycle, while the heat exchanges occur at constant pressure.

We can evaluate the performance of such a cycle using the T–s diagram. As the heat exchanges occur at constant pressure the entropy changes during these processes can be written as

$$s_2 - s_3 = s_1 - s_4 = C_p \ln (T_2/T_3) = C_p \ln (T_1/T_4)$$

assuming that C_p is essentially constant over the temperature range in question. Thus T_4 is given by

$$T_4 = T_1 T_3 / T_2 \tag{16.4}$$

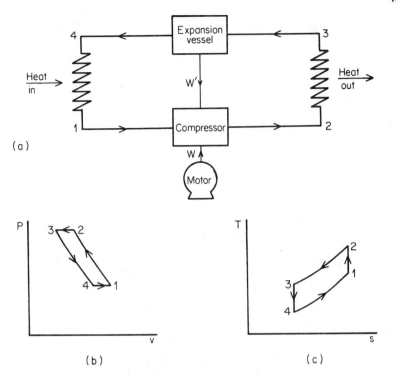

Figure 16.6. Reversed Joule cycle: (a) schematic, (b) pressure–volume diagram, (c) temperature–entropy diagram

The coefficient of performance is given by the ratio of the heat delivered, $C_p(T_2 - T_3)$, to the work done. This latter quantity is itself the difference between the work done by the compressor, $C_p(T_2 - T_1)$, and the work done by the gas in expanding, $C_p(T_3 - T_4)$. Thus the coefficient of performance is

$$CP = \frac{C_p(T_2 - T_3)}{C_p(T_2 - T_1) - C_p(T_3 - T_4)}$$

$$= \frac{T_2 - T_3}{(T_2 - T_3) - (T_1 - T_4)}$$

$$= 1 \bigg/ \left(1 - \frac{T_1 - T_4}{T_2 - T_3}\right)$$

(We shall omit the subscript and assume that CP refers to heating unless otherwise stated.) Using equation (16.4) this reduces to

$$CP = 1/(1 - T_1/T_2) \tag{16.5}$$

Now as the expansion and compression are adiabatic, they can be described by the familiar law: $P^{1-\gamma} T = \text{constant}$. Then if R denotes the pressure ratio between the heat exchangers,

440

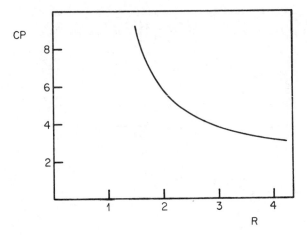

Figure 16.7. Coefficient of performance, CP, as a function of pressure ratio, R, for a heat pump using air ($\gamma = 1\cdot4$) and operating on the reversed Joule cycle

$$T_3/T_4 = T_2/T_1 = R^{(\gamma-1)/\gamma}$$

and equation (16.5) may be rewritten as

$$CP = 1 \Big/ \left(1 - (1/R)^{\frac{\gamma-1}{\gamma}} \right)$$

The value of γ for air is approximately $1\cdot4$ and using this value leads to the variation of CP with R shown in Figure 16.7. It is clear that the best coefficients of performance can be achieved when the pressure ratio, and hence the temperature difference, is low. This may be compared with the result for the reversed Carnot cycle. The difference in this case is that for a given value of R, the coefficient of performance is independent of the particular temperature range employed.

The final refrigeration cycle to be considered is the absorption cycle. As our main concern is with the heat pump, and as the absorption cycle is not well suited to this application, we shall only give a brief description for the sake of completeness. Absorption cycles use a vaporizable liquid, for example ammonia, as a refrigerant and a second liquid, for example water, as an absorbent for the first. A schematic diagram of the flow processes involving both the water and the ammonia is shown in Figure 16.8. The right-hand side of the diagram is identical to the condenser–evaporator arrangement of a device utilizing the vapour compression cycle, except that there are no longer a compressor and a throttling valve linking them. Instead, heat energy is used to drive an absorption–generator unit which effectively acts as a compressor. The ammonia circulates completely around the circuit, that is from the evaporator through the absorber and generator to the condenser, whereas the water only circulates between absorber and generator.

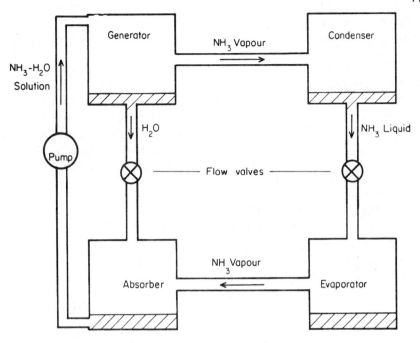

Figure 16.8. Absorption refrigeration cycle

In the condenser the ammonia loses heat and condenses. The liquid ammonia is passed through a flow valve to the evaporator. Here the pressure is sufficiently low for evaporation to occur, the latent heat being extracted from the evaporator surroundings. The ammonia vapour enters the absorber where it is absorbed by the water with an accompanying release of the heat of solution. The concentrated solution can now be passed easily to the high-pressure region of the generator using a pump. (As the absorption process has effectively 'compressed' the volume of vapour this is not a compressor, but merely a circulating device requiring little energy.) In the generator the water and ammonia are again separated using heat energy. The water flows back to the absorber via a flow control valve and the ammonia vapour passes directly to the condenser. Thus the absorption refrigeration cycle is essentially a heat engine and a vapour compression cycle combined.

The majority of heat pump installations have utilized the vapour compression cycle. Consequently, we shall tend to concentrate on this type of heat pump for the bulk of this chapter. Nevertheless, many of the points raised in discussion are also applicable to air heat pumps, working on the reversed Joule cycle, and these will be more fully treated in Section 16.5.

16.2 Sources of Heat

So far we have failed to comment on the low-temperature source of heat required in heat pump installations. In fact there is a number of suitable sources and the atmosphere, the earth and various water supplies have all been exploited in the past. Moreover, there is increasing interest in the use of the Sun's energy as a primary source of heat but this has still to be developed.

The atmosphere is undoubtedly the most commonly employed source. Air is fanned over the evaporator coils which are usually finned to provide efficient heat transfer. A temperature difference of 20°F between the atmosphere and the refrigerant in the evaporator is typical. Air is an eminently suitable source of heat in that it is continuously available at all locations and leads to equipment of moderate size as well as installation and operating costs that are low. In addition, the design of air-source heat pumps can be adapted to facilitate factory production of standard units.

A heat pump is at its most efficient when the temperature difference between the evaporator and the condenser is small. This is immediately obvious in the case of a reversed Carnot heat pump, equation (16.2). Therefore, high values of the coefficient of performance can be achieved by using a source of heat whose temperature is as close as possible to the temperature of the space to be heated. For most of the heating season the air temperature is reasonable in this respect. However, it is somewhat variable and this leads to considerable problems in attempting to estimate the required capacity for the installation. If a heat pump is used that can meet the heating load under the most severe winter conditions, then it will be operating at considerably less than full power for most of the time. This represents a waste of capital expenditure. As a compromise it is common to use auxiliary heating during the worst conditions. Another disadvantage of using air as a heat source is the formation of frost on the evaporator heat exchanger coils with a consequent reduction in the efficiency of heat transfer. This can be circumvented by periodic defrosting of the coils but still entails an overall reduction in the actual coefficient of performance: the energy used to defrost the coils increases the total energy requirement of the system.

Water is also a very useful source of heat. If anything it is preferable to air in that the temperature of a given supply fluctuates much less violently than does the atmospheric temperature but, against this, it is not so universally available. The main problem with using any type of water supply as a heat source is scaling and corrosion of the heat exchanger coils but this has been overcome in many other engineering plants. There is a number of types of water supply to be considered. Firstly, there are the natural sources of sea water, surface water (rivers and lakes) and well water. Secondly, there is towns water and, thirdly, there is waste water from industrial processes.

It will be recalled that towns water was actually used as an auxiliary source of heat in the first application of heat pumps but this is an extremely expensive supply today. Surface water, either in rivers or lakes, is an adequate source, especially as its temperature is unlikely to reach freezing point in all but the coldest conditions. Well water is even more suitable as its temperature is usually higher than on the surface during the winter months, and its temperature variation is small. Needless to say, though, the availability of well water is limited. Sea water may well prove to be an excellent source for a country such as the United Kingdom which has so much coastline but scaling and corrosion problems are particularly severe in this case. However, the most promising source for the future may well be waste water from industrial processes. This is often at temperatures well above atmospheric which raises the possibility of extremely high coefficients of performance. Obviously the availability of such waste water is strictly limited but this should be a case of the heat pump going to the heat source rather than the other way round. Industrial consumption of electrical power could be reduced if the waste heat in this water were extracted and recycled (Chapter 15).

It is feasible to extract heat from the earth using evaporator coils which are buried underground but this has not been widely used for a number of reasons. Thus it is difficult to estimate the heat-transfer characteristics of a particular type of soil and, once the system is installed, repair work on the buried coils is not simple. The ground temperature slowly decreases as the pump operates which in turn reduces the performance and, as the ground will become frozen hard in the course of the heating season, considerable damage may be caused to gardens!

Considerable effort is now being expended on research into the use of solar energy as a source of heat for heat pumps. When available, solar energy can provide heat at a significantly higher temperature than other sources so co-efficients of performance would be expected to be high. However, solar energy is an extremely intermittent source and can obviously provide no heating during the cold winter evenings. Nevertheless, it is still worthwhile to investigate its use in conjunction with other sources of heat or in association with a heat-storage system inside the building.

Solar source heat pumps may be either direct or indirect. In the direct type, the evaporator coils are embedded in the solar collector, for example a darkened flat plate insulated by a glass cover (Chapter 12). Without the glass cover, the evaporator will actually extract heat from both the air and the sunlight! In an indirect solar heat pump, a fluid such as air or water is circulated through the solar collector, is warmed by the solar energy, and is then used as the heat supply for a normal air-source or water-source heat pump. When air is used as the fluid, it is a simple matter to allow for periods of low solar radiation flux. By opening the air circuit to the atmosphere, the pump can use the atmosphere as a heat source, with the solar collector acting as a preheater. Alternatively, in strong sunlight the air circuit is closed to exclude the outdoor air with the result that all the heat is derived from the solar energy.

The source used and the medium employed to transmit heat to the space concerned enable a simple classification of heat-pump types to be drawn up. This classification is displayed in Table 16.1 and some of the advantages and disadvantages of each type are also listed. The table also refers to the use of each heat pump in reverse mode, that is to say, as an air conditioner. Now the source of heat is the air inside the building and a suitable external heat

Table 16.1 Classification of heat pumps

Source	Availability	Advantages	Disadvantages	Suitability as heat sink in reverse mode	Types of pump
Air	Excellent	Low installation and running costs	Large temperature variation; frosting of outdoor coils	Good	Air-to-air Air-to-water
River water Sea water	Reasonable Good in United Kingdom		All water supplies give problems of scaling and corrosion; towns water has very high operating costs		
Well water Towns water	Rare	High temperature		Good except for waste water	Water-to-air Water-to-water
Waste water	Usually available for industrial consumers	High temperature (All have small temperature variation)			
Earth	Excellent		Temperature of source falls as pump operates; evaporator coils inaccessible	Good	Earth-to-air Earth-to-water
Solar energy	Intermittent and unpredictable	High-temperature source			Direct/indirect solar source heat pump

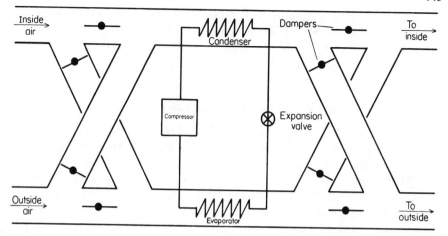

Figure 16.9. Air-to-air heat pump and air-conditioner. The dampers are shown in heating mode position—to convert to cooling mode, all dampers are rotated through 90°

sink must be found for the heat extracted. All of the heat sources mentioned above can also be used as heat sinks, with the obvious exception of solar heat. Waste water is not a particularly suitable heat sink but, as the main aim of building heat pumps to work with waste water as the heat source is to reduce the amount of waste heat emitted, this is not a great drawback. Nor is the earth a particularly satisfactory heat sink, but, if it has been used as a heat source throughout the winter, it may be in a condition of permafrost, and will, therefore, be adequate as a heat sink for air-conditioning during the summer.

The switch from heat-pump mode to air-conditioning mode can be easily brought about by reversing the flow of refrigerant in the basic cycle. (Of course, the piping and valves must be arranged to ensure that refrigerant always passes through the compressor and the expansion valve in the same direction.) In the case of an air-to-air heat pump the mode switching can be accomplished in an alternative manner, as shown in Figure 16.9. The heat-pump assembly is a compact unit to which both external and internal air are drawn along a system of ducts. In the heating mode, the external air passes over the evaporator coils and the internal air over the condenser coils. When cooling is required, the air supplies are redirected by means of dampers so that the internal air now passes over the evaporator and the external air over the condenser. Thus the role of the heat exchanger coils remains unchanged after the switchover.

If a heat pump is to be used in reverse mode during the summer, it is especially important to ensure that the design balances the requirements of summer and winter operation. In the case of an air-source heat pump we have already referred to the reduction in performance as the external air temperature drops. This means that the system design for a given heating load is more complex than is the case in, say, an oil-fired heating system. If the heat pump is designed with too low a balance point (the external temperature at which the pump capacity exactly meets the heating requirements) then there will be a considerable excess of cooling capacity in summer. Once again, the optimum course to steer leads to a balance point which provides enough capacity for all but the coldest weather, while the deficit in these extreme conditions can be made good by auxiliary heating supplies.

16.3 Design Considerations (Vapour Compression Pumps)

It is not our intention to enter into a detailed description of the various components used in vapour compression heat pumps; this can be found in the trade journals and refrigeration engineering handbooks. (See, for example, the ASHRAE handbooks referenced at the end of this chapter.) We shall content ourselves with a brief mention of the more salient design considerations. In the United Kingdom the climatic conditions are such that we need only consider the heating role of the heat pump, except in the very large installations where air conditioning may be necessary in summer. In the United States and elsewhere, however, the heat pump may also act as an air conditioner on the domestic scale, and so we shall deal with this more general case of a dual role heat pump.

Load Estimation

In order to arrive at specifications for the compressor and other components it is essential that the design heating and cooling loads should be known. As has been mentioned before, it is customary to specify a heat pump capable of meeting the cooling load and then to use auxiliary heating during severe winter weather. From the point of view of capital investment it is uneconomic to oversize the cooling load. In large installations it is, of course, important to take into account ventilation and also the possibility of transferring heat internally from exceptionally warm areas, for example rooms containing a large amount of electrical equipment, to cooler areas.

Auxiliary Heating

This is best located in the circuit of the heat-transporting medium (for example, water or air) after the condenser heat exchanger. Thus the condenser can operate at a lower temperature which leads to a better coefficient of performance for the heat pump, although the overall performance is, of course, reduced by the auxiliary power consumed.

Water Temperature

When water is used as the heat-transporting medium in central heating systems, it is generally believed that high water temperatures are necessary for efficient space heating. In fact a temperature of 120°F (49°C) will suffice if care is taken to eliminate draughts and to design the radiators correctly. From the control viewpoint a low water temperature is desirable as it reduces the temperature range through which the throttling valve must operate. Incidentally, for a domestic hot water supply it is possible to use a heat exchanger in the refrigerant discharge line from the compressor where the refrigerant gas is at a high superheat temperature.

Heat Storage

Every heating system incorporates thermal storage to some extent in that the walls and fittings of any building absorb heat from and discharge heat to the air space. This can be aided by using materials of high thermal capacity thus giving a reduction in the heating capacity required to cope with peak load conditions. The most common method is to use warm water storage tanks. The water in these tanks can be maintained at a temperature greater than that of the space being heated, in which case it is used directly for heating purposes whenever the need arises, or it can be kept at a lower temperature and used as an intermittent heat source during periods of peak load.

Compressors

There are two main types of compressors, those with balanced compression and those with free compression. The former class include *reciprocating* and *rotary compressors* while the latter includes *centrifugal compressors*. Centrifugal compressors are not particularly suited to small heat pump installations. They are best used for handling large quantities of fluid and tend to surge under low load conditions. They are utilized in large installations with water as the heat source; a suitable storage system can help by smoothing the load.

Reciprocating compressors are generally used in heat pumps of up to 100 ton capacity. They have an intrinsic inefficiency resulting from the clearance between the piston and the cylinder head in top dead centre position. This clearance volume of gas which remains in the cylinder after discharge will re-expand to a larger volume when the pressure falls to the inlet pressure. This means that the compressor does not admit a full cylinder of fresh refrigerant from the condenser. This inefficiency can be characterized by a quantity known as the 'volumetric efficiency' of the compressor. Referring to Figure 16.10, this can be written as

$$\eta_v = (V_a - V_d)/(V_a - V_c)$$
$$= 1 + C - C(v_1/v_2)$$

where C, the clearance, is defined to be $V_c/(V_a - V_c)$, and v_1 and v_2 are the specific volumes of the vapour before and after compression.

The volumetric efficiency has an important effect on the overall capacity of the heat pump. Suppose that we have a single cylinder reciprocating compressor running at N rpm. Then if D is the cylinder bore and L is the stroke (in feet), the refrigerant pumping rate is given by

448

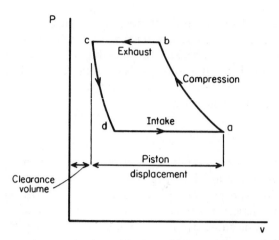

Figure 16.10. Behaviour of piston compressor. (Point b is equivalent to point 2 in the vapour compression cycle of Figure 16.3)

$$m = \pi D^2 L N \eta_v / (4 v_1) \text{ lb per min}$$

and the heating capacity by

$$Q = m(h_2 - h_3)/200 \text{ tons}$$

where we use the notation of Figure 16.3. These relations also show that v_1 is an extremely important factor when selecting a particular refrigerant (Section 16.4).

Reciprocating compressors usually have clearance volumes of 5 per cent, but lower clearance volumes can be useful in low-temperature applications. Rotary compressors are similar to the reciprocating type but have lower clearance volumes and, consequently, higher volumetric efficiencies. In the range 10 to 120°F they have approximately one-quarter greater capacity than medium clearance reciprocating compressors, but, against this, the power demand is greater in reverse mode operation.

If an oversized compressor is used to cope with low source temperatures then for most of the year it will be under-used and, therefore, a waste in terms of capital investment. We have seen that one way of overcoming this problem is by the use of auxiliary heating. Another alternative is to use *staged compression*. For example, if the outside air temperature is 10°F, we might use one compressor to pump from 10 to 40°F and then compress to 120°F using a second machine. In normal conditions the compressors are arranged to run in parallel between, say, 40 and 120°F. They are also arranged in parallel during the cooling cycle. It is common to use rotary compressors as the first stage in staged compression systems, paired with a reciprocating compressor.

It is important that no liquid refrigerant leaks into the compressor for it can lead to excessive wear as a result of injured valves or diluted lubricant. The liquid can gain access after a restart when liquid accumulated in the evaporator is passed directly into the compressor. Alternatively 'floodback' may occur, that is the continuous return of a fraction of the liquid refrigerant as a result of wrong control settings or the use of an excessively large refrigerant charge. To protect against this it is important to use the minimum charge and the correct setting for the liquid feed to the evaporator, as well as isolating the compressor during shutdown.

Expansion Valves

The most common type of expansion valve is the thermostatic one which controls the flow rate of the liquid refrigerant entering the evaporator by monitoring the superheat of the gas leaving the same. This ensures that the entire evaporator is active and that no liquid refrigerant is returned to the compressor. Constant pressure expansion valves can also be used. The evaporator outlet pressure controls the mass flow rate of liquid to balance the capacity of the compressor pump.

Heat Exchangers

In an ideal heat exchanger the temperature of the fluid giving up heat is equal to the temperature of the fluid accepting heat. In addition the pressure drop through the exchanger is zero. However, the size of a given heat exchanger is limited by considerations of cost and available space and the limitation on size leads to failings in these ideal restraints. A non-zero temperature difference between the two fluids involved in the heat exchange means that the range of temperature experienced by the refrigerant is increased and consequently that the heat pump's coefficient of performance is reduced. Similarly, a pressure drop, indicative of mechanical losses, means that more work must be expended to drive the refrigerant around the system and this again reduces the coefficient of performance. Thus the design engineer must balance the saving in capital cost inherent in small exchangers against the poorer performance which results from their use.

With an air-source heat pump air must be fanned over the external heat exchanger in order to provide efficient heat transfer (the same applies to the internal exchangers if air is used as the heat transporting medium). This also entails a reduction in the performance of the pump as the power required to drive the fans must be included in the figure for the total input power.

Defrost Control

When an air-source heat pump is used under conditions of low temperature and high humidity, frost will accumulate on the evaporator heat exchanger surface. If such a build-up becomes severe, then it must be removed in order that efficient heat transfer can be maintained. The frequency of defrosting depends, of course, on the prevailing air conditions, reaching a maximum at a dew-point temperature of 32°F. It is not usually necessary to defrost for dew-point temperatures of below 10°F or above 50°F. There may well be condensation at higher temperatures than this but the condensate will not freeze as the surface temperature of the refrigerant coils will be above 32°F.

There are a number of methods which can be used to ensure automatic defrosting. The most common of these is to switch to the cooling cycle for a short period of time. Alternatively, electrical resistance heaters can be used to warm the evaporator directly, or hot water can be sprayed over the evaporator. The control of the defrost cycle can be achieved in a number of ways. The simplest is an arbitrary time switch which initiates the defrost mechanism every 30 minutes or so when the outside air temperatures is sufficiently low to warrant this. The temperature differential between the refrigerant and the air will increase during frost formation and this can provide an alternative control for defrosting. Similarly, an air pressure switch can be employed. The deposit of frost increases the air resistance across the evaporator coils so a pressure differential switch can be used to activate the defrost cycle. In all these cases the defrosting can be terminated by a control which monitors the pressure or temperature of the liquid refrigerant in the outdoor coils.

Exhaust Heat

In large buildings the air must be extracted to ensure efficient distribution of heat and adequate ventilation. If it is arranged that the extraction can be performed at one or a small number of central points, then the warm spent air can be used as an additional heat source for the heat pump. This is most simply effected in the so-called 'run-around' system. An anti-freeze solution such as ethylene glycol is pumped between heat exchanger coils located in the inlet air duct and the exhaust air duct. Thus the glycol is warmed by the exhaust air and is used to preheat the inlet air to the heat pump. Frosting of the coils in the exhaust duct can occur but the defrost procedure is simply a matter of switching off the glycol pump.

Liquid Subcooling Coils

Earlier in this chapter we described how the vapour compression cycle could be improved if the liquid refrigerant was subcooled prior to throttling. Warm liquid from the liquid refrigerant receiver is allowed to flow through a subcooling coil where it preheats the ventilation air and is itself cooled. Lowering the refrigerant temperature increases the heating effect, as shown in Figure 16.4. It is perfectly feasible to achieve capacity increases of up to 20 per cent with this technique and, furthermore, the capacity increase is greatest when the outdoor temperature is low, that is when the pump is at its least efficient.

16.4 Refrigerants

The properties of the refrigerant used in a vapour compression heat pump are the dominant factors in controlling its performance. As we shall see, the choice is determined in principle by the thermodynamic properties of the refrigerant, that is to say, its capacity for transferring heat, but there are many other factors which must be taken into account. Obviously cost and availability are of paramount importance from the marketing point of view, as are the flammability and toxicity of the refrigerant when the safety aspects are considered. Other properties such as viscosity, surface tension and density also affect performance.

The design engineer is, as usual, confronted with the task of deciding between these somewhat conflicting choices. Thus it is desirable to have as high a pressure in the evaporator as possible but at the same time a low condensing pressure is preferable in order that the compression ratio of the compressor should be low. Similarly, a refrigerant of low viscosity and surface tension will have good flow properties but these same properties will inhibit the formation of droplets during condensation which is advantageous for heat transfer. In order to economize on power it is also desirable that the refrigerant has a high latent heat so that more heat can be transferred for a given amount of refrigerant. As a rough guide to this requirement, Trouton's rule, an empirical law, may be used. This states that the latent heat per mole at the boiling point is proportional to the temperature at the boiling point.

There is now a multitude of refrigerants in existence, and we can divide them into a number of classes by their chemical nature. Firstly, we have the inorganic materials of which the most important examples are ammonia, carbon dioxide and sulphur dioxide. There are many other examples, of course; hydrogen, argon, oxygen and nitrogen can all be used, as can water and air! Secondly, there are the saturated aliphatic hydrocarbons such as methane and ethane, and, thirdly, there are the unsaturated hyrocarbons such as ethylene. The fourth group is probably the most important class of modern refrigerants, the halogenated organic compounds. There are numerous examples of this group, some based on the saturated hydrocarbons (for example, trichlorofluoromethane), some based on the unsaturated hydrocarbons (for example, tetrafluoroethylene) and still others based on cyclic organic compounds (for example, octacyclofluorobutane). Table 16.2 lists some of the physical properties of various representatives of these groupings, together with their assigned refrigerant number.

One reason for the modern tendency to use halogenated organics as refrigerants is their comparative safety. Of the most popular inorganics, ammonia forms an explosive mixture with air at concentrations of about one-fifth by volume and is, moreover, toxic. Thus a concentration of one per cent of the gas can be fatal within one hour. Sulphur dioxide, although non-flammable, is even more toxic, a similar concentration being fatal within minutes. The hydrocarbons are all prone to explosions but are otherwise harmless to life. Generally speaking, though, the halogenated organic compounds are very safe. They are in the main non-flammable (there are exceptions such as methyl chloride). There are some such as carbon tetrachloride and chloroform whose toxicity approaches that of ammonia but chlorodifluoromethane and trichlorofluoro-

Table 16.2 Physical properties of refrigerants

Number	Chemical name	Formula	Molecular weight	Boiling point (°F)	Freezing point (°F)	Critical temperature (°F)	Critical pressure (psia)
717	Ammonia	NH_3	17	− 28	− 108	271	1657
744	Carbon dioxide	CO_2	44	− 109	− 70	88	1071
764	Sulphur dioxide	SO_2	64	14	− 104	315	1143
50	Methane	CH_4	16	− 259	− 297	− 116	673
170	Ethane	$CH_3.CH_3$	30	− 127	− 278	90	708
1150	Ethylene	$CH_2:CH_2$	28	− 155	− 272	49	732
11	Trichlorofluoromethane	CCl_3F	137	75	− 168	388	639
12	Dichlorodifluoromethane	CCl_2F_2	121	− 22	− 252	234	597
13	Chlorotrifluoromethane	$CClF_3$	104	− 115	− 294	84	561
13B1	Bromotrifluoromethane	$CBrF_3$	149	− 72	− 270	153	575
21	Dichlorofluoromethane	$CHCl_2F$	103	48	− 211	353	750
22	Chlorodifluoromethane	$CHClF_2$	86	− 41	− 256	205	722
40	Methyl chloride	CH_3Cl	50	− 11	− 144	289	969
114	Dichlorotetrafluoroethane	$CClF_2 \cdot CClF_2$	171	38	− 137	294	474
115	Chloropentafluoroethane	$CClF_2.CF_3$	154	− 38	− 147	176	458
1114	Tetrafluoroethylene	$CF_2:CF_2$	100				
C318	Octafluorocyclobutane	C_4F_8	200	21·5	− 42·5	239·5	395
502	Azeotrope of 22 and 115 (48·8 per cent of 22 by weight)		112	− 50		179	619

methane are as safe as carbon dioxide. Safest of all are dichlorodifluoromethane and bromotrifluoromethane which remain non-toxic in concentrations of up to 20 per cent. It is ironic that the safest refrigerants should be based on the noxious halogen gases!

The chemical properties of refrigerants must also be taken into account when deciding upon the materials to be used in the construction of the heat pump. Sulphur dioxide will, of course, dissolve in any water present to form sulphurous acid which can attack iron or steel. Ammonia is also corrosive and cannot be used with copper or any copper alloy. Once again the halogenated refrigerants are the best choice as they can generally be used with most common metals. Metals do tend to promote the thermal decomposition of halogenated compounds but this is not usually a problem under normal operating conditions.

The halogenated organic refrigerants, marketed under the trade name 'Freon' by Du Pont de Nemours International S.A., are now the standard refrigerants used in most refrigeration and heat-pump systems. The most common ones are R 12, R 22, R 114 and R 502. R 21 has been more or less replaced by R 114: they have similar thermodynamic properties but R 114 has a larger number of fluorine atoms per molecule and is, therefore, more stable. (C 318 is even more stable but is not often used because of its high cost.) R 502 is also tending to replace R 22 and is particularly recommended by Du Pont for heat-pump operation. To see why this should be so it is necessary to examine the thermo-dynmic properties of these refrigerants.

The most convenient way of displaying these properties is by means of a

Critical volume (ft³ lb⁻¹)	Liquid heat capacity at 30°F (Btu lb⁻¹ F⁻¹)	C_p/C_v at 100°F	Dielectric constant	Liquid viscosity at (centipoises)				Vapour viscosity at			
				0°F	40°F	80°F	120°F	0°F	40°F	80°F	120°F
0·068	1·10		15·5		0·230	0·210				0·0093	
0·035	0·62		1·59		0·095	0·064				0·0140	
0·031			15·6		0·350	0·290				0·0120	
0·099											
0·076											
0·073											
0·029	0·21	1·14	1·92	0·677	0·517	0·417	0·349	0·0096	0·0103	0·0110	0·0116
0·029	0·22	1·14	1·74	0·335	0·286	0·255	0·232	0·0113	0·0119	0·0126	0·0132
0·028	0·29	1·14	2·3								
0·022											
0·031	0·24	1·17	4·88	0·484	0·397	0·337	0·294	0·0101	0·0108	0·0115	0·0121
0·031	0·28	1·18	1·68	0·291	0·256	0·232	0·214	0·0113	0·0122	0·0130	0·0137
0·043	0·37			0·298	0·263	0·237	0·217	0·0094	0·0101	0·0108	0·0115
0·028	0·23	1·09	1·83	0·598	0·454	0·366	0·307	0·0102	0·0109	0·0116	0·0122
0·026											
0·03											
0·029											

pressure–enthalpy (P–h) diagram. Such P–h charts for R 22 and R 502 are shown in Figure 16.11. To compare refrigerants we shall choose a cycle in which the refrigerant is pumped between 10 and 120°F. This is obviously an extreme case for heat-pump operation in a country such as the United Kingdom in that a lower temperature of 10°F would correspond to an air temperature of, say, 20°F, while an upper temperature of 120°F is more than adequate for an air-to-air heat pump and is suitable for an air-to-water pump.

The use of the P–h chart to analyse this cycle is illustrated in Figure 16.12. The values of enthalpy around the cycle can be read off directly as h_1, h_2 and h_3. The condensing and evaporating pressures, P_{cond} and P_{evap}, are also available directly and hence the compression ratio, c.r., can be calculated. The net heating and cooling effects (NHE and NRE) per unit weight are $h_2 - h_3$ and $h_1 - h_3$. The temperature at point 2 corresponds to the compressor discharge temperature, T_{dis}, and the specific volume, v_1, can be obtained from the constant volume curves. The results of such an analysis are listed in Table 16.3 for R 22, R 114, C 318 and R 502 together with the results for ammonia (R 717) for comparison. We can also allow for superheating by extending the line 41 until it reaches the isotherm corresponding to the degree of superheat specified and then completing the cycle. (It should be noted that no data are given for R 114 or C 318 for the case of a cycle involving no superheat. This is because such a cycle would involve wet compression: the line 12 in the diagram would lie partly within the mixed liquid–gas region of the P–h diagram.) Results for 60°F superheating are included in Table 16.3. The final set of data presented refers to the required compressor displacement for one ton of refrigeration. This can be calculated from the net refrigeration effect (NRE) and the specific volume according to

$$\text{displacement} = 200\, v_1/(NRE)\ \text{ft}^3\ \text{min}^{-1}$$

454

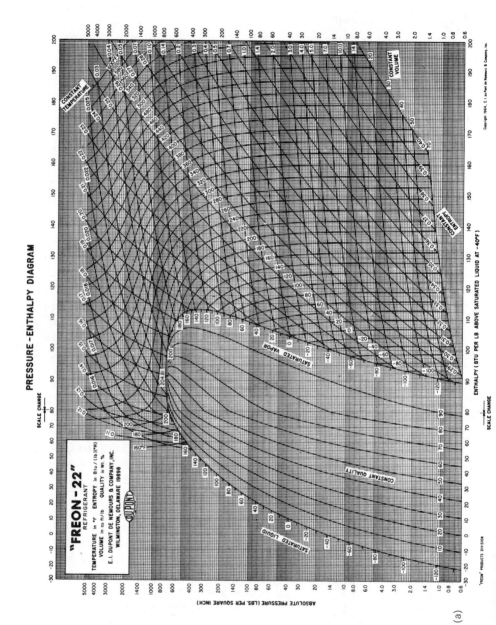

PRESSURE-ENTHALPY DIAGRAM

(a)

455

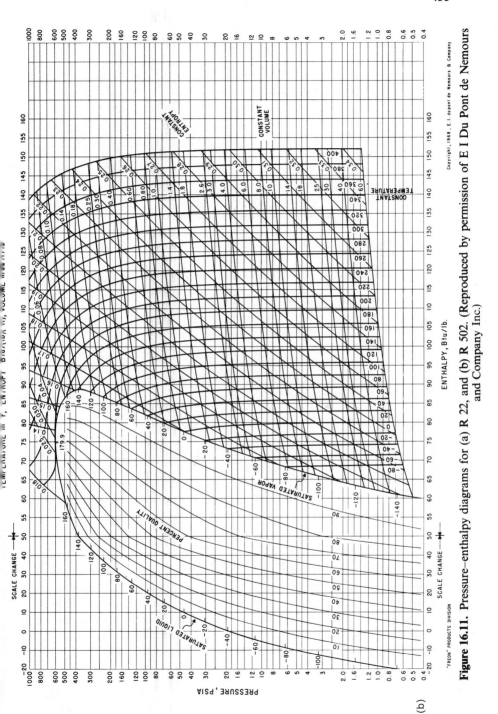

Figure 16.11. Pressure–enthalpy diagrams for (a) R 22, and (b) R 502. (Reproduced by permission of E I Du Pont de Nemours and Company Inc.)

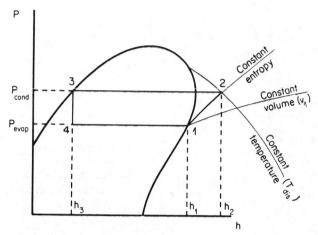

Figure 16.12. Illustrating the use of pressure–enthalpy diagrams

There are a number of important points to be made in connection with Table 16.3.

Coefficient of Performance

For the refrigerants chosen there is remarkably little variation in *CP*. That for C 318 is slightly below the other values. R 502 has a somewhat low CP when no superheating is used in the cycle but when 60°F superheat is allowed it improves considerably.

Pressure

Low operating pressures are desirable in reciprocating compressors, on which point R 114 and C 318 score heavily. However, their evaporating pressures are less than atmospheric (15 psia) which means that air and water could leak into the system. Leakage from the compressor will be worse at high compression ratios so the low ratio of R 502 is advantageous in this respect.

Temperature

It has been found that many of the maintenance problems associated with heat pumps can be attributed to the high refrigerant temperatures at the compressor outlet. High discharge temperatures in hermetic (sealed) compressors can lead to decomposition of the lubricant circulating with the refrigerant and may also affect the insulation of the motor windings. The low discharge temperatures of C 318 and R 502 are, therefore, to be preferred.

Specific Volume

The significance of v_1 has already been stressed in Section 16.3. A low value of v_1, such as in R 502, is a tremendous asset to a refrigerant. The most significant

Table 16.3 Refrigerant performance in heat pump cycle from 10 to 120°F

	R 717		R 22		R 114	C 318	R 502	
Superheat (°F)	—	60	—	60	60[a]	60[a]	—	60
Enthalpy (Btu lb⁻¹) h_3	179	179	46	46	37	40	44	44
h_2	748	793	123.5	135.5	92.5	84	91	102
h_1	615	643	105.5	113.5	79.5	72.5	78	88
C.P.	4.3	4.1	4.3	4.1	4.3	3.8	3.6	4.2
P_{cond} (psia)	286		277		61	92	300	
P_{evap}	38.5		46		7.4	11.5	55	
c.r.	7.4		6.0		8.2	8.0	5.5	
NHE (Btu lb⁻¹)	569	614	77.5	89.5	55.5	44	47	58
NRE	436	464	59.5	67.5	42.5	32.5	34	44
T_{dis} (°F)	283	355	165	224	145	127	135	185
v_1 (ft³ lb⁻¹)	7.3	8.2	1.15	1.3	4.5	2.4	0.75	0.85
NHE/v_1 (Btu ft⁻³)	78	75	67	69	12	18	63	68
Compressor displacement for 1 ton refrigeration (ft³ min⁻¹)	2.6	2.65	3.0	2.9	16.7	11.1	3.2	2.9

[a] No figures are given for R 114 or C 318 in the absence of superheating as this would involve wet compression.

458

quantity, however, is the net heating effect per unit volume (NHE/v_1) or, alternatively, the compressor displacement needed for one ton of refrigeration. The best values of NHE/v_1 are obtained with R 22 and R 502 and, correspondingly, they require compressors with the smallest capacity.

Ammonia

It should be noted that ammonia, R 717, has a good coefficient of performance and a high value of NHE/v_1 but, against this, it requires a high compression ratio and gives a high discharge temperature. The essential reason for not using ammonia in domestic applications is one of safety.

It is clear, then, that R 22 and R 502 are the refrigerants most suited to heat pump operation. R 22 has indeed been the most commonly used refrigerant but R 502 is now replacing it in many applications. Some of the reasons for this have already been mentioned but it is perhaps a useful exercise to spend some time on a direct comparison of the two. In favour of R 502 we have the reduced maintenance requirements (as a result of the lower compressor discharge temperature), the higher flow rate which facilitates control, and a capacity which increases relative to that of R 22 as the operating temperature is decreased. Thus, although the performance of R 22 is comparable to that of R 502 in the cycle described above, more extreme conditions will tend to favour R 502. Moreover, the compression ratio needed for R 502 is less than that required for R 22. The relative performance of the two refrigerants is illustrated in Figure 16.13.

Naturally, the use of R 502 does have some disadvantages. It is more expensive than R 22 (currently by about 60 per cent) but cost of refrigerant is only a small fraction of the installation and running costs of a heat pump. When R 502 is used in hermetic compressors it suffers from poor oil miscibility but this can be overcome by the use of synthetic oils and is not, in any case, severe at normal operating temperatures. Finally, in reverse mode (cooling) operation, R 22 does have a higher capacity.

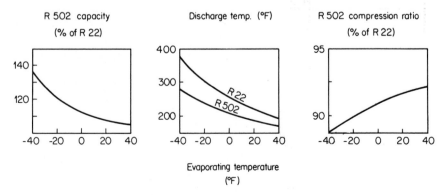

Figure 16.13. Relative performances of R 502 and R 22. (Reproduced by permission of E I Du Point de Nemours and Company Inc.)

In conclusion, it would appear that R 502 is the most suitable refrigerant for use in heat pump-installations in the United Kingdom and Northern Europe.

16.5 Air Heat Pumps

In Section 16.1 we presented an idealized treatment of air heat pumps on the assumption that their operating cycles could be represented by the reversed Joule cycle. The relationship between coefficient of performance and pressure ratio, R, was shown in Figure 16.7. We must now consider how the performance will be affected by inefficiencies in the compressor and in the expansion vessel. In practice the expansion efficiency, η_t, can be made higher than the compression efficiency, η_c, and both can be higher than 0·85 in modern equipment.

It can be shown that the coefficient of performance when such inefficiencies are accounted for is given by the expression

$$CP = \frac{t + \eta_c \dfrac{t - r}{r - 1}}{t - \eta_t \eta_c}$$

where $r = R^{(\gamma - 1/\gamma)}$, $t = T_2/T_3 = T_1/T_4 = V_2/V_3 = V_1/V_4$. (Here the notation is the same as that of Figure 16.6.) The resulting behaviour of the coefficient of performance with pressure ratio is shown in Figure 16.14 for various values of η_c and η_t. The effect of these efficiencies on performance is striking: introducing 10 per cent inefficiencies of compression and expansion degrades the performance by approximately 50 per cent. Also shown in Figure 16.14 is the variation of the coefficient of performance with the temperature ratio, t, for a fixed value of R. It is clear, then, that the best performance can be obtained when R is as low as possible and t as high as possible. However, the situation is not as simple as this in that t and R are linked and depend on the particular temperature range of operation.

Let us now turn our attention to design considerations for heat pumps

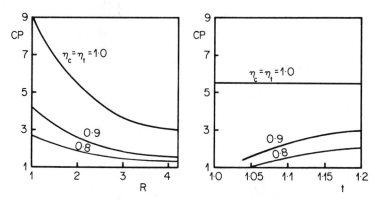

Figure 16.14. Coefficient of performance of an air heat pump as a function of (a) pressure ratio, R, and (b) temperature ratio, t

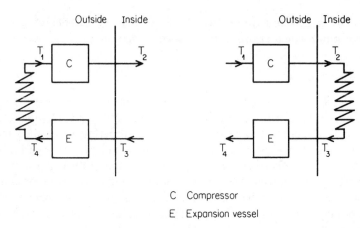

C Compressor

E Expansion vessel

Figure 16.15. Possible arrangements for air heat pumps. (Temperature notation as for Figure 16.6

operating on this cycle. Obviously the main engineering task is to provide for efficient heat transfer from the working substance at the higher pressure and to the working substance at the lower pressure. (This is analogous to the condenser and evaporator arrangement of a vapour compression heat pump.) Two possible arrangements for an air heat pump are illustrated in Figure 16.15. By comparison with the vapour compression heat pumps described previously, it should be noted that in both cases the air circuit is open and only one heat exchanger is required. Designs (a) and (b) differ in that the pressure in the heat exchanger is below atmospheric pressure in the first case but is above atmospheric pressure in the second case. This means that the compressor supplies warm air directly into the building to be heated in design (a). The heat exchanger in this case transfers heat between air flows at atmospheric pressure and below atmospheric pressure. Consequently it must be of larger size for a given heating load than the exchanger of design (b) where the transfer takes place at higher pressure.

A simple reversal of the air flow around these circuits will, of course, permit these designs to be converted into coolers. In addition, design (a) can be modified to act as an air conditioner in this reverse mode operation, that is to purify and humidify the air as well as cool it. This is easily achieved by installing a filter in the inlet stream to the expansion vessel, E. and by arranging that the air enters the building from the compressor via a water spray. Design (b) would involve a separate ventilation flow for air conditioning.

16.6 The Case for the Heat Pump

We began this chapter by pointing out that the heat pump was fairly extensively used in North America but had been singularly unsuccessful in breaking into the market for heating appliances in the United Kingdom and much of Northern Europe. Installations in the United Kingdom are famous through their rarity value. The Norwich Town Hall heat pump uses river water as a heat source to provide a heating capacity of 90 kW with a coefficient of performance of 3·2. The Royal Festival Hall, London, is also heated in this way using the River Thames as the heat source. Gas is used to power the twin centrifugal compressors which provide heating capacities of 2500 kW at a performance figure of 5·1.

Much of the reticence in advocating heat-pump systems undoubtedly stems from a basic mistrust of a relatively new system with such extravagant claims for its performance. Many of the major design problems with heat pumps have only been overcome within the last 15 years and the heat pump's early reputation for unreliability has been long remembered in some quarters. Perhaps, then, it is worth emphazing its attractiveness both economically and environmentally.

From the economic point of view, the main argument in favour of the heat pump is its extremely efficient use of electrical power. A coefficient of performance of four is easily attainable in moderate conditions and a seasonal performance factor (SPF) of three is likely. It can be argued, then, that the efficiency of fuel use in electrical space heating should never be quoted without a three to one improvement implicit in the use of a heat pump. It is true that the heat pump represents a greater initial investment but the long-term economic advantages are obvious.

There are many severe critics of any use of electrical power for space heating. They argue that the efficiency of electricity generation is low (the most modern power stations have efficiencies of 40 per cent but an average figure of 30 per cent is perhaps more accurate) and that space heating should be achieved by the combustion of fossil fuels on the premises. This overlooks a number of factors. Firstly, power stations burn low grade fuel (Chapters 5 and 6) whereas the oil used in domestic central heating systems has been considerably refined. Secondly, the efficiency of oil and gas combustion on the domestic scale may not be as high as is commonly believed. Thus the Westinghouse Energy Utilisation Project (1973) gives the average efficiency of a gas furnace between services as not more than 47 per cent. A gas furnace is considerably more efficient than an oil-fired boiler so it is clear that we must question the standard quotation of 70 per cent for the efficiency of an oil-fired system. This means that electrical heating is barely less efficient than these other systems, and at any rate depends on fuel of a much lower grade. (This has the advantage of releasing the higher grade fuels for the more essential uses of transport.) It should be clear from these figures that electrical heating which employs a heat pump is by far the most efficient scheme.

It is, of course, possible to power the heat pump by other means. Suppose that an internal combustion engine is used to drive a heat pump with a coefficient of performance of four. If we use a 40 per cent efficient diesel engine the amount of heat delivered by the pump will be 160 per cent of the calorific value of the fuel. In addition the exhaust gases from the diesel can be passed through a heat exchanger with, say, 80 per cent transfer efficiency to give a further 48 per cent (0·8 × 60) of the fuel's heat content towards space heating. The grand total of heat delivered is, therefore, just over 200 per cent of the fuel's calorific value! However, it must be admitted that electrical power is more attractive to the consumer and that internal combustion engines require more maintenance than electric motors.

The heat pump is environmentally desirable on two counts: the reduction of power input to the environment which its use entails and its potential for handling waste heat. The former point can be illustrated by the example of a typical house requiring 15 kW of heating in winter. Consider meeting this heating load in three ways, by electrical resistance heating, by an electrically driven heat pump and by an oil-fired boiler. Electrical resistance heating is more or less 100 per cent efficient so the local heat loss to the environment will be 15 kW. However, the efficiency with which the electricity is produced is at best 40 per cent so the overall environmental heating is 40 kW. By using a heat pump the power consumed locally is reduced to 4 kW (assuming a coefficient of performance of about four) which corresponds to an overall environmental heating of 10 kW when the waste heat from the power station is allowed for. Thus there is a four to one advantage in favour of the heat pump. The oil-fired heating system, even at the traditionally accepted figure of 70 per cent, would warm the environment at a rate of 21·5 kW, twice as much as the heat pump. The heat pump can moreover be used to tap the waste heat from industrial plant. The high source temperatures lead to a high coefficient of performance and the use of the heat pump cools the effluent which is ecologically desirable.

This, then, is the case for the heat pump. It is assuredly a strong one, and one that should be acted upon by all the bodies responsible for power generation.

References

Haldane, T. G. N., *J. Inst. Elec. Eng.*, **68**, 666 (1930).
Westinghouse, Energy Utilisation Project, Technical Report 10–30–73, 1973.

Further Reading

Ambrose, E. R., *Heat Pumps and Electrical Heating*, John Wiley and Sons, New York, 1966.
ASHRAE Guide and Data Book: Handbook of Fundamentals 1967; Applications Volume 1968; Equipment Volume 1969; Systems Volume 1970, (American Society of Heating, Refrigeration and Air-Conditioning Engineers).
Davies, S. J., *Heat Pumps and Thermal Compressors*, Constable, London, 1950.

17 Transport of Fuel

Up to now we have considered mainly the technologies of fuel and power, without considering how the product is to reach the consumer. There have been exceptions, the most obvious of which is Chapter 7, where it was noted from the outset that electricity is basically a means of energy transmission. We have also considered thermal transmission in Chapter 15, and pipeline transmission of hydrogen as an adjunct to the 'hydrogen economy' in Chapter 14. However, we have not discussed the transport of fuel, and we have not made any attempt to compare the various methods of energy transmission. In the present chapter we attempt to remedy these omissions, and, as usual, we begin with an appraisal of the technology involved.

17.1 Sea Transport of Oil and Gas

The specialized oil tanker is by no means new; the first real tanker, with a capacity of 3000 tons, was launched in 1886. There have been spectacular increases in the size of tankers in recent years; the most recently completed vessels have been in excess of 300,000 tons deadweight, and the 200,000 ton vessel is now by no means uncommon. This, together with the spectacular accidents which can occur with vessels of this size, has led to an increased public awareness of large oil tankers and their problems.

The reason for this increase in size can be understood quite easily. Until 1967 the Suez Canal provided a very short and convenient route from the major oil producers of the Middle East to Western Europe where the oil was consumed. The maximum size of vessel which could use the Canal fully laden was about 50,000 tons. Furthermore, until about 1950 much of the crude oil produced in the Middle East was refined near the source of production, and the vessels engaged in the trade were called upon to carry both crude oil and refined products. It was, and for the most part still is, uneconomic to carry very large quantities of a single refined product, and this, combined with the restriction of the Suez Canal, kept vessels to around 25,000–50,000 tons.

During the 1950's new refineries were set up in the consuming countries, partly as a result of the nationalization by Iran of the Anglo–Iranian Oil Company, and the consequent separation of transport routes for crude oil and refined products led the oil companies to consider rather larger vessels for the carriage of crude.

The Six Day War of 1967 led to the closure of the Suez Canal, and faced oil shippers with two problems. Firstly, the journey around the Cape of Good Hope increased the cost of transporting the crude oil, and secondly, the increased time taken for the journey reduced the number of voyages which could be made per year by each vessel, so there was a shortage of shipping capacity. Clearly there was a need for new vessels, and for vessels of lower operating cost. Tanker designers, freed from extraneous limitations on vessel size, were able to put technical considerations first and to design much larger vessels. These led to economies, for the following reasons:

(1) The capital cost per ton of capacity tends to be less for larger vessels, so that they represent a better investment than does an equivalent capacity made up of smaller vessels. The important proviso here is that they must be kept busy; clearly two small vessels are more adaptable than one large one, and thus the larger vessel can make its benefits felt most strongly only when the market for oil is stable or expanding.

(2) The complement of crew is not much greater for a large vessel than for a smaller one, especially as most modern vessels have a high degree of automation. A reduction in the number of crew per unit capacity leads to a direct saving in salaries, but perhaps more important, it saves space in the vessel, since fewer men need less accommodation and services. The space thus saved can be used for cargo. There is a further point that crew-men may sometimes be difficult to find, since the Merchant Navy is not the most favoured choice of profession.

(3) All other factors being equal, the speed at which a ship of given capacity and given power can move through the water is proportional to $(\text{length})^{1/2}$ (Barkla, 1964). There is thus an advantage of higher speed to be gained in the use of a long hull, which by reducing the time for a voyage makes a vessel more productive. Alternatively, a given speed can be attained with a reduced engine power, which leads to fuel economy.

Despite the foregoing considerations, there are other factors involved which tend to limit the size of vessels. The most obvious is the engineering problem of manufacturing vessels over 300 metres long able to withstand the severe stresses caused by heavy seas. Such problems do not seem to be the limiting factor, however; it seems that geographical constraints are likely to be more important. For example, there are plans to widen and deepen the Suez Canal to encourage its use by larger vessels. Present plans would allow 200,000 tonners to use the canal on their southward journey when they are carrying ballast rather than a full cargo, and even though they would have to go via the Cape when laden, the saving of time for the round trip would be attractive, especially at today's high interest rates. Very large vessels would still be used for other routes, but for the trade between Western Europe and the Middle East a size of 200,000 tons may become the normal maximum. Natural waterways provide another constraint, particularly the relatively confined and very busy English Channel by which vessels reach European ports such as Rotterdam and Canvey Island (the alternative route via the north coast of Scotland, as well as being much further, is hazardous in winter). Furthermore, the docking of very large vessels requires deep water at a reasonable distance from the shore, and for environmental reasons it is desirable for this to be adjacent to existing industrial complexes. Few sites offer such a combination.

The constraint which has received most publicity, although it is actually less well-founded than those mentioned previously, is the problem of accidents. When a very large vessel (such as the well-known Torrey Canyon) suffers a serious accident, the scale of damage caused by the released oil can be immense. In fact, the probability of a large vessel suffering an accident is no greater than that of a smaller vessel, and since one large vessel replaces several smaller ones, the actual number of accidents should be less. On statistical grounds the annual pollution caused by accidents should not depend on the size of the vessels being used, but only on the amount of oil being carried. This argument, although oversimplified, is nevertheless not unreasonable, but it does assume that the safety and manoeuvrability of large vessels is kept at an adequate level. This brings us to a study of safety factors in the design of large tankers.

Stopping a large tanker is by no means easy. Ships are not usually equipped with brakes, and the usual way of stopping in emergency is to run the engines at full speed astern. A large tanker is a very efficient vessel, requiring only relatively small engines, but because the engines are small, their ability to stop the vessel is rather limited. A typical 200,000-ton vessel, travelling at 16 knots, takes about 16 minutes to stop completely, and travels about 2·5 miles in so doing. While this is acceptable in open water, it is less satisfactory in confined spaces. It would be quite uneconomic to provide the vessel with greatly enlarged engines which would normally never be used, and an alternative system using brakes would require considerable development work. Braking a large vessel is not at all easy, mainly because of the very large forces involved. Nevertheless, various designs are being considered for braking devices which apply acceptable forces to the ship's hull. Parachutes with diameters of about three metres, slung over the side and trailing in the water, have been tested in Japan, and have been successful in reducing the stopping distance to half its previous value.

Tied up with the problem of stopping is the problem of steerage way, that is the minimum speed which a vessel needs to maintain good steering. This can be quite high, as much as 10 knots in some cases, but it can be reduced by attention to hydrodynamic design in the vicinity of the rudder. An alternative possibility is the use of small propellers mounted near the bow to provide sideways force for manoeuvring; this is done already in some specialized ferry boats working in small harbours, but their use in very large vessels of high efficiency presents serious problems.

A quite different safety problem is the danger of explosion when vapour from the cargo mixes with air. This is not likely to happen when the tanks are full, but during unloading, and especially during the cleaning of tanks (Lawes, 1970) there is a significant risk. It is possible to let nitrogen or inert gas into the tanks as they are unloaded, and this can reduce the risk appreciably.

The oil spills which result from accident are spectacular, but they are appreciably less than can result from the cumulative effects of careless tank cleaning operations (Chapter 15). Cleaning the tanks is essential in a vessel carrying refined oil products to avoid contamination of the new cargo by the previous one, but it is also necessary to a limited extent in crude-oil carriers because some maintenance of the empty tanks is often done during the return journey and it would be difficult to undertake such work without first removing the crude oil. There is also the need to carry ballast in the form of sea-water when the ship is not carrying cargo, and if this were carried in oily tanks it would become contaminated and when discharged at the port prior to loading of cargo it would contaminate the shore. Consequently a method of cleaning the tanks is needed, and for environmental reasons it must not put out any oil while near land, and preferably should not release oil on the high seas either. This is usually achieved by washing the tanks with water from high-pressure sprays, and collecting the dirty water in a slop tank where the oil, having a density less than water, will float. The water is then discharged, and the oil is retained in the ship.

Most of the foregoing account has been applicable only to crude-oil tankers, for as we noted at the beginning of this section, it is unusual to transport huge quantities of a single refined product. This is partly for the simple reason that few areas require a quarter of a million tons of one product, say gasoline, at one delivery, and also because many of the refined products are highly dangerous; gasoline is again the obvious example. As we saw earlier, it has been common since the 1950's to site refineries in the consuming countries, but in the aftermath of the oil price crisis, the oil producers have found themselves with a large surplus of capital, and it is likely that some of this may be invested in refining capacity in their own territories. This may lead to an increase in demand for refined product carriers, and thus we may yet see a decrease in the average size of oil tankers, especially with the reopening of the Suez Canal.

One very specialized example of a refined product carrier is the liquefied methane tanker. The shipping of liquid methane is now well established (Sivewright, 1972) despite its low boiling temperature of 113 K (at one atmosphere). The first really large-scale operation of this type began in the late 1950's when the North Thames Gas Board investigated the possibility of bringing natural gas from Algeria to Britain, and ordered two special vessels from Vickers in Barrow and Harland and Wolff in Belfast. The amount of natural gas carried in this way is still vastly less than the traffic in other petroleum products, but with the shortage of natural gas in the United States it is likely to grow quite rapidly.

Shipping a cargo of very low specific gravity (0·45), low temperature and great inflammability is not straightforward. The main problem is adequate insulation, which is necessary for two reasons. Firstly it reduces boil-off, which in a good design is as little as 0·25 per cent per day and which is used to fuel the ship's boilers. Secondly, it prevents contact with the ship's hull, which would suffer brittle fracture if allowed to get cold. A secondary problem, which arises from the low density of the cargo, is the design of tanks which make use of as much hull space as is consistent with safety.

17.2 Pipeline Transport of Oil and Gas

Crude oil is reasonably fluid at ordinary temperatures, and it can be transported by pipeline relatively easily. For transport between the wells and the tanker terminal, and at the other end of the journey from the tanker terminal to the refinery, pipelines are of course the only practicable means. They can also be used in favourable circumstances for the whole of the journey from the wells to the refinery, so replacing transport by ship.

The economics of pipeline operation are quite different from those of batch transport systems such as tankers. They are totally inflexible in that they cannot be rerouted to new sources or new consumers without major engineering, and as they involve great capital investment they are only economic when used at or near their capacity, so thay cannot even be used to supply variable demand. Because of these problems, pipelines are only useful when both the supply and the market can be relied upon.

Over a similar route, transport by pipeline is rather more expensive than transport by sea, but there are some circumstances where large ships are at a disadvantage, and in these cases it may be advantageous to consider installing a pipeline to supplement if not to replace ships. For example, because of the great distance involved in going round the Cape of Good Hope and the limitations of the Suez Canal, it is economic to run a pipeline from the Middle East to the Mediterranean to supply some of the Western European demand.

As well as crude oil, it is possible to carry refined products by pipeline, and even to carry several different products in the same pipeline in batches. It might appear to be necessary to separate the batches by solid plugs, and this is indeed done in some cases, but if the flow of liquid is kept turbulent, the mixing between batches is very slight. In any case, some degree of mixing can usually be tolerated.

Gas can be transported by pipeline either in the gaseous or liquid state. Liquid is easier to pump, and because of its higher density a greater fuel value can be transported by a given pipeline. There is a further advantage to pumping a liquid rather than a gas in that there is a more favourable relationship between pressure and distance in the case of a liquid. A liquid flowing in a streamlined manner in a pipe flows at a rate V (in volume per unit time) given by Poiseuille's formula

$$V = \frac{P\pi r^4}{8\, l\eta}$$

where P is the pressure drop along a length l of pipe, r is the pipe radius and η is the liquid viscosity. Consequently, the pressure in the pipe will be a linear function of distance. Admittedly, the flow of oil in a practical pipeline is unlikely to be truly streamlined, and indeed in the case of product pipelines it is desirable, as noted above, to avoid streamline flow. However, it is possible to apply a modified form of Poiseuille's formula in the case of turbulent flow. By contrast, a gas is compressible, and it is not possible to express flow rate solely in terms of volume flow. Instead it is the mass flow rate which is important, and by suitable choice of units it is possible to express mass in terms of the product of pressure and volume, pV, provided the temperature is constant. Poiseuille's formula is then modified to

$$p_1 V_1 = \frac{(p_1^2 - p_2^2)\pi r^4}{16\, l\eta} \text{ (Newman and Searle, 1948)}$$

where p_1 and p_2 are the pressures at the start and end of a pipe of length l. From this it is evident that there is a dependence of flow rate on the square of the pressure, and consequently the pressure in the pipe will be a parabolic function of distance. The correct siting of pumping stations is thus much more important in a gas pipeline than in a liquid pipeline.

For the above reasons it is desirable to transport in the liquid state, but it is not always practicable to do so. Propane and butane have critical temperatures of 369 and 407 K, respectively, and since these are both well above the 'normal' atmospheric temperature, they can be liquefied by compression alone. Pipeline transmission as liquid is thus relatively straightforward, provided that suitable valves can be used to keep that pipe under the necessary pressure. Methane is less easy to handle, for its critical temperature is 171 K, well below normal temperatures, and in order to remain liquid it must be cooled below this temperature. Even then, the pressure required to retain the liquid state is 46 bars, and to produce liquid at normal atmospheric pressure requires refrigeration to 113 K. Neither of these prospects is likely to appeal to the pipeline engineer, and consequently it is normal to handle methane as a gas.

Pipes for large pipelines are made from rolled steel plate, joined by electric arc welding. If possible they are buried, partly for aesthetic reasons and partly to avoid problems of thermal expansion. Pumping stations are necessary at intervals, depending on the pipe diameter, but typically every 80 km. Gas pipelines are rather more demanding in pumping requirements, and more frequent stations are needed.

Submarine pipelines are constructed in a similar way to land pipelines, but the construction is done on special barges, and the pipe is lowered continuously from the barge to the sea bed. Once laid it can be buried in a trench created by a water-jetting machine which runs along the pipe blasting away the sea bed beneath it.

Corrosion is a problem in both submarine and land pipelines. It can be reduced by coating the outer surface with bitumen, and attaching sacrificial anodes of zinc to the pipe at suitable intervals.

17.3 Transport of Coal

As it is a solid, coal is much less amenable to large-scale transport than are the fluid hydrocarbons. Sea transport is, of course, perfectly feasible, although as we noted in Chapter 5, the major consumers of coal are also the major producers, and the volume of international trade in coal is much less than in the case of oil. The vessels engaged in this trade are for the most part small, of the order of 1,000 tons.

On land there is at present no alternative to rail and road transport, both of which are comparatively expensive, although the use of automatic handling equipment helps to reduce manpower, and equally important, demurrage charges (the charge made by the haulage company for delaying its vehicles while waiting to unload). In a few cases, such as where a power station is adjacent to a coalmine, conveyors can be used successfully, but this ties the power station to using coal from that mine, and can be unfortunate if unforeseen problems cause increases in mining costs or decreases in supply. Some experimental work has been done on the pipeline transport of pulverized coal as a slurry in water, but erosion has proved severe, and removal of water from the slurry has presented problems. For all of these reasons it is desirable to minimize as far as possible the distance over which coal is transported, and the siting of large coal-fired plant such as power stations in close proximity to mines is likely to continue.

Brown coal and peat have a lower calorific value than hard coal, and it is highly desirable in such cases to minimize the distances involved in transport; this is especially true of milled peat, where as we saw in Chapter 5, about 50 per cent of the weight is moisture. Power stations burning these low-grade fuels are thus invariably sited at the mine or bog. In cases where transport is unavoidable, it is desirable to reduce both the water content and the volume by briquetting prior to transport.

The possibilities of inland water transport for heavy cargo such as coal have been raised occasionally, notably by conservation societies. The applicability of canals is limited by the small size of vessels which can negotiate many of the older waterways and this is particularly true in Britain. In Europe the availability of natural waterways such as the Rhine gives canal transport certain advantages, but clearly it is not universally applicable.

17.4 Transmission of Energy

An alternative to transport of primary fuel is conversion to a secondary fuel such as a gas, or to a form of energy such as electricity or heat. Such a conversion is bound to be less than completely efficient, and must, therefore, involve some waste of fuel, but as we have seen in earlier chapters there may well be a net saving owing to the greater efficiency of large-scale central plant. Moreover, there may be environmental advantages; for example, whatever objections one may have to a high-voltage transmission line, it is undoubtedly less unfortunate than the possible alternative of a stream of heavy trucks carrying coal. Several different energy transmission systems have already been discussed in earlier chapters; electricity in Chapter 7, hydrogen in Chapter 14 and thermal in Chapter 15.

Many authors have attempted to present comparisons between various alternative means of energy transmission. Unfortunately it is exceptionally easy in this type of calculation to end up with an answer favourable to one's own viewpoint; an example of such a case is to be found in Chapter 15 where, in order to compare the economics of heat and electricity transmission, the cited author has taken the electricity transmission system as a 50 kV underground cable, which is a rather unlikely and expensive choice. In view of this difficulty, the present authors intend to avoid the issue. Let it suffice to say that the problem is undoubtedly susceptible to proper economic analysis *in a given situation*, but the conclusion reached may be inapplicable in different circumstances.

References

Barkla, H. M., *Bulletin Inst. Physics*, **15**, 57 (1964).
Lawes, G., *New Scientist*, 19 March, 1970.
Newman, F. H. and V. H. L. Searle, *The General Properties of Matter*, 4th ed., Edward Arnold, London, 1948.
Sivewright, S., *New Scientist*, 15 June, 1972.

Further Reading

Goss, R. O., *Studies in Maritime Economics*, Cambridge University Press, Cambridge, 1968.
Hubbard, M., *The Economics of Transporting Oil to and within Europe*, Maclaren, London, 1967.
Lowson, M. H., (Ed.), *Our Industry—Petroleum*, British Petroleum Company, Ltd., London, 1970.

18 Economic and Social Aspects of Fuel and Power

No account of the energy situation would be complete without some reference to people and to money. The basic reasons for the existence of any industry are that people need its product and that money can be made by supplying it. Economics is of particular importance in the energy industries, for the choice between different sources of power is often based largely on present and predicted costs of producing the power required. Social consideration are of similarly great importance, for the large scale of investment, the environmental impact and the number of jobs involved in the energy industries are such that they affect everyone in the community.

The authors do not claim any expertise in the social sciences or economics, and the reader should bear in mind that what follows is not necessarily authoritative. It is, rather, a scientist's view of some selected topics, chosen because the authors find them interesting or relevant to their own experience.

18.1 Coal

In most of the older industrial nations of the West, the coal industry has suffered a continuous stream of problems for several decades. Several authors have made detailed studies of coal mining in particular countries, including Henderson (1958) on the American coal industry and Simpson (1966) on the British coal industry. One of the most pressing difficulties has been matching supply and demand. There was an acute shortage of coal in Britain and throughout Europe in the post-war years from 1945 until about 1956. In the case of the United Kingdom, the government attempted to remedy the situation by rationing domestic coal (which encouraged consumers to use unrationed electricity and gas, both of which were coal-based) and forbade miners to leave the industry for other jobs (which is unlikely to have improved productivity). Increased mechanization was not as helpful as it might have been, since the biggest shortfall of supply was in large coal for domestic use and railways, while mechanized coal cutting tends to produce small coal suitable for power stations. To remedy the shortage of large coal, opencast mining was expanded, and a substantial amount of both large and small coal was imported. During 1954 the government authorized the conversion of 16 power stations to oil firing, and in 1955 a major programme of nuclear power stations was proposed, both measures being designed to help solve the coal shortage. From 1956 onward, however, the demand for coal fell, not for the traditional reason of an economic recession, in fact there was a boom at the time, but because consumers chose oil instead. Despite the Suez crisis, which temporarily drew attention to the problems of reliance on Middle Eastern oil, the coal industry continued to lose to its competitors. The shortage of coal quickly became a surplus, especially of small coal, and the government hastily modified its oil-conversion programme for power stations and slowed down the nuclear programme. Yet this was not enough, and it became necessary for the Coal Board to curtail output. To some extent this was done by phasing out a few high-cost pits, reducing overtime and by the method of manpower reduction described by the inelegant phrase 'natural wastage', but these measures were not enough. A possible solution would have been to close a larger number of high-cost pits, with the benefit of making the industry more competitive, but unfortunately most of these pits were located in areas where mining was the only major industry, and massive social upheaval would have been caused by the resulting unemployment. A more acceptable solution, and the one actually adopted, was a cut-back in opencast mining, for the workers involved were not usually traditional miners and were, therefore, less likely to suffer as a result (or so it was thought) and further, because the productivity of opencast was better than underground mining, it was possible to reduce output by the required amount without creating quite so much unemployment. Unfortunately, opencast mining was cheaper, precisely because of this higher productivity, and the decision inevitably had unfortunate effects on the Coal Board's financial position. The situation continued in this form, exacerbated by the arrival of natural gas and nuclear electricity and helped only partially by a government policy of encouraging the growth of alternative industries in some of the threatened mining areas, until the latest crisis. Now the Coal Board and the British Government are proposing to expand the mining industry, new seams are to be developed, and the future of the work-force seems secure, but with the rather sorry history of coal in recent years, one is tempted to wonder if this is merely another short-term phase. The major problem is that it is extremely difficult to expand or contract underground mining rapidly. To open a new working requires heavy capital investment and several years of preparation, and since mining is not an attractive industry to those who have not been brought up with it, it can be difficult to attract enough men when necessary. It is not an easy decision to close a face at a time of contraction, because the prospect of making miners redundant is unwelcome, and further, flood damage may make subsequent reopening impossible.

The cyclic nature of demand is not the only problem facing the industry. Perhaps more difficult are problems associated with its work-force. Mining is a labour-intensive industry, particularly in those countries where most of the coal is obtained underground. This is illustrated by figures issued by the National Coal Board (1974) which indicate that wages and related costs accounted for 54 per cent of the cost of coal in 1972–1973. Inevitably, therefore, industrial disputes are a source of difficulty, for any increase in wages causes a significant rise in price. Furthermore, strikes can cause loss of production for far longer than their immediate duration. A related problem, absenteeism, has caused disquiet from time to time. Many of these labour problems arise from the undoubtedly unpleasant conditions of work, and they may perhaps become of less significance as truly remote control of cutting and loading machinery becomes more widespread. Another labour problem, which is sometimes forgotten by those outside the industry, is health and safety. Underground mining has always had a relatively poor record on both these counts, but major efforts have been made to improve the situation. The success of these efforts can be judged by National Coal Board figures which are reproduced in Table 18.1. The improvement is substantial, but it is still considerably more dangerous than, say, university teaching.

Table 18.1 Health and Safety figures for underground mining

Diagnoses by pneumoconiosis medical panels

Year	1954–1958	1959–1963	1964–1968	1969	1970
Rate per 1000 men	6·1	4·8	2·2	2·0	2·7

Serious accidents and deaths

Year	1947	1957	1967	1971
Number of accidents	2446	1900	1000	598
Number of deaths	618	390	140	92

(Figures derived from National Coal Board (1972). Reproduced by permission of the National Coal Board.)

18.2 North Sea Oil

One of the most significant areas of oil exploration in the World at present is the North Sea. Initial estimates of output from the British sector were guardedly small, but current figures seem likely to turn Britain into an oil exporter in the foreseeable future. This has inevitably involved a considerable upheaval in those parts of the country most directly affected, particularly the east coast of Scotland, and the rural west coast north of the Clyde. The problem has been most acutely felt on the west coast, because there has been very little industrial development in these areas, and there has been much concern about the effect on natural beauty and upon the life-style of the people.

The concern about natural beauty is certainly justified, as all large scale industrial construction is unsightly. However, it may be possible to confine certain 'movable' developments to areas of existing industrial activity such as the Clyde. Of course, not all development is 'movable'; for example, a shore terminal for a submarine pipeline must be as near as possible to the producing wells to minimize the very high capital cost involved, but such developments are probably in a minority. Spreading out the development has other advantages in that it places less localized demands on industrial infrastructure such as roads and power supplies, and it also enables a larger section of the population to benefit from increased employment.

The concern about the effect on people's life-style is probably rather less well justified, for if development is properly planned, the local population may well benefit from the increasing availability of sophisticated employment. Areas such as the west coast of Scotland have suffered for centuries from a lack of skilled or professional employment, and it has long been traditional for young men to leave for the industrial south, a trend which has removed much native talent from the areas, as well as distorting the sex ratio. On the other hand, excessive development in a rural area will generate shortages of building land and housing, forcing up prices, and will overtax the existing social services such as education and health. This argues not for a ban on development, but rather a careful involvement of government in planning and licensing.

To some extent it is necessary to distinguish between capital-intensive industries such as oil extraction and refining, and labour-intensive industries such as the construction of oil rigs. Labour-intensive industries are normally regarded as desirable in areas of high unemployment; however, in the present case this conventional viewpoint may not be applicable. The actual number of underemployed and unemployed local people in any one rural area is probably insufficient to meet the labour requirements of a major labour-intensive industry, and it becomes necessary to 'import' workers from beyond daily travelling distance, and to accommodate them in sometimes less than ideal temporary living accommodation. This clearly presents social problems. A capital-intensive industry, in contrast, employs relatively few people, and often needs highly-trained and well-paid staff. It is true that the majority of such staff are likely not to be local people, but there seems to be a tendency for former emigrés to return in response to the demand, and thus many of the staff may not be complete strangers to the district. A more tangible benefit to the community arises through the arrival of new money, for the employees are likely to spend a substantial amount of their income in the local shops and on local service industries. Such spending can bring substantial benefits to the existing members of the community without requiring any significant change in their life-style. One can argue, therefore, that North Sea oil is likely to bring more benefit to the local community when the current phase of exploration has moderated, and the wells are quietly producing.

One should also distinguish between these new industries and those which are extensions of traditional ones. A good example of the latter is servicing of offshore drilling rigs, which involves small ships carrying not only 'deck and engine stores' but also fuel for both machinery and men. This can be achieved by extension of existing ports and shipping facilities.

18.3 Peat and Rural Development

In developing native peat resources, the Irish peat authority, Bord na Mona, has been principally concerned with providing an economic competitor to imported fuels such as coal and oil. A second important concern has been the development of a new industry in parts of the country which have suffered for centuries from depopulation and from lack of satisfactory employment other than small-scale farming. The programme seems to have been particularly successful in this respect, as has been demonstrated by the work of Bristow and Fell (1971), who have quantified the costs and benefits in considerable detail.

The labour force is about 6000 (Bord na Mona, 1972) and the majority are employed at or near the bog areas. Housing was a major problem in the early stages, and some rather unsatisfactory temporary accommodation was erected, but this has now been superseded by small housing estates, most of which are located at existing centres of population. The exploitation of the bogs for fuel is expected to continue for several decades, and training schemes are in operation to ensure a continued supply of trained personnel. There are normally about 400 trainees, some of whom stay in the industry, but others leave to take up other employment, by which mechanism the peat industry operates as a means of technical education.

One problem which has still to be solved is that not all the inhabitants of the bog villages who wish to work can find jobs in the peat industry. This is particularly true of the young women, for whom the number of jobs is rather small. This type of problem arises frequently in one-industry towns, and it is normally solved by commuting, but this is ruled out by the remoteness of the bog areas. The government is trying, with limited success, to induce other industries into the areas.

One obvious problem for the future is what happens when the peat is exhausted. It has been found that provided a few feet of peat is left in place at the bottom of a bog, it is possible in most cases to grow crops of suitable types. The most successful crop appears to be grass for cattle-grazing, although in some favourable cases vegetables have been grown successfully, and in some less favourable cases it has still been possible to resort to coniferous forestry. It is too early to say if all the crop-growing ventures will prove successful. However, there is the prospect of a permanent industry to replace peat as a source of income for the community, and there is the further benefit that the area of the nation's productive agricultural land is increased.

18.4 Costs of Electricity Plants

In comparing different type of electricity generating plant, one must make proper allowance for the differences in capital (installation) costs. In a careful theoretical study, Turvey (1968) has presented mathematical analyses of the problem, but for those readers who prefer to avoid mathematical economics, the situation can be outlined as follows. Conventional fossil-fuel generating stations are cheap to install, but have relatively high fuel costs. Nuclear stations have a capital cost per kilowatt of installed capacity which is roughly twice as high as fossil-fuelled stations, but their fuel costs are much lower. 'Natural' power stations such as hydro and tidal installations have even higher capital costs, but have fuel costs which are effectively zero. Solar collectors, at the present state of technology, are exceptionally expensive to install, but again the fuel cost is effectively zero. There are, of course, standard accounting procedures which are able to incorporate these factors, and while these may differ in detail between public and private industries, the general principles are the same. For example, the British nationalized fuel and power industries have used discounted cash-flow techniques in their budgeting for several years. However, it must be remembered that many of the rules by which the accounting is done are to some extent arbitrary. In the British nuclear power programme, a decision was taken to give the stations an economic life of only 20 years, compared with the 30 years given to conventional plant; the 20-year figure may or may not prove correct, but the arbitrary decision to choose it gave nuclear stations a significant handicap compared with fossil stations. The interest rates which the government requires the nationalized energy industries to earn on their capital investment are presumably related to prevailing commercial rates, but the exact figure is still arbitrary, and it is always open to the government to write off some of the debt (as it has done on occasion with the railway network) whenever it wishes to offer assistance. For these and other reasons it is extremely difficult to arrive at a 'correct' costing when comparing different types of plant, and it is even more difficult when attempting to compare, say, electricity with gas. This is not to say that the comparison should not be attempted, but rather that having made it one should treat it with reserve. In the long term it is possible that social cost–benefit analysis may well be more useful than conventional accounting techniques.

18.5 Fuel Saving

There are important social as well as economic considerations in reducing energy demand. It may appear to be desirable, for example, to restrict fuel consumption as much as possible, in order to save foreign exchange, but this may generate severe unemployment in all industries whose existence relies on use of fuel. The social and economic disruption caused in this way could easily outweigh any benefits and would in any case be politically unacceptable in a democracy. Conversely, there may be political considerations which force governments to introduce fuel-saving measures which are in fact more trouble than they are worth. An example of such a case is the vehicle speed restrictions introduced in December 1974 by the United Kingdom government. An official report (CPRS, 1974) estimated the saving in oil imports resulting from an *overall* speed limit of 50 instead of 70 miles per hour to be as little as 0·3 million tons per year, and the effect of the measures actually adopted, in which only non-motorway roads were restricted, while the motorway speed limit was kept at 70 mph, is inevitably even less effective than that. The reason for its introduction is probably its psychological effect, and also its likely effect on accidents, rather than any real expectation of fuel saving.

A further addition to the complexity of the problem is provided by regional differences. It may appear obvious to the inhabitant of more urban parts of the World, such as the south-east of England or the eastern seaboard of the United States, that significant savings in fuel and important improvements in the environment could result from restrictions in the use of private automobiles, together with greater use of public transport. It is indeed obvious, but it is even more obvious that in order to achieve any significant benefits there must be a good and reasonably well utilized public transport system. Such a system already exists in metropolitan areas, but in rural areas it does not. Even if it did, it would not necessarily be economic because of the low population density, and it might not even result in saving of fuel, for a bus travelling empty apart from the driver has a fuel consumption per passenger-mile of infinity! It is important for any nationally-imposed measures to be fair to all regions, and measures designed to restrict private transport are unlikely to achieve this aim.

18.6 Conclusions

To summarize, it is the authors' view that there is more to energy policy than supplying the demand as cheaply as possible, and more to it than promoting fuel-saving policies. As well as the consumers' interests, it is possible to take into account the interests of the producers as well, and to use the siting of major projects as a useful form of regional development. Badly planned projects can damage the quality of life, and lead to the familiar criticism of large-scale technology, but if care is taken it is perfectly possible for great benefits to accrue to society quite apart from the primary aim of producing power.

References

Bord na Mona, *The Moving Bog*, Bord na Mona, Dublin, 1972.

Bristow, J. A. and C. F. Fell, *Bord na Mona—a cost-benefit study*, Institute of Public Administration, Dublin, 1971.

Central Policy Review Staff, *Energy Conservation*, H. M. S. O., London, 1974.

Henderson, J. M., *The Efficiency of the Coal Industry*, Harvard University Press, Cambridge, Mass., 1958.

National Coal Board, *Black Diamonds Silver Anniversary*, National Coal Board, London, 1972.

National Coal Board, *Facts and Figures*, National Coal Board, London, 1974.

Simpson, E. S., *Coal and the Power Industries in Postwar Britain*, Longmans, London, 1966.

Turvey, R., *Optimal Pricing and Investment in Electricity Supply*, Allen and Unwin, London, 1968.

Further Reading

There is a wealth of literature on economics of fuel and power, and a limited amount on social aspects. The reader is recommended to read the books by Bristow and Fell, Henderson, Simpson and Turvey listed in the references, but this list is by no means exhaustive. A particularly interesting treatment may be found in:—MacKay, D. I., and G. A. MacKay, *The Political Economy of North Sea Oil*, Martin Robertson, London, 1975.

19 Energy Policy

Throughout this book we have tended to concentrate on the practical aspects of energy supply and use. Apart from a few instances, we have made little reference to policy decisions. The single fact that underlines all future energy policies is that our existing pattern of energy generation and use is possible for only a relatively short period. Oil and natural gas supplies are limited and may last for only a few decades. Although it may be possible to extend their use for a century or more through further discoveries and by increased energy conservation measures, we must operate on the premise that the 'age of oil' will soon be over. Coal will be a primary energy source for considerably longer, while nuclear fission will be only a very short-lived contributor unless the breeder reactor is fully developed. Unfortunately, the undoubted hazards of handling plutonium make the security and even the morality of fast breeder reactor development problematical.

The extreme short-term developments can be clearly seen. Governments will attempt more and more to control the exploitation of the oil reserves within their territorial boundaries. This has already begun with the large government involvement in the OPEC countries and the fiscal controls imposed by governments with a claim on North Sea oil and gas fields. Meanwhile there will be attempts to reactivate and expand coal mining in those countries where it has been allowed to run down. Nuclear power programmes will be expanded and some breeder reactors will be brought into service—the Dounreay Prototype Fast Breeder Reactor, the United Kingdom's first commercial breeder power reactor, was commissioned on March 3rd, 1974. In the longer term, beyond 1985 or so, forecasting becomes more difficult. It is complicated by concern centred on the unpleasant side effects of the nuclear fission programme and by uncertainties over the development of a viable nuclear fusion power system. Undoubtedly, however, coal will continue to provide the basis of power production for some time to come. Reserves will probably outlast the entire third millenium, though several estimates show coal becoming scarce by the year 2500.

The current high price of oil is the factor most likely to lead to a change of attitude on the part of industrialized nations. The huge import bills that must be faced if oil imports are not reduced may provide the incentives for governments to encourage research into other possibilities. Those governments with the courage to invest large sums in somewhat imaginative schemes may well prove to be the wise ones in the long run. Success is close at hand and the rewards are colossal. There are many other slightly less spectacular power sources which ought to be developed. In principle, solar power is comparable in capacity to fusion power, and the technology for tapping it is making significant progress. On a less widespread scale it may be possible to further reduce our dependence on fossil and nuclear fuels by developing secondary natural resources such as water power. In Chapter 13 we saw that most of the available hydropower capacity is already being exploited in the developed countries. Nevertheless, there is vast scope for such exploitation in Africa and in South America. The technology exists for making use of tidal power and this should be encouraged wherever there is a suitable location. Wind power will probably only be capable of small-scale power generation, but the associated wave power is deemed by the United Kingdom Government to be worthy of intensive study. Hand in hand with this exploitation of new power sources must go the development of still more efficient generation based on traditional fuels.

The most significant reappraisal of policy may well concern our use—or misuse—of energy and our attitude to waste. Waste material should be recycled as much as is economically possible. Refuse can be burnt both to provide heat and to generate electricity, though it may well prove to be the case that paper, for example, will be more important for its organic chemical content than its heat content. The heat pump will become an important heating device with blown-air heating systems taking advantage of the high figures of merit for small temperature differentials becoming more common. Efforts to save fossil fuels must be encouraged, if only because this provides breathing space for discussion as to the best methods of replacing them as power sources.

While on the subject of waste, it is worth noting that the electricity industry has been remarkably unpopular of late among a small but vocal section of the populace who have cited the relatively low thermal efficiencies of power stations in their argument against the use of electricity and the existence of a public electricity supply. Further, there have recently been numerous examples of the exploitation of the power network as a means of furthering the ends of those engaged in industrial and political disputes, leading to power cuts or voltage reductions. This has had the inevitable effect of accentuating the ill-feeling against the supply authorities. Among the more reasonable reactions to the problem has been the upsurge in demand for standby generators for industries where even a few hours loss of supply can be serious. Among the less reasonable reactions has been a romantic attraction to 'self-sufficiency'. In view of this, it is worth spending a little time examining if there should be a public electricity supply, and studying its advantages and disadvantages.

We have repeatedly emphasized throughout this book that the efficiency of electricity generation is mainly limited by thermodynamic constraints, rather than by the incompetence of the generating authorities. In any case, the efficiencies of other technologies is often not much better, and sometimes significantly worse (see Chapter 16), and further, they tend to require higher-grade fuel. Since the major objection to electricity has been its use in space heating, it is worth recalling that by means of a heat pump even this objection can be overcome. Thus there is no necessity to object to electric space heating, and certainly not to its other applications in which it is essential.

The most pressing reason for a *centralized* electricity generation and transmission system is a statistical one, namely the saving in plant and associated capital investment which is possible because all consumers do not need electricity at maximum demand at the same time. The second reason is a question of convenience in the widest sense of the term. Any electricity system requires some attention, and since at present the only really feasible types of generator are mechanical, the amount required is considerable. By locating as much of the equipment as possible in central generating stations, it is possible to employ a few specialist engineers to look after the generators, while the inherently more reliable transmission equipment can be serviced by mobile teams of specialists, all of which results in much saving of time and expense compared with individual servicing of small plants. Further, electricity plants of the size required by the average household are noisy and space consuming, and not the sort of thing one would wish to have in the garage or garden shed. Finally, and perhaps most important of all, the electricity supply is of such great importance to the household that a failure for an extended period is highly inconvenient. A failure of supply is much less likely in a mains supply than in a private supply and, moreover, when it occurs, it is the responsibility of someone else to put it right, which comes as a tremendous relief to the average householder.

The foregoing argument has presented the case for a certain amount of centralization, perhaps as far as an individual town or county, but does not account for the need for national transmission systems. Why, then, has the electricity system developed into its present structure, with giant power plants of 1000 MW and more linked nationally and internationally? The first reason is related to the nature of the fuels used for electricity generation. Small plants use refined oil, usually gas oil, which is expensive in relation to its energy content. To deal effectively with the cheapest fuels, such as low-grade coal, residual oil and nuclear fuels, large plants are essential. The second reason is related to thermal efficiency; large steam turbines are invariably more efficient than small ones. It might be argued that the importance of this can be overemphasized, for the 40 per cent efficiency of a large turbine can be reached by a diesel engined plant of much smaller size. However, the diesel plant requires more maintenance, is noisier and requires (usually) a more expensive fuel. The third reason is the dramatic fall in capital cost per unit generating capacity which occurs when plant is enlarged. This economy of scale is most noticeable in nuclear stations, and since the capital cost is rather higher than for fossil-fuel plant, it is of great importance to instal the largest possible size of nuclear reactor. A fourth reason, of particular importance for coal-fired plant, is related to the location of the generator close to the fuel supply. It is considerably cheaper, as well as environmentally desirable and more economical in labour, to transmit electricity to consumers in the city from power stations located close to the coal mine, than to transport coal from the mine to power stations close to the city. The same is in general true for oil-fired stations, though to a lesser extent because transport of oil is usually cheaper than that of coal. For nuclear stations the cost of fuel transport is negligible, but it is still deemed desirable to locate the power station away from populated areas in order to allay public fears about safety.

This is the case for large power plants, and for their siting at places other than centres of consumption, thus necessitating transmission systems from the plant to the consumers, but there is not yet a convincing case for interconnection of generating plants on a national and international grid system. To understand this point we must examine what are possibly the most fundamental reasons of all for the present structure of the electricity industry, one primarily economic and one primarily technical though with an economic spin-off.

The generating stations in a network are used in order of running cost. If one imagines two adjacent transmission systems which are not interconnected but which might easily be so, it is likely that at any moment the cost of generating a small extra amount of power in either of the two systems will be different, so there would be a net saving of cost if the two systems were interconnected and that small extra amount of power generated in the one with the lower cost at that time. The effect may appear trivial, and indeed it must be admitted that the saving per unit is not large. However, such is the scale of the industry that very large savings can accrue in a very small time, soon offsetting the cost of the link between the systems.

In order to be ready for an expected or an unexpected increase in demand, a certain amount of generating plant must be rotating and synchronized to the mains but not generating, or at least not generating at its full capacity, so as to be ready when required. Similarly, to guard against unforeseen plant faults, enough spare capacity must be instantly available to cope with the failure of a generating set, and ideally there should be enough available to replace the capacity of the largest generator currently in use, in case that particular one should fail. Plant held ready in this way is called *spinning reserve*, and the amount needed in a modern power system can be very large indeed. Returning to our mythical example of the two systems which are not interconnected but could easily become so, it is clear that each must have sufficient spinning reserve to cope with its own emergencies. Once interconnected, however, the laws of probability can be relied upon to suggest that disaster is not likely to strike both systems simultaneously so they can dispense with some of their spinning reserve and rely on each other. Since the installation of, say, 200 MW of power transmission capacity between two systems is invariably much cheaper than the construction of an extra 200 MW of generating plant, there is much to be gained economically as well as technically from the interconnection of the two systems.

The foregoing argument makes clear that at the present stage of development of the electricity industry there is little likelihood of a change from the established structure of large central power plants heavily interconnected. It is perhaps worth examining the circumstances in which it might be feasible to have a less centralized arrangement. A small neighbourhood plant, possibly allied to district heating, does fulfil the statistical need for randomization of demand, provided it is big enough, and it also meets the requirement of convenience. It fails the spinning reserve requirement, however, and it is at an economic disadvantage while large plants are less expensive to install. Only if the installation cost of a new technology were sufficiently small would it be economic to install spare capacity rather than to interconnect. (The authors reject any suggestion that any new technology might be sufficiently reliable to obviate the need for spare capacity; such a suggestion would depart too far from human experience!) An alternative way of circumventing the present structure is to install domestic-size units, possibly powered from natural sources. This is emotionally attractive but fails the tests of acceptability. There is no possibility of randomizing demand and it is inconvenient. Only if a system of electricity supply can be devised which is sufficiently cheap in capital cost that it can be afforded in a size capable of coping with maximum household demand, and which is reliable, can the back-garden power station become acceptable.

In our discussion of power generation and economy in the production and use of energy, we must never allow ourselves to forget that the 'energy crisis' is a meaningless concept to most of the World's population. It is only a small percentage of the total population that notices its effects. While people in industrialized societies worry about having enough energy to drive their cars, washing machines, deep-freezers and dish-washers, the rest of the World has to worry about finding its next meal, or if there will be enough rain for next year's crops. Of grave concern at the moment is the nearly static and extremely low per capita energy consumption of some countries, frequently areas of large population, for example India, where the population growth negates the possible beneficial effects of higher energy consumption. If it is accepted that there is a correlation between a country's per capita energy consumption and gross national product, and between gross national product and individual well-being, then it is clear that countries with a low GNP would stand to gain from an increased energy input. This is nothing more than a statement that increased automation leads to increased productivity. Therefore, if we feel that an improvement in man's standard of living is worth pursuing on a world-wide basis, then we must find ways of increasing the energy input to the developing countries.

The overriding conclusion which may be drawn from this book is that in our own lifetimes, and probably in the lifetimes of our children, there will be no energy crisis in the true sense of the term. There will undoubtedly be some adjustment in life styles and fuel use, but there will probably be no fundamental change. What must be considered, though, is the future of our grandchildren and great-grandchildren. For them the situation does begin to assume a serious aspect. Do we care enough about them to take the necessary steps now? We could buy our childrens' cheap power by using breeder reactors, but at what cost to later generations? We could build solar electricity stations with today's technology but the economic cost is high. How do we balance these two factors? These are the sort of decisions that must be faced if we are to leave our descendants both a fairly clean World to live in, and enough energy supplies to enable them to live in it comfortably.

It would be very easy to write about the future and paint a rose-tinted picture of great wealth based on the cheap and plentiful supply of power from fusion reactors. It would be equally easy to assume the fashionable gloomy posture and predict that we are all doomed. These are the two extremes. The former was much favoured in the science fiction writings of the late nineteen fifties, while the latter has been much favoured in the pseudo-ecological writings of the early nineteen seventies. The truth will obviously be neither of these.

The technology for maintaining most of our power supplies and for economizing in their use already exists. It remains only for governments and individuals to take their courage in one hand, and their chequebooks in the other, and apply it.

Appendix 1 Units

Although SI (Système Internationale) units are now accepted as standard throughout the scientific world, other units are still in use, especially in the United Kingdom and the United States. It is hoped that this appendix will serve to explain the units used throughout this text to all readers, whatever their scientific or engineering background. To facilitate this, some helpful conversion factors are included.

Basic SI Units

Length	metre	(m)
Mass	kilogram	(kg)
	tonne	(1000 kg)
Time	second	(s)
Electric current	ampere	(A)
Absolute temperature	kelvin	(K)

Prefixes:

$$G \times 10^9$$
$$M \times 10^6$$
$$k \times 10^3$$
$$c \times 10^{-2}$$
$$m \times 10^{-3}$$
$$\mu \times 10^{-6}$$
$$n \times 10^{-9}$$

Basic British Units

Length	foot (ft)	1 ft $= 0.305$ m
Mass	pound (lb)	1 lb $= 0.454$ kg
	[ton (2240 lb)*	1 ton $= 1.017$ tonnes]
Time	second (s)	

* The United States of America (short) ton is 2000 lb.

Temperature

There are two temperature scales in everyday use, the metric *Centigrade* scale and the British *Fahrenheit* scale:

$$0°C = 32°F \qquad\qquad 100°C = 212°F$$

The corresponding absolute scales of temperature are the *Kelvin* and *Rankine* scales. These are related to the above by:

$$* \; 273°K = 0°C \qquad\qquad 491°R = 32°F$$

* It is now customary to write this simply as 273 K.

Energy units

The basic SI unit of energy is the *Joule* (J). When a force of 1 newton (1 N), capable of giving a mass of 1 kg an acceleration of 1 m s^{-2}, moves a body of mass 1 kg a distance 1 m, the work done is 1 J.

A power output of 1 watt (W) is achieved when work is done at the rate of 1 J s^{-1}.

It is sometimes convenient, especially when dealing with electricity supply and consumption, to use the units watt-hour and kilowatt-hour (Wh, kWh) for energy. Thus,

$$1 \; Wh = 3·6 \; kJ \qquad\qquad 1 \; kWh = 3·6 \; MJ$$

The units *calorie* and *kilocalorie* are still common in chemical applications. They are mainly used to refer to thermal energy; the calorie is defined as the amount of heat required to raise the temperature of 1 gram of water through 1 °C. The relationship between calories and joules is

$$1 \; cal = 4·2 \; J$$

In the British system of units, the unit of heat energy is the *British Thermal Unit* (Btu). This is the amount of heat required to raise the temperature of 1 lb of water through 1 °F. Thus,

$$1 \; Btu = 252 \; cals = 1055 \; J$$

One *therm* is defined as 100,000 Btu.

The British unit of power is either the *horse-power* (hp) or the Btu hr^{-1}.

$$1 \; hp = 0·75 \; kW \qquad\qquad 1 \; Btu \; hr^{-1} = 0·293 \; W$$

In fuel technology it is often the case that the calorific value of fuels such as oil is quoted in terms of *million tons of coal equivalent* (mtce). For example, one million tons of oil represents 1·69 mtce.

$$1 \; mtce = 2·55 \times 10^{13} \; Btu$$

Oil is, of course, also measured in terms of *barrels*. A barrel, defined as 42–US gallons, has an energy content of $5·62 \times 10^6$ Btu.

In some specialized applications the *electron volt* (ev) is used as the unit of energy. This is the energy acquired by an electron when it is accelerated through a potential difference of 1 volt.

$$1 \; eV = 1·6 \times 10^{-19} \; J$$

Finally, for convenience and because simple mathematical relationships to energy exist, non-energy units are frequently used to represent energy:

(1) Mass through the relativistic transformation $E = mc^2$, where c is the velocity of light (3×10^8 m s^{-1}). This is of particular importance in nuclear physics, where the standard unit of mass is the *atomic mass unit* (amu).

$$1 \text{ kg} = 8.99 \times 10^{16} \text{ J}$$
$$1 \text{ amu} = 1.66 \times 10^{-27} \text{ kg} = 1.49 \times 10^{-10} \text{ J}$$

(2) Temperature through the Boltzmann relation $E = (3/2)kT$. As $k = 1.38 \times 10^{-23}$ J K^{-1}, a room temperature of 300 K leads to an energy of 6.2×10^{-21} J, or approximately 1/25 eV.

(3) Frequency and wavelength through Planck's relationship $E = h\nu$, $c = \nu\lambda$, where $h = 6.6 \times 10^{-34}$ J s. Thus light of wavelength 1000 nm is transmitted in quanta of energy 1.25 eV.

Other Derived Units

Frequency	hertz	(Hz)	s^{-1}
Force	newton	(N)	m kg s^{-2}
Charge	coulomb	(C)	s A
Electric potential	volt	(V)	W A^{-1}
Magnetic flux	weber	(Wb)	V s
Magnetic flux density	tesla	(T)	Wb m^{-2}
Inductance	henry	(H)	Wb A^{-1}
Pressure	pascal	(Pa)	N m^{-2}

(other common units of pressure are the *psi*— pounds per square inch—and the *bar*.

$$1 \text{ bar} = 10^5 \text{ Pa}$$

Standard atmosphere = 1 bar or 15 psi)

Appendix 2 Thermodynamics and Heat Engines

It is not intended that this appendix should present a complete treatment of the theory and practice of thermodynamics; for this the reader is referred to the standard textbooks. Rather, a summary of thermodynamics terminology will be given followed by brief discussions of the basic heat engine cycles and their theoretical efficiencies.

Terminology

Thermodynamics is governed by several laws, the most important of which for general purposes are the First and Second. The First Law states that the increase in the internal energy U of a system is equal to the sum of Q, the heat input to the system, and W, the mechanical work done on it. For small inputs dW and dQ, this can be written as

$$dU = dQ + dW$$

The Second Law states that it is not possible for a self-acting machine, unaided by an external agency, to make heat pass from one body to another at a higher temperature. From an analysis of cyclic processes, it is possible to introduce a quantity S, the entropy, which is defined by $dS = dQ/T$. That is the change in entropy during a process at temperature T is equal to the heat input divided by T. In a cyclic process this equation must be integrated to give the overall entropy change. If the working fluid of the machine is always in equilibrium with the surroundings, the cycle is reversible and the change is zero. If there are losses this will not be true, and an alternative statement of the Second Law is that the entropy of the universe must increase.

It is well known that the pressure p, the volume V and the temperature T of a gas are related ($pV = nkT$ for an ideal gas). These quantities specify the *state* of the gas, or more generally of the fluid. In this they can be regarded as thermodynamic coordinates. From these coordinates and a knowledge of the heat or mechanical work supplied to a fluid, various functions of state can be defined. Entropy and internal energy have already been introduced, and three other important functions are:

Enthalpy	$H = U + pV$
Helmholtz free energy	$F = U - TS$
Gibbs free energy	$G = H - TS$

Changes in the state of the fluid can be viewed as being either *adiabatic* (isentropic), in which no heat is exchanged between the fluid and its surroundings, or *isothermal*, in which the temperature of the fluid remains constant. A general change of state may be a combination of these two.

It is now possible to represent changes in the state of a fluid graphically in a variety of ways. The classic representation is the p–V diagram, on which area has the dimensions of work, but others are frequently either more appropriate or more informative. It must be emphasized that all these curves of state are complementary, and that specific quantities (quantity per unit mass) are generally used. Two important examples are the T–s, temperature–entropy, and the p–h, pressure–enthalpy curves. On the T–s diagram, area has the dimensions of heat content per unit mass, while the p–h diagram allows enthalpy changes to be read directly off the abscissa.

Heat Engines

A heat engine is a device in which either heat is transferred from a high-temperature source to a low-temperature sink, doing mechanical work in the process, or mechanical work is supplied to effect a transfer of heat from the low temperature to the high temperature. The fundamental machine of this type is an idealized one proposed by Carnot.

The Carnot Cycle

Consider a fixed mass of a perfect gas contained in a cylinder that can be thermally isolated from its surroundings. The cylinder contains a piston which is free to slide and there is a heat source at temperature T_1 and a sink at temperature T_2. The starting point is taken as point A in Figure A2.1 and the

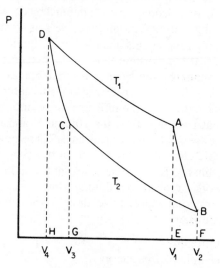

Figure A2.1. Carnot cycle

temperature and volume are T_1 and V_1, respectively. From A the gas is allowed to expand adiabatically to B and, as no heat is supplied, the temperature must drop to T_2 when its volume is V_2. The cylinder is now connected to the sink and is compressed isothermally to V_3 when it is further compressed adiabatically to T_1 and V_4 and then, after connection to the source, isothermal expansion will bring the gas back to A.

During this cycle work is done by the surroundings on the gas, and vice versa. Along DA the gas does work on the surroundings represented by the area DAEH; during the adiabatic expansion AB, it does work ABFE. On the other hand, during the two compressions, work BCGF and GCDH is done by the surroundings on the gas. The net work delivered by the engine is therefore the area ABCD. Since at the end of the cycle, the gas has returned to its original state, this work cannot have been done at the expense of internal energy, and we deduce that the heat Q_1 absorbed from the source during the expansion DA must be greater than that, Q_2, rejected during the compression BC, and that the difference, $Q_1 - Q_2$, is the mechanical work delivered. The efficiency of the engine is defined as

$$\frac{\text{Work delivered}}{\text{Heat absorbed}} = (Q_1 - Q_2)/Q_1$$

The ratio of expansion and compression is determined by considering the adiabatic changes. During the adiabatic expansion from A to B, the volume change is given by

$$T_1/T_2 = (V_2/V_1)^{\gamma - 1}$$

During the adiabatic compression from C to D we have,

$$T_1/T_2 = (V_3/V_4)^{\gamma - 1}$$

Hence

$$(V_2/V_1)^{\gamma - 1} = (V_3/V_4)^{\gamma - 1}$$

or

$$V_2/V_1 = V_3/V_4$$

This means that the ratio of isothermal expansion is the same as that of isothermal compression. If this ratio is r, then the heat taken in at temperature T_1 is $RT_1 \ln r$, and that rejected at T_2 is $RT_2 \ln r$. Consequently the efficiency of the engine is

$$\eta = \frac{RT_1 \ln r - RT_2 \ln r}{RT_1 \ln r} = \frac{T_1 - T_2}{T_1}$$

This particular derivation applies only to a perfect gas working fluid. It can be shown that the result holds for any working fluid in a Carnot engine. All practicable heat engines will have lower efficiencies as the Carnot engine represents an ideal—unattainable in practice.

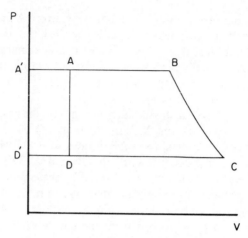

Figure A2.2. Rankine cycle, P–V diagram

The Rankine Cycle

This is more appropriate for many heat engine applications which involve condensation of the working fluid. It is also an idealized cycle, however, and is therefore unattainable in practice. As with the Carnot cycle, it does give the upper efficiency limit for such systems. The basic assumption is that there is a separate condenser and a feed heater in the boiler. In the cycle shown in Figure A2.2, A'B represents the admission of high pressure working fluid in the vapour state; BC is the expansion line (adiabatic) and CD', the exhaust to the condenser. The cycle can also be shown on a $T - s$ diagram as in Figure A2.3. The line CE is the saturated vapour line representing dry saturated working fluid (fully evaporated), AB is the saturated liquid line representing complete condensation and Z is the critical point. Initially liquid working fluid is at temperature T_2 (Point A) and is heated to T_1 at constant pressure along the line AB. The gain in entropy is $FG = C_p \ln (T_1/T_2)$. The fluid is next fully vaporized at temperature T_1, along BC, leading to a gain in entropy of $GK = L_1/T_1$, where L_1 is the latent heat of evaporation at temperature T_1. The fluid is now expanded adiabatically along CD to the original temperature T_2 and then condensed back to A once more. The total work done during the cycle is given by the area ABCD and the heat supplied by FABCK. The efficiency is therefore

$$\eta = \frac{\text{Area ABCD}}{\text{Area FABCK}} = \frac{\text{FABG} - \text{FALG} + \text{LBCD}}{\text{FABG} + \text{GBCK}}$$

$$= \frac{C_p(T_1 - T_2) - C_p T_2 \ln(T_1/T_2) + (L_1/T_1)(T_1 - T_2)}{C_p(T_1 - T_2) + L_1}$$

If the fluid is incompletely vaporized so that it has a dryness fraction q_1

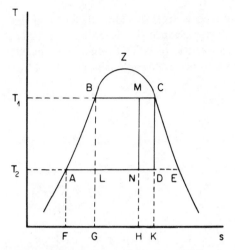

Figure A2.3. Rankine cycle, T–s diagram

at the temperature T_1, then the cycle is limited by MN rather than CD and the dryness fraction is $q_1 = BM/BC$. This leads to an efficiency

$$\eta_w = \frac{C_p(T_1 - T_2) - C_p T_2 \ln(T_1/T_2) + q_1 L_1(T_1 - T_2)/T_1}{C_p(T_1 - T_2) + q_1 L_1}$$

At any point in the cycle, the dryness fraction can be determined by the equivalent ratio. For example, at D it is $q = AD/AE$.

The other possibility is that the working fluid becomes superheated. In this case the cycle is as shown in Figure A2.4, and contains an extra section. It is now ABCMNA. The increase in work due to superheating is DCMN and

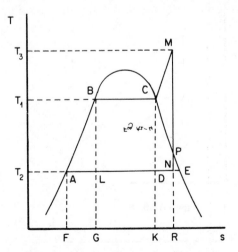

Figure A2.4. Rankine cycle with superheat

the extra work supplied is KCMR. The efficiency of this part of the cycle is higher because the temperature T_3 is greater and hence the area DCMN represents a greater fraction of KCMR than the efficiency η of the saturated cycle. The total work done in the cycle is now ABCMN and is given by

$$W = C_p^l(T_1 - T_2) - C_p^l T_2 \ln(T_1/T_2) + L_1(T_1 - T_2)/T_1 + C_p^v(T_3 - T_1) -$$
$$C_p^v T_2 \ln(T_3/T_1)$$

C_p^l and C_p^v represent the specific heats at constant pressure in the liquid and vapour phases.

The heat supplied is FABCMN = FABCK + KCMR

$$= L_1 + C_p^l(T_1 - T_2) + C_p^v(T_3 - T_1)$$

making the efficiency

$$\eta_s = \frac{(C_p^l + L_1/T_1)(T_1 - T_2) - C_p^l T_2 \ln(T_1/T_2) + C_p^v(T_3 - T_1 - T_2\ln(T_3/T_1))}{L_1 + C_p^l(T_1 - T_2) + C_p^v(T_3 - T_1)}$$

The evaluation of these efficiencies is clearly complicated and an alternative approach used in practice is to express the efficiencies in terms of enthalpies H and to obtain values of H from steam tables or their equivalent (see Chapter 7).

None of these cycles is exactly attainable as true adiabatic compression or expansion is unrealizable in practice.

The Stirling Cycle

This cycle replaces the piston displacements of Carnot's engine by constant volume flow through a regenerator or vessel packed loosely with heat absorbing material. The P–v diagram is shown in Figure A2.5, and the cycle is as follows: Air at T_1 expands isothermally along AB through a volume ratio r, taking

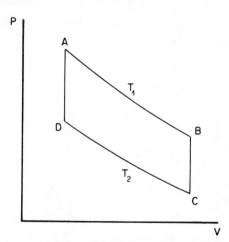

Figure A2.5. Stirling cycle

in heat equivalent to the work done, $RT_1\ln r$. This air then passes at constant volume through the regenerator, its temperature falling to T_2 and the pressure undergoing a corresponding decrease. The amount of heat stored in the regenerator is $C_v(T_1 - T_2)$. The air is now brought into contact with the cold sink at temperature T_2 and is compressed isothermally by the engine piston to its original volume, rejecting heat $RT_2\ln r$. Finally the air is passed through the regenerator in the reverse direction, its temperature rising to T_1 again. The amount of heat taken in from the regenerator is $eC_v(T_1 - T_2)$ where e is the efficiency of the regenerator. The efficiency of the cycle is therefore given by

$$\text{Heat supplied} = RT_1\ln r + C_v(T_1 - T_2)(1 - e)$$

$$\text{Heat rejected} = RT_2\ln r + C_v(T_1 - T_2)(1 - e)$$

making

$$\eta = \frac{R(T_1 - T_2)\ln r}{RT_1\ln r + C_v(T_1 - T_2)(1 - e)}$$

If the regenerator is ideal so that $e = 1$, this reduces to $(T_1 - T_2)/T_1$.

All of the cycles discussed so far rely on an external source and sink for the supply and rejection of heat. By contrast the internal combustion engine derives the heat supply from ignition of the fuel, while the rejection is accomplished partly through the use of a coolant, and partly by the rejection of the exhaust gases at the end of the working stroke. Only two cycles will be considered.

The Otto Cycle

This follows basically the same cycle as the Stirling engine with the addition of an exhaust stroke DE and a pumping stroke ED, both at nearly atmospheric pressure. It is therefore a constant volume cycle as shown in Figure A2.6. It

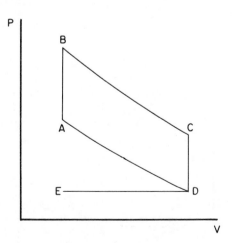

Figure A2.6. Otto cycle

is used for gas and petrol engines. A four-stroke cycle is as follows. Starting at E, the piston is withdrawn sucking an air–fuel mixture into the cylinder. At D, the inlet valve is closed and the mixture is compressed to A by moving the piston back along the cylinder once more. At this point the fuel is ignited and the pressure rapidly increases to B when the piston is driven out to C and the cylinder volume has reached its maximum value once more. The exhaust valve is opened and the pressure drops to about atmospheric at D when the piston is driven in again to E, expelling the spent gases. The exhaust valve is closed and the cycle restarts. Unfortunately the air–fuel mixture is not well behaved and the compression and expansion strokes are far from ideal. This shows itself in the value of γ that must be used in $pv^\gamma = $ constant to attain any sort of agreement with experiment. If the engine were perfect and working on the corresponding air cycle, its efficiency would be given (as shown in Chapter 6) by

$$\eta = 1 - (1/r)^{\gamma - 1}$$

where r is the compression ratio. In practice efficiencies are much lower than this and γ is different during the compression and expansion strokes.

The Diesel Engine

This is a constant pressure cycle engine in which the toe of the cycle is omitted to make the compression and expansion strokes equal, as shown in Figure A2.7. In addition the fuel must be injected and exhausted. Air is taken into the cylinder along FA, and is compressed along AB by movement of the piston. At B the fuel is injected in a fine spray into the compressed and hot air, and ignites. This fuel injection is continued for the portion of the stroke BC, C being known as the *cut-off*. Thereafter more or less adiabatic expansion occurs

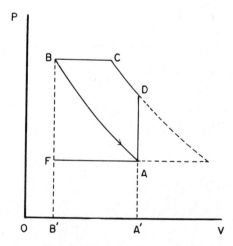

Figure A2.7. Diesel engine cycle

to D followed by exhaust of the spent gases to A and then to F. The heat received is $C_p(T_C - T_B)$. The heat rejected is $C_v(T_D - T_A)$ and the efficiency is

$$\eta = 1 - \frac{C_v(T_D - T_A)}{C_v(T_C - T_B)}$$

Since all the temperatures can be expressed in terms of T_A:

$$T_B = T_A r^{\gamma - 1}; \ T_C = T_B(V_C/V_B) = T_B\beta = \beta T_A r^{\gamma - 1}; \ T_D = T_C(V_C/V_D)^{\gamma - 1} =$$
$$T_A r^{\gamma - 1}(\beta/r)^{\gamma - 1},$$

where r is the compression ratio OA'/OB', we can write the efficiency as

$$1 - (1/r)^{\gamma - 1}(\beta^\gamma - 1)/(\gamma(\beta - 1))$$

This efficiency is dependent on both the compression ratio r and the volume ratio β. For a given compression ratio, the higher the temperature at C after combustion, the greater will be the value of β and since this must be greater than unity, the efficiency will decrease as the maximum temperature increases.

Index